Effective Use of Social Media in Public Health

Effective Use of Social Media in Public Health

Edited by

Kavita Batra
Department of Medical Education, Kirk Kerkorian School of Medicine at UNLV, Las Vegas, NV, United States;
Office of Research, Kirk Kerkorian School of Medicine at UNLV, Las Vegas, NV, United States

Manoj Sharma
Department of Social and Behavioral Health, School of Public Health, University of Nevada, Las Vegas (UNLV), NV, United States;
Department of Internal Medicine, Kirk Kerkorian School of Medicine at UNLV, Las Vegas, NV, United States

Academic Press is an imprint of Elsevier
125 London Wall, London EC2Y 5AS, United Kingdom
525 B Street, Suite 1650, San Diego, CA 92101, United States
50 Hampshire Street, 5th Floor, Cambridge, MA 02139, United States
The Boulevard, Langford Lane, Kidlington, Oxford OX5 1GB, United Kingdom

Copyright © 2023 Elsevier Inc. All rights reserved.

No part of this publication may be reproduced or transmitted in any form or by any means, electronic or mechanical, including photocopying, recording, or any information storage and retrieval system, without permission in writing from the publisher. Details on how to seek permission, further information about the Publisher's permissions policies and our arrangements with organizations such as the Copyright Clearance Center and the Copyright Licensing Agency, can be found at our website: www.elsevier.com/permissions.

This book and the individual contributions contained in it are protected under copyright by the Publisher (other than as may be noted herein).

Notices

Knowledge and best practice in this field are constantly changing. As new research and experience broaden our understanding, changes in research methods, professional practices, or medical treatment may become necessary.

Practitioners and researchers must always rely on their own experience and knowledge in evaluating and using any information, methods, compounds, or experiments described herein. In using such information or methods they should be mindful of their own safety and the safety of others, including parties for whom they have a professional responsibility.

To the fullest extent of the law, neither the Publisher nor the authors, contributors, or editors, assume any liability for any injury and/or damage to persons or property as a matter of products liability, negligence or otherwise, or from any use or operation of any methods, products, instructions, or ideas contained in the material herein.

ISBN: 978-0-323-95630-7

For information on all Academic Press publications visit our website at https://www.elsevier.com/books-and-journals

Publisher: Stacy Masucci
Acquisitions Editor: Elizabeth A. Brown
Editorial Project Manager: Timothy J. Bennett
Production Project Manager: Fahmida Sultana
Cover Designer: Vicky Pearson Esser

Typeset by TNQ Technologies

"To my father, who always taught me to never give up. I know that he is looking down on me from heaven and smiling, proud of the person that I have become."
*<div align="right">**Kavita Batra (KB)**</div>*

"To my beloved mother in heaven, who always believed in me. I wouldn't be the person I am today without her."
*<div align="right">**Manoj Sharma (MS)**</div>*

Contents

List of contributors	xvii
About the contributors	xix
Foreword	xxix
Preface	xxxi
Acknowledgments	xxxiii

Section I
Introduction

1. Social media and types with their current applications in public health and healthcare
Tung Sung Tseng and Gabrielle Gonzalez

Social media	3
Role of social media in public health	4
COVID-19 and social media publications	7
Social media and the dangers of misinformation	8
Types of social media	9
Social media and public health interventions	11
Social media and social ecological model	13
Social media and youth	14
Ethical issues in social media	15
Challenges and opportunities of social media	16
Chapter summary	18
Questions for discussion	18
Websites to explore	19
Key terms—definitions	19
References	20

2. Gender and age-specific use of social media
Jody L. Vogelzang

Introduction	23
Social media growth and access	23
Age-specific use of social media	24
Children	24
Adolescence	26

Adults	28
Vulnerability of adults in social media	29
Older adults	30
Older adult and social media during COVID	31
The impact of gender on social media use	32
Stereotypical access and use of social media	32
Cisgender	33
Chapter summary	37
Discussion questions	37
Definition of key terms	37
Websites to explore	38
References	39

Section II
Social media use and mental health outcomes in diverse populations

3. Social media and young adults
Bronwyn MacFarlane and Jason Kushner

Social media and young adults	45
Social media usage among young adults	46
Empirical research concerns about social media use among young adults	46
Key terms defined	48
Mental health and counseling trends and issues	49
Summary	58
Questions for discussion	59
Website to explore	59
Pew Research Center	59
Activity for the reader	59
Key terms used in the chapter	59
References	60
Further reading	63

4. Social media: utility versus addiction
Ram Lakhan, Bidhu Sharma and Manoj Sharma

Introduction	65
Prevalence of social media use	67
Social media during the COVID-19 pandemic	68
The epigenetic effect is a future concern	69
Addiction to social media	70
The theoretical basis of social media addiction	70
Summary	86
Conclusion	87
Questions	88

	Definition of key terms	88
	Website to explore	91
	Substance Abuse and Mental Health Services Administration	91
	Popular social networking websites/applications	91
	References	91

5. Social media use among older adults and their challenges
Ram Lakhan, Bidhu Sharma and Manoj Sharma

Introduction	99
Growing older population and its consequences	100
Role of social support and social connectedness	101
Social media	104
Research-based evidence of benefits of social media usage among older adults	105
Policy approaches to increase social connectedness	106
Challenges associated with social media for older adults	106
Social media literacy-related challenges among older adults	107
Physical challenges related to social media use among older adults	108
Financial challenges related to social media among older adults	108
Psychological challenges associated with social media use among older adults	109
Summary	113
Conclusion	116
Questions	116
Definition of key terms	117
Websites to explore	118
References	119

6. Social media, diversity, equity, and inclusion
Tiffiny R. Jones and Sely-Ann Headley Johnson

Identifying diversity, equity, and inclusion	125
How does health equity, diversity, and inclusion affect mental health?	127
Barriers to mental health service utilization	128
Linking social media, mental health, and diversity equity and inclusion	134
Implications for public health practice	135
Chapter summary	136
Definition of key terms	138
Review or discussion questions	138
Websites to explore	139
References	139
Further reading	142

Section III
Social media and global exposure to research

7. Ethical, privacy, and confidentiality issues in the use and application of social media
Amar Kanekar and Joseph Otundo

Social media and its varied applications	145
Role of ethics in social media usage	146
Role of privacy in social media usage	148
Case study questions	153
Case study questions	154
Role of confidentiality in social media usage	155
Review of literature	157
Major ethical issues related to social media	157
Confidentiality	158
Can ethics and confidentiality Coexist?	158
Trust	158
Application of privacy and confidentiality in public health studies using social media	159
Chapter summary	161
Questions for discussion	162
Important terms defined	162
Websites to explore	163
Centers for Disease Control and Prevention, social media tools, guidelines, and best practices	163
Ethical dilemmas of social media and how to navigate them from Norwegian Business School	163
Internet safety rules while using social media for teens	163
Issues in ethics: ethical use of social media	163
Protecting student privacy on social media	164
Public health guide to social media 101	164
Social media research: public health versus privacy	164
Social media and web 2.0 policy: US Department of Commerce	164
Theme issue: social media, ethics, and COVID-19 misinformation	164
References	165
Further reading	167

8. Applications of social media research in quantitative and mixed methods research
Rose Marie Ward, Mai-Ly N. Steers, Akanksha Das, Shannon Speed and Rachel B. Geyer

Introduction	169
Applications of social media research	170
Platform choice and across platforms use	170
Posting across multiple platforms	170
Self vs. other generated content	171

Relationships with others	172
Engagement (likes, comments, and reposting)	173
Content (images, videos, and text)	174
Reach (followers, friends of friends, viral, and influencers)	177
Motivations for posting	178
Change over time	179
Experimental	180
Real-time	181
Self-report vs. actual content	182
Social media use measures	183
Challenges for quantitative and mixed methods research	184
Future directions for quantitative and mixed methods social media research	185
Conclusion	186
Questions for discussion	186
Websites to explore	186
Facebook application using social network analysis	186
Social network analysis—social circles of Facebook	186
Definition of key terms	187
References	187

9. Applications of social media in qualitative research in diverse public health areas
Geetanjali C. Achrekar and Kavita Batra

Introduction	193
Social media and qualitative research	194
Methods/tools	196
Applications of social media in diverse public health areas	198
Application of social media by health care professionals	199
Social media and public	201
Social media and public health organizations	203
Pitfalls of social media	205
Guidelines for ethical use of social media in public health	206
Conclusions and future directions	208
Questions for discussion	208
Websites to explore	208
Key terms	209
References	210
Further reading	215

10. Role of social media in research publicity and visibility
Sely-Ann Headley Johnson and Tiffiny R. Jones

Introduction	217
Research dissemination patterns	217
Novel publication metric trends	218

Using social media for health information	219
Social media as dissemination avenues	220
Visual abstracts	220
Challenges and advantages of disseminating research on social media	221
Effect of social media on health outcomes in spite of research findings	222
Linking health literacy and media literacy	223
Engaging policy makers through social media	224
Disseminating research on social media and the rise of enterprise social media	224
Implications for public health practice	225
Conclusions	226
Chapter summary	226
Definition of key terms	227
Review or discussion questions	227
Websites to explore	227
References	228
Further reading	230

Section IV
Social media, public health communication, and pedagogy

11. Social media and policy campaigns
Gayle Walter

Introduction	233
Overview of social media in advocacy and policy change	234
Social media use by members of Congress	234
Benefits of using social media in policy development	238
Examples of use	239
Epidemiology and surveillance	239
Health promotion messaging	239
Public health policy	242
Public opinion	243
Social media influence in crises	245
The dialogic theory and engagement framework	245
Future directions	247
Conclusions	247
Questions	248
Websites to explore	248
Centers for Disease Control and Prevention—CDC social media tools, guidelines, and best practices	248
The Community Toolbox—using social media for digital advocacy	248
U.S. Department of Health and Human Services—social media policies	249

	Definitions of key terms	249
	References	249
	Further reading	251

12. Social media and Infodemiology—use of social media monitoring in emergency preparedness
Kavita Batra, Ravi Batra and Manoj Sharma

Introduction	254
Information Epidemiology or Infodemiology	254
Uses of Infodemiology on historical context	255
Research landscape of Infodemiology	255
The expanded framework of Infodemiology	256
Social media and the COVID-19 pandemic	257
Infodemic: A new front of challenge in the COVID-19 battle (mis vs. dis vs. mal)	257
Fake news triangle and meta-framework of fake news impact	259
COVID-19: A pandemic of publications	261
"Infodemic knowledge" and public health at risk	262
Social media: A double-edged sword	264
Efforts to combat "infodemic"	265
Strategies to flatten the "Infodemic curve"	266
Conclusions and future directions	267
Questions for discussion	268
Websites to explore and a list of credible sources	268
Key terms with definitions	269
References	270
Further reading	275

13. Social media, online learning, and its application in public health
Janea Snyder

Social media	277
Health communication	280
Health equity guiding principles for inclusive communication	281
Skill building discussion questions activity	283
Social marketing and public health	283
Health promotion, prevention, and psychosocial health	284
Chapter summary	288
Chapter key terms	288
Chapter review questions	289
Websites to explore to learn more	289
CDC social media tools, guidelines, and best practices	289
Suggested activities to the readers	289
CDC public health media library	289

Suggested activities to the readers	289
Quality matters	289
Suggested activities to the readers	290
American public health association	290
Suggested activities to the readers	290
Texthelp	290
Suggested activities to the readers	290
References	290
Further reading	292

Section V
Social media in healthcare

14. Application of social media in designing and implementing effective healthcare programs
Priyanka Saluja, Vishakha Grover, Suraj Arora, Kavita Batra and Jashanpreet Kaur

Introduction	295
Role of social media in healthcare	296
Role of social media in health promotion	297
Role of social media in circulating health facts and ceasing the wrong information	299
Role of social media in research	301
Role of social media in professional development	301
Role of social media in improving doctor–patient communication and health services	301
Popular social media platforms for doctors	301
Benefits of using social media by healthcare professionals	303
Professional recommendations for using social media	305
Limitations and challenges	305
Conclusions	307
Chapter summary	308
Questions for discussion	308
Websites to explore and a list of credible sources	309
Definition of key terms	310
Health awareness	310
Health promotion	310
Infodemiology	311
Professional networking	311
Social media	311
Social media analytics	311
References	311

15. Role of social media in telemedicine
Rasika Manori Jayasinghe and Ruwan Duminda Jayasinghe

Telemedicine	317
Social media	319
Social media in healthcare	321
Role of smartphones in using social media in health	323
Use of social media in telemedicine	324
Guidelines for social media use in telemedicine	324
Use of social media by different specialties/branches in healthcare	325
Different uses of social media in healthcare	326
mHealth and telemedicine	326
Telemedicine, social media, and curriculum	329
Telepharmacy	330
Positive and negative points of using social media in telemedicine	331
Recommendations for using social media in telemedicine	332
Summary	334
Questions for discussion	334
Websites to explore	335
References	336
Further reading	338

Section VI
Epilogue

16. Innovative uses of social media in public health and future applications
Manoj Sharma

The growing use of social media among public health and healthcare professionals	341
The upward trend of the use of social media in public health research	343
Social media and crowdsourcing	343
Artificial intelligence and social media	343
Rise in the use of social media in teaching, training, and workforce development of public health and healthcare professionals	344
Increasing use of social media in disease surveillance	344
Rising use of social media in public health campaigns and messaging	345
Greater use of social media by local health departments	345

Social media surveillance in K-12 settings	345
Counteracting threats from the use of social media	346
Expanding the use of social media for health education and health promotion interventions	346
Use of social media for equity and social justice	346
Use of social media for quality assessment and safety of healthcare services	346
Use of social media in policy change interventions	347
Conclusion	347
References	347
Further reading	349
Index	351

List of contributors

Geetanjali C. Achrekar, GVM's College of Commerce and Economics, Ponda, Goa, India

Suraj Arora, Department of Restorative Dental Sciences, College of Dentistry, King Khalid University, Abha, Saudi Arabia

Kavita Batra, Department of Medical Education, Kirk Kerkorian School of Medicine at UNLV, University of Nevada, Las Vegas, NV, United States; Office of Research, Kirk Kerkorian School of Medicine at UNLV, University of Nevada, Las Vegas, NV, United States

Ravi Batra, Department of Information Technology, Coforge Ltd., Atlanta, GA, United States; Social and Behavioral Health, School of Public Health, University of Nevada, Las Vegas, NV, United States

Akanksha Das, Psychology Department, Miami University, Oxford, OH, United States

Rachel B. Geyer, Psychology Department, Miami University, Oxford, OH, United States

Gabrielle Gonzalez, Behavioral and Community Health Sciences, School of Public Health, Louisiana State University Health Sciences, New Orleans, LA, United States

Vishakha Grover, Department of Periodontology and Oral Implantology, Dr. H.S.J. Institute of Dental Sciences, Panjab University, Chandigarh, Punjab, India

Sely-Ann Headley Johnson, Wayne State University, Detroit, MI, United States

Rasika Manori Jayasinghe, Faculty of Dental Sciences, University of Peradeniya, Kandy, Sri Lanka

Ruwan Duminda Jayasinghe, Faculty of Dental Sciences, University of Peradeniya, Kandy, Sri Lanka

Tiffiny R. Jones, Wayne State University, Detroit, MI, United States

Amar Kanekar, School of Counseling, Human Performance and Rehabilitation, College of Business, Health and Human Services, University of Arkansas, Little Rock, AR, United States

Jashanpreet Kaur, Dr. H.S.J. Institute of Dental Sciences, Panjab University, Chandigarh, Punjab, India

Jason Kushner, University of Arkansas at Little Rock, College of Business, Health and Human Services, School of Counseling, Human Performance, and Rehabilitation, Little Rock, AR, United States

Ram Lakhan, Department of Health and Human Performance, Berea College, Berea, KY, United States

Bronwyn MacFarlane, Department of Educational Leadership Studies, Curriculum, and Special Education, College of Education and Behavioral Science, Arkansas State University, Jonesboro, AR, United States

Joseph Otundo, School of Counseling, Human Performance and Rehabilitation, College of Business, Health and Human Services, University of Arkansas, Little Rock, AR, United States

Priyanka Saluja, Department of Dentistry, University of Alberta, Edmonton, AB, Canada

Bidhu Sharma, School of Arts and Science, University of Louisville, Louisville, KY, United States

Manoj Sharma, Department of Social and Behavioral Health, School of Public Health, University of Nevada, Las Vegas, NV, United States; Department of Internal Medicine, Kirk Kerkorian School of Medicine at UNLV, University of Nevada, Las Vegas, NV, United States

Janea Snyder, Health Education and Promotion, University of Arkansas at Little Rock, Little Rock, AR, United States

Shannon Speed, NIDA/NIAAA IRP—NIH, Translational Addiction Medicine Branch Lab, CPN Section, Biomedical Research Center, Baltimore, MD, United States

Mai-Ly N. Steers, School of Nursing, Duquesne University, Pittsburgh, PA, United States

Tung Sung Tseng, Behavioral and Community Health Sciences, School of Public Health, Louisiana State University Health Sciences, New Orleans, LA, United States

Jody L. Vogelzang, Grand Valley State University, Allendale, MI, United States

Gayle Walter, Department of Health and Human Physiology, University of Iowa, Iowa City, IA, United States

Rose Marie Ward, Department of Psychology, University of Cincinnati, Cincinnati, OH, United States

About the contributors

Dr. Geetanjali C Achrekar, MA, Ph.D., is an Associate Professor in Economics at the GVM's College of Commerce and Economics, Goa (India), and has teaching and research experience of over 2 decades. Dr. Achrekar mentors several doctoral level candidates and her areas of interest include but are not limited to the economic growth and development, health, and environmental economics. Dr. Achrekar has successfully completed a postdoctoral project related to the coastal sustainability and is currently leading a government-sponsored project on tribal welfare schemes in Goa. She serves as a key member to the Departmental Research Committees of Goa University and is an external referee for the Ph.D. theses of some state universities in Maharashtra, India.

Dr. Suraj Arora, BDS, MDS, serves as an Assistant Professor at College of Dentistry, King Khalid University, Abha, Saudi Arabia. He received his master's degree in Conservative Dentistry and Endodontics from BHU, Varanasi, India. Mr. Arora has 13 years of academic and research experience gained by working in reputed dental institutes. His research interests include endodontic microbiology, endodontic anatomic variations, effect of COVID-19, dental composites, and nano dentistry. He has published multiple peer-reviewed articles on the effect of plant extracts on endodontic microbiota, nonsurgical endodontic management of anomalous teeth, biomaterials, and the impact of COVID-19 among dentists. He has worked in the following projects: "Evaluation and Comparison of Antibacterial Activity against Oral Pathogenic Microorganism of Different Plant Extracts: A microbiological study" and Use of CBCT in assessing anatomical structures in dental implants and endodontic procedures. He is the Lifetime Member of the Federation of Operative Dentistry of India. Dr. Arora has a patent "Effectiveness of different systems for root canal preparation in eliminating enterococcus faecalis" registered in the Office of the Controller General of Patents, Designs, and Trademarks under the Government of India.

Dr. Kavita Batra, PhD, MPH, BDS, FRSPH, serves as an Assistant Professor and Senior Biostatistician with the Kirk Kerkorian School of Medicine at UNLV. Dr. Batra began her career as a dental surgeon in India and received her master's and doctorate degrees in Public Health from the University of Nevada, Las Vegas, USA. As the Co-Vice Chair of the research seminar planning committee for the office of research and director of internal medicine resident research within the school of medicine, Dr. Batra provides statistical and research mentoring to faculty, residents, fellows, and medical students. Her research interests include, but are not limited to, maternal and child health, COVID-19, social determinants of health, vaccine hesitancy, clinical research, and evidence synthesis. Dr. Batra has published multiple peer-reviewed articles and presented her work at several state, national, and international public health conferences. She has an extensive experience in quantitative and qualitative research. Dr. Batra serves on the topical advisory panel of several journals, including *Dentistry*, *Vaccines*, and *Healthcare*. Dr. Batra is the President-elect for the Delta Omega-Delta Theta Chapter at UNLV, a Grad Alumni Ambassador, a key member to the Sexual Misconduct Taskforce, Nevada System of Higher Education, and a statewide secretary to the Nevada Public Health Association.

Ravi Batra, MCA, is a doctoral scholar within Social and Behavioral Health Department, School of Public Health, University of Nevada, Las Vegas. Mr. Batra has 18+ years of experience in data analytics, coding, database development, and quality assurance (through automation). Since the inception of the COVID-19 pandemic, Mr. Batra has coauthored several publications, which are already making quite an impact toward the evidence synthesis. As a part of Mr. Batra's dissertation, he intends to utilize his advanced coding skills to query large datasets for investigating the social determinants of health and resulting disparities prevailing among our communities.

Akanksha Das, MA, is a clinical psychology doctoral student at Miami University. Broadly, her research interests center on understanding factors (individual and structural) that support conditions for equitable well-being. To that end, she has applied a multimethod approach (e.g., psychophysiology and self-report, as well as social attitudes, and policies) with the hope of understanding how we can leverage structural and individual factors to promote mental health equity among minoritized communities.

Rachel Geyer, MA, is a clinical psychology doctoral student at Miami University. Her research interests include examining transdiagnostic factors that influence both anxiety and substance use (e.g., anxiety sensitivity and distress tolerance). Rachel also conducts research exploring substance use motives and examining alcohol use patterns and problems (e.g., blackouts) across undergraduate and graduate students.

Gabrielle Gonzalez is a Ph.D. candidate at the School of Public Health at Louisiana State University Health Sciences in New Orleans, LA. She has a special interest in childhood and adolescent obesity. She received her master's in Behavioral and Community Health Sciences from the School of Public Health at Louisiana State University Health Sciences and her Bachelor of Science in Biological Sciences from Louisiana State University in Baton Rouge, LA.

Dr. Vishakha Grover is currently working as an Associate Professor in Department of Periodontology at Dr. Harvansh Singh Judge Institute of Dental

Sciences, Panjab University, Chandigarh. Dr. Grover received her Ph.D. in Periodontology from SGT University, Gurugram. Her research interests include oral health—related quality of life, periodontal microbiology, oral fluid—based diagnostics, dental implants, perio-systemic relationship, and genetics. She has published more than 80 scientific publications in various quality scientific indexed national and international dentistry and allied medical journals with significant impact. Dr. Grover serves as reviewer and an editorial board member to several national and international scientific journals. She has received the grant from the Department of Science and Technology, Govt. of India, for her research projects on exploring perio-systemic relationship in Indian population and antimicrobial peptides in oral hygiene applications. Dr. Grover has been awarded a special mention "excellence in research" award in 2016 from a national level organization honoring science and technology awards to most eminent Indian researchers. She has been associated with the Indian Society of Periodontology as an active executive committee member of ISP for the past 11 consecutive years and contributed in various roles as Joint Secretary, National Coordinator for annual rapid revision activity for postgraduates, and Convenor for the activities for ISP women wing. Currently, Dr. Grover is serving as the National Scientific Program Convener for Indian Society of Periodontology.

Dr. RM Jayasinghe is a Professor in the Department of Prosthetic Dentistry, Faculty of Dental Sciences, University of Peradeniya, Sri Lanka, for over 7 years. She has experience working as a consultant in Restorative Dentistry at multiple government hospitals such as Teaching Hospital, Kurunegala, and National Dental Institute Colombo, Sri Lanka, and School of Dentistry and Oral Health, Griffith University, Australia. She obtained MS in Restorative Dentistry from the Postgraduate Institute of Medicine, the University of Colombo in 2009. She has widely published on dental education, prosthetic dentistry, restorative dentistry, and implant dentistry.

Dr. RD Jayasinghe is the Chair Professor of Oral Medicine and Periodontology and a Specialist in OMF Surgery at the Department of Oral Medicine and Periodontology, Faculty of Dental Sciences, University of Peradeniya, Sri Lanka. He is an Adjunct Faculty in the Department of Oral Medicine and Periodontology, Saveetha Dental College and Hospitals and Saveetha Institute of Medical and Technical Sciences, Chennai, India, and serves as the Director of Centre for Research in Oral Cancer, Faculty of Dental Sciences, the University of Peradeniya, and the Director, Operational Technical Secretariat (OTS), University of Peradeniya, AHEAD Operations. He has widely published in the field of dental education, oral cancer, and oral potentially malignant disorders and has addressed different forums in the field of oral cancer.

Dr. Sely-Ann Headley Johnson, Ph.D., has experience teaching over 15 courses (undergraduate and graduate) in Statistics, Health Education, Public Health, and Food Safety. Her research interests include health education ethics, hybrid violence mitigation, food safety, and unintentional drug overdose. Dr. Headley Johnson is currently an Epidemiologist at the CDC Foundation serving at the Indiana Department of Health, in Indianapolis, IN. Dr. Headley Johnson has a Master of Public Health degree with a concentration in Epidemiology and earned a Ph.D. in Health Education from the University of Toledo in 2020.

Dr. Tiffiny Jones, MPA, MPH, Ph.D., is an Assistant Professor at Wayne State University in Detroit, Michigan and an advocate for public health. Not only does she have extensive teaching experience, she has also taught a variety of courses including courses in theory, statistics, public health, health education, women and gender studies, and Africana/Black studies. She enjoys most teaching courses that directly apply to real-world problems and allows her students to critically work through real-world solutions. She has over 10 years of experience working in the public sector (including public health and nonprofit). Her primary research areas are maternal/child health (specifically among Black women), and health outcomes of racism and racist practices. She also has research interests in reproductive health, child/maternal health, food and research justice, health disparities, and community-based participatory research. She has a Master of Public Health and Master of Public Affairs from the University of Missouri and a Ph.D. in Health Education from the University of Toledo.

Dr. Amar Kanekar is a Professor and Graduate Program Coordinator for Health Education and Health Promotion at the University of Arkansas at Little Rock. His 16 years of teaching experience involves more than 30 different courses (undergraduate and graduate) in the areas of public health, health education, and health promotion. Recipient of numerous teaching awards, his pedagogical techniques involve online—distance learning, hybrid, and face-to-face courses with web enhancement using instructional technology for synchronous and asynchronous interaction. His research areas of interest focus on adolescent health, measurement in health education, global health, online and hybrid pedagogy, and health behavior interventions. He has published more than 75 publications in refereed and nonrefereed venues and has more than 120 presentations at the local, state, national, and international levels. Dr. Kanekar currently serves as the Interim Undergraduate Coordinator and the Graduate Coordinator for the health education/promotion program within the School of Counseling, Human Performance, and Rehabilitation at the University of Arkansas at Little Rock. He currently serves nationally on the Society for Public Health Education Ethics Committee and Editorial Board for the *American Journal of Public Health*. He additionally serves on the Review Board of the *Frontiers in Nutrition* Journal and Section Editor for the journal *HealthCare*.

Jashanpreet Kaur, BDS, received her bachelor's degree in Dentistry from Dr. Harvansh Singh Judge Institute of Dental Sciences, Panjab University,

Chandigarh, India. Her research interests include public oral health, social factors affecting oral health, and oral health education.

Dr. Jason D. Kushner, Ph.D., is a Professor of Counselor Education at the University of Arkansas at Little Rock. A Certified School Counselor and Licensed Professional Counselor, Dr. Kushner's professional profile includes positions as a high school English teacher, school counselor, college counselor, and outpatient mental health counselor. During his time as a school counselor, Dr. Kushner coordinated dropout prevention interventions for high school students while teaching adult basic education in his spare time. Since his appointment in higher education, Dr. Kushner has developed over 15 fully online and hybrid courses incorporating engaging technology including social media to provide statewide coverage and access to graduate programs in counseling. As a mental health counselor, Dr. Kushner has worked extensively with adolescents and adults around issues of depression, anxiety, mood disorders, and developmental concerns. A frequent presenter on issues related to school and mental health counseling, Dr. Kushner's research agenda features presentations and publications in the areas of multicultural counseling, teaching modalities in counseling, and innovative counseling practices.

Dr. Ram Lakhan, DrPH, is a tenure track Assistant Professor in the Department of Health and Human Performance at Berea College in Kentucky. He has a therapeutic degree in Intellectual Disabilities from India. He has worked extensively for people with intellectual disabilities, mental illness, and their family members in the remotest areas of central India. He has implemented large-scale community-based mental health programs, obtained service grants, and conducted research. He has published over 30 data-based peer-reviewed papers in academic journals and 5 book chapters in the textbooks of prestigious publishers including Springer and Oxford University Press. His scholarly work revolves around mental health issues in people with intellectual disabilities, and his application of behavior change theories in improving health outcomes for special populations is highly acknowledged among academicians and researchers.

Dr. Bronwyn MacFarlane, Ph.D., Arkansas State University doctoral faculty in educational leadership, has experience evaluating and designing educational programming and curriculum. As a professor of gifted education for 13 years, Dr. MacFarlane was honored as one of the top three professors at the University of Arkansas at Little Rock in 2021 where she published four books including *Specialized Schools for High-Ability Learners* (2018) and *STEM Education for High-Ability Learners* (2016). She taught 32 graduate course topics, authored/coauthored 25 articles, 95 reports, 25 book chapters, and delivered over 150 presentations. She earned her doctorate with dual specializations in gifted and K-12 education administration at the College of William and Mary after teaching across K-12 content areas and coaching. Leadership roles included Associate Dean of the UALR College of Education and Health Professions; Academic Dean of the Summer Institute for Gifted at

Princeton University; NAGC Chair of STEM and Counseling Networks; journal guest editor; national columnist of *The Curriculum Corner* in Teaching for High Potential; and recognition with the 2018—2019 Early Leader Award by the National Association for Gifted Children. She serves as the nationally elected board member on the Executive Committee for the World Council for Gifted and Talented Children. Her fifth forthcoming book is *Social and Emotional Learning for Advanced Children in Early Childhood: Birth to 8*. As an educational leader, Dr. MacFarlane is regularly invited to speak to audiences about talent development, perseverance around barriers, and improving organizational/program design and culture.

Dr. Joseph O Otundo is an Assistant Professor of Health Education and Promotion at the University of Arkansas at Little Rock since 2018. Dr. Otundo is involved in teaching, research, and community service. He earned his bachelor's degree from Kenyatta University, Master's degree in Health Studies from Southeastern Louisiana University in 2013, and Ph.D. in Kinesiology from Louisiana State University in 2017. From 2017, he was a Visiting Professor at Southeastern Louisiana University. His research interests include pedagogy in health education and promotion, physical activity, and mental health. Dr. Otundo is an Author and Coauthor of more than 12 articles in peer-reviewed journals and more than 20 conference contributions. He has also written and coauthored four book chapters. In addition, he reviews research articles for several journals. Dr. Otundo has presented talks and conducted workshops both locally and internationally. On pedagogy, he has developed an integrated model of interest theory and self-determination theory in physical activity. In addition, he has explored instructional design and social support as they relate to physical activity. In his research design, Dr. Otundo uses both the traditional methods for data collection as well as modern technologies such as social media platforms.

Priyanka Saluja, BDS, is in her second year of the MSc Dentistry Program at the University of Alberta in Edmonton, Alberta, Canada. She received her bachelor's degree in Dentistry from BFUHS, Punjab, India. Dr. Saluja has 12 years of academic and research experience gained by working in reputed dental institutes. Her research interests include dental public health, oral health of minority population, social determinants of health, role of fluoride concentration on dental fluorosis, effect of plant extracts on endodontic microbiology, and effect of COVID-19 on dentists. She has published multiple peer-reviewed articles in the reputed journals. She is also working on the design of dental pulp testers used for vitality testing of the teeth to gain its complete advantage.

Bidhu Sharma is a biology major undergraduate student at the University of Louisville. He has a special interest for writing on contemporary issues from public health points of view. While he was a high school student, he published his observations related to physical activities during the COVID-19 pandemic and science-based papers on the impact of social environments on physical and mental well-being in his local newspapers. He has also published science-based articles in health magazines and is an alum of the Kentucky Governor's Scholars Program. Social media and its implication on education is one of his current focuses. Pursuing this interest, he requested to become a part in writing the chapters in this textbook. He provided a student's perspective in writing both chapters in this textbook in addition to literature search, making diagrams, and writing the manuscript.

Dr. Manoj Sharma, MBBS, Ph.D., MCHES®, is a Public Health Physician and Educator with a medical degree from the University of Delhi and a doctorate in Preventive Medicine from the Ohio State University. He is currently a tenured Full Professor and Chair of the Social and Behavioral Health program at the University of Nevada, Las Vegas in the School of Public Health and an Adjunct Professor in the Department of Internal Medicine at the Kirk Kerkorian School of Medicine at UNLV. In his career, spanning over 35 years, he has trained/taught over 6000 health professionals. He has worked for local health departments; state health departments/agencies; federal government agencies; nonprofit agencies; professional organizations; and international agencies including governments of other nations in his career. He is ranked in the top one percentile of global research scientists from 176 fields by Elsevier. He has been awarded several prestigious honors including American Public Health Association's Mentoring Award, ICTHP Impact Award, J. Mayhew Derryberry Award, and William R. Gemma Distinguished Alumnus Award at the Ohio State University among others. His research interests are in developing and evaluating theory-based health behavior change interventions, obesity prevention, stress-coping, community-based participatory research, and integrative mind—body—spirit interventions.

Dr. Janea Snyder is an Associate Professor and Certified Health Education Specialist (CHES) at the University of Arkansas at Little Rock in Health Education and Promotion. She completed her doctorate at Texas Woman's University in Denton, Texas, where she majored in Health Studies. Dr. Snyder has served as a state representative for the American Heart Association's Southwest Affiliate Health Equity Committee, a board member for the Arkansas Single Parent Scholarship of Pulaski County, and a board member for the University District Development Corporation of Little Rock, Arkansas. She serves as a host for Community Development Minute, an educational program for KUAR Radio in which she shares informative health education public service announcements. She also serves as an Arkansas Coalition for Obesity Prevention (ArCOP) grant recipient representative for Growing Healthy Communities for the University District Community of Little Rock, Arkansas. She has serves as a reviewer for various manuscripts for higher education and public health education—related journals and has publications in both respected professions. Her service in higher education has also included serving as a graduate faculty consultant for Walden University and the

University of the People. Her research interests are health disparities, heart disease prevention, online education, teaching and service learning, comprehensive sex education, obesity prevention, and community health.

Dr. Shannon Speed, Ph.D., is a Postdoctoral Fellow in the Translational Addiction Medicine Branch (TAMB) Lab with the National Institutes of Health (NIH) Intramural Research Program. She works primarily in the Clinical Psychoneuroendocrinology and Neuropsychopharmacology Section of TAMB, which is a joint National Institute on Drug Abuse and National Institute on Alcoholism and Alcohol Abuse laboratory. Shannon's research examines the intersection between addictive disorders (alcohol and other substances) and neuroendocrine systems, gut−brain axis, and other peripheral−central pathways, with the goal of identifying new targets and treatments with special emphasis on work in the field of drunkorexia.

Dr. Mai-Ly Nguyen Steers, Ph.D., is an Assistant Professor at Duquesne University in the School of Nursing. As an applied social psychologist, she has developed two parallel research trajectories in the addictions field: (1) examining psychosocial factors, particularly social norms, in relation to drinking, and (2) exploring the influences of social media on health and well-being.

Dr. Tung Sung Tseng is an Associate Professor with tenure and the Charles L. Brown, MD, Endowed Professor of Health Promotion and Disease Prevention in Behavioral and Community Health Sciences (BCHS) at the School of Public Health within the Louisiana State University Health Sciences Center in New Orleans. He is a Certified Health Education Specialist (CHES) and Master Certified Health Education Specialist (MCHES). Dr. Tseng's research has focused on tobacco, cancer, and obesity-related health disparities. As PI or co-Investigator on several university- and NIH-funded grants, he has been developing effective intervention approaches and collaborated with researchers to identify mediation and interaction effect between a genotype and a social or behavioral risk factor. He has more than 80 peer-review publications and 150 professional conference presentations. Dr. Tseng has substantial experience in professional leadership and community service activities. He chaired the Ethics Committee of the Society for Public Health Education (SOPHE) and the Genomic Forum of the American Public Health Association (APHA) and was a coconvener for the International Comparisons of Healthy Aging Interest Group, the Gerontological Society of America (GSA). Dr. Tseng also serves as an editor, associate editor, reviewer, or member of the editor/review board for 30 more professional journals.

Dr. Gayle Walter, Ph.D., MCHES®, is currently an Associate Professor of Instruction in the Department of Health and Human Physiology at the University of Iowa. She earned her Ph.D. in Public Health and her Master of Public Health (MPH) from Walden University with an emphasis on community health promotion and education. Her undergraduate degree is in health services administration. She has been teaching for several years in both graduate and undergraduate programs in public health, health care administration, and health services delivery. Her subject matter areas of expertise include the U.S. healthcare system, the social determinants of health, and cultural competency. She formerly served on the Board of Trustees for the Society for Public Health Education (SOPHE) and was President of the Iowa Society for Public Health Education (IASOPHE). She also currently serves as the Chair of her department's Diversity, Equity, and Inclusion (DEI) Committee.

Dr. Rose Marie Ward, Ph.D., is the Vice Provost for Graduate Education, Dean of the Graduate School, and Professor of Psychology at the University of Cincinnati. Her research examines college student alcohol consumption with respect to social media, alcohol-induced blackouts, and drunkorexia. In addition, she explores alcohol consumption and sexual assault across diverse identities.

Dr. Jody L. Vogelzang, a nationally recognized speaker, researcher, registered dietitian, and health education specialist, is an expert in community health. She defines herself as a research practitioner and has spent the last 3 decades teaching in higher education. Her peers have formally recognized her for excellence in professional practice (Excellence in Public Health Nutrition) and service to the profession (Medallion Award). She holds a bachelor's degree in Dietetics from Michigan State University, a Master of Science in Health Science from Grand Valley State University, a Master of Arts in Biology from Miami University, and a Ph.D. in Health Services specializing in community health from Walden University. She is active in various professional organizations, serves as the Board President for a social service organization in Grand Rapids, Board President for a local school board, Board Secretary for an international foundation in Haiti, and volunteers with other organizations working toward food justice and refugee resettlement.

Foreword

The growth in the use of social media is staggering. According to Statistica, 70% of the US population and 49% of the world population are active social media users. The average, cumulative time spent on social media is about 2 hours a day. Social media presents an underdeveloped opportunity to promote public health. Social media is a powerful tool for virtually all interests in society. The rapid rise of social media has contributed to both good and bad effects on communities. A recent example would be the global COVID-19 pandemic. Public health efforts to promote testing, masks, and vaccines were disseminated with unprecedented speed. Just as quickly social media became a source of disinformation contributing to the polarization of the public. The fog of conflicting information can drive out evidence-based knowledge. Vaccine hesitancy and general distrust of health mandates grew among the public.

A core principle of public health is consumer participation. Participation engages community members in identifying goals and methods to achieve health goals. Social media is the sword that cuts both ways. Social media provide an unprecedented opportunity for marginalized people to bring attention to issues they face. At the same time, social media opens the floodgates to collective outrage without reflection. In the background, algorithms are tailoring the information that shapes public perception. Community participatory models have developed methods of reaching out to engage the community to build ownership and commitment to public health efforts. A goal of public health programs is to build enduring public support for conditions that support personal and public health. Community participatory models can be translated into social media technologies to provide unprecedented reach into the community. Social media is likely to possess unique features contributing to the effectiveness of community participatory models to promote health.

I worry about the experience of social media in creating a passive citizen. Social media use can isolate people from health challenges looming. Clever algorithms can render people a tool of social media. Endless streams of information replace knowledge, clicks on links replace carefully considered decisions, contacts replace interacting with social networks, and posting on social media replaces personal actions. Social media was constructed to serve purposes other than personal and community health. The role of social media in a health-promoting lifestyle should be studied.

I am optimistic that public health professionals can use social media to empower people. Part of this effort will be to ensure consumers have the skills to discern evidence-based health information from anecdotes. We can encourage consumers to use social media to engage in collective actions to promote public health. Social media may be a powerful tool to promote social networking and engagement to enhance personal health. This book is timely. It overviews many ways in which the power of social media can be used to serve the public interest and promote public health.

I commend the authors and editors of this book for bringing to the readers current information on this important and timely subject. This book targets an important void in the existing literature. I sincerely hope public health professionals consider the recommendations of this book. I am confident that social media can contribute to the professional practice of health promotion.

R. L. Petosa, The Ohio State University, Columbus, OH, United States

Preface

We have great pleasure in placing this timely compendium, *Effective Use of Social Media in Public Health* in your hands that can be used as a textbook for undergraduate as well as graduate-level courses in public health. Practitioners in public health and healthcare will also find this book useful as they navigate various intricacies of social media in their disciplines. This book is aimed at filling a void in these fields for such a book. We ardently hope that the readers will like it and provide us with their support.

Social media has gained prominence in all walks of life including public health. The applications continue to amaze us as we progress further with technological advancements. There is growing utilization of social media in public health and healthcare at the global level as it has increased the speed of communication between professionals, and this will gain even more importance in years to come. All these advances will make the information provided in this book relevant and useful to the readers.

This work is a compilation of chapters from various leading public health professionals who have used and continue to use social media in their practice or research. Hence, this book provides a pragmatic perspective on the applications of social media in public health and its subspecialties including biostatistics, epidemiology, health promotion, environmental health, occupational health, health policy, and health management. The readers will find this eclectic approach of varying perspectives from different authors refreshing and enriching.

This is a comprehensive textbook, which discusses diverse topics and subtopics in public health having at its core a focus on social media uses, applications, and management in diverse scenarios. In this book, we have discussed types of social media and their current applications in public health and healthcare, gender and age-specific use of social media, implications of social media related to mental health outcomes among diverse populations, applications of social media in public health research, the relation of social media with public health communication and pedagogy, and future directions. The coverage we think is extensive for the reader and covers almost all aspects related to the applications of social media in public health and healthcare.

This book has utilized some excellent pedagogical features that will aid learning especially for students. Some of these features are as follows:

- **Learning objectives** for each chapter.
- **Questions for discussion** at the end of each chapter.
- **Websites to explore** at the end of each chapter with discussion or activities to complete.
- **Up-to-date examples** from the current literature
- A list of **important terms defined** in the chapter.
- A comprehensive list of **references** for further reading on the issues discussed in the chapter

It is also envisaged that this book will be useful for those preparing for the Certified in Public Health (CPH) exam administered by the National Board of Public Health Examiners (NBPHE). Also, those preparing for the Certified in Health Education Specialist (CHES)® exam and master Certified in Health Education Specialist (MCHES)® exam administered by the National Commission on Health Education Credentialing (NCHEC) will find this material particularly useful to prepare for implementation, evaluation and research, and communication responsibilities.

For the instructors adopting this book, we have prepared a set of PowerPoint slides that can be obtained from the publisher and modified as needed to present material in the course. We welcome diverse and interdisciplinary faculty who teach public health–related courses (related to social media–based applications) at undergraduate- and graduate-level courses across the United States and international educational institutions to adopt this book. We would be delighted to receive feedback for improving this book further from the instructors, students, practitioners, and any other users of this book.

Spring 2023

Dr. Kavita Batra, BDS, MPH, PhD, FRSPH

Dr. Manoj Sharma, MBBS, PhD, MCHES®

Editors

Acknowledgments

Contributing to a book project as editors and authors is more rewarding than we imagined at the time of establishing this project. We especially want to thank the individuals who helped us to achieve our project goals through their support, motivation, and positive energy. First, we would like to acknowledge Dr. Amar Kanekar (Professor, University of Arkansas at Little Rock, Arkansas, USA) for his sincere efforts in shaping our book proposal and recruiting potential contributors.

We greatly appreciate Drs. Rick L. Petosa, Matthew Asare, Jerome Kotecki, and Traci Hayes for reviewing our book proposal and for sharing their keen insight with us.

Next, we would like to acknowledge Ms. Elizabeth Brown (Senior Acquisitions Editor, Elsevier), Mr. Timothy Bennett (Editorial Project Manager, Elsevier), and their dedicated team members to help us on this project. All of them work tirelessly to ensure that the project is continuously monitored and all queries from the contributing authors are addressed in a timely manner. A special thanks to the Elsevier publishers for turning authors' ideas into a meaningful compendium.

We are forever indebted to our family, friends, and students, who always encouraged us to become the better versions of ourselves each day and stood by us whenever we needed them the most. Next, we are grateful to our teachers, who shared the gift of their time and knowledge to mentor us.

We want to praise and thank God, the Almighty, most of all, who has granted us with countless blessing, knowledge, and opportunities to lead this book project. Without God, we wouldn't be able to do any of this.

Section I

Introduction

Chapter 1

Social media and types with their current applications in public health and healthcare

Tung Sung Tseng and Gabrielle Gonzalez
Behavioral and Community Health Sciences, School of Public Health, Louisiana State University Health Sciences, New Orleans, LA, United States

Learning Objectives

- Introduce the role of social media or web-based interventions that have been widely applied in several areas of public health.
- Discuss the types of social media with their current applications as powerful and efficient tools.
- Discuss ethical issues in social media.
- Discuss the role of the *social ecological model* as an ecological perspective on social media use in public health.
- Propose the role of COVID-19 in increasing social media-based public health publications.
- Discuss the challenges and opportunities of using social media in public health research.

Social media

Social media has transformed since its creation to be a platform that is not only used to communicate with friends, family, and other members of society but also to promote consumerism, share thoughts and ideas, as well as share pertinent information related to the public's interest (Kietzmann et al., 2011). Social media is defined as "forms of electronic communication (such as websites for social networking and microblogging) through which users create online communities to share information, ideas, personal messages, and other content (as videos)" (Merriam-Webster, 2022). Additionally, social media can also be viewed as "computer-mediated tools that allow people/peers to create, share, or exchange information, ideas, and pictures/videos in virtual communities and networks" (Mazer et al., 2007). Social media (such as Facebook, LinkedIn, Twitter, Instagram, Line, WeChat, Chat rooms, Blogs, Wikis, YouTube, Flickr, and additional social networking sites) differs from traditional or industrial media in many ways including quality, reach, frequency, usability, immediacy, and permanence (Apuke, 2016). Social media allows users to share, create, and exchange information and ideas throughout multiple communities and networks. Healthcare information is readily available on different social media platforms and can influence the masses.

Three-quarters of US adults use social media, and 50% report that information spread on social media influences their healthcare-related decisions (Giustini et al., 2018). According to the Pew Research Center's social media factsheet of 2021, 72% of the public used some type of social media (Pew Research Center, 2021). Young adults are early adopters of social media and have the highest usage (Pew Research Center, 2021). However, older adults have been steadily increasing their social media presence. In 2021, the percentage of US adults that use at least one type of social media was 69% white, 77% Black, and 80% Hispanic (Pew Research Center, 2021). YouTube (81%) and Facebook (69%) reign as the most used social media sites when sampling the general population (Pew Research Center, 2021). US adults incorporate social media usage into their daily lives with seven-in-ten Facebook users visiting the platform at least once per day (Pew Research Center, 2021).

With the observed high usage of social media in the population, it is imperative that public health and healthcare professionals benefit from this readily available method to improve the health of the population. According to a 2018 report, the percentage of US adults who use at least one social media site increased from 5% in 2005 to 69% in 2018 (Pew Research Center Internet & Technology, 2018). In 2020, Facebook Messenger and Instagram had about 1.2 billion monthly active users (Pew Research Center, 2021). Social media has been prevalent to public health development for over a decade and has aided health professionals and patients in promoting healthier lifestyles and coping skills (Giustini et al., 2018). It is an available method to disseminate information to the public rather than the costly and ineffective method of journal publications (Gatewood et al., 2020). Research published in professional journals is often hard due to fees and individuals that are not educated in proper literature searches. Social media presents a cost-effective way to inform the public about health issues, disease outbreaks, and responses to public health issues (Gatewood et al., 2020).

Role of social media in public health

Social media was first used as a platform to share health information in the early 2000s. Empirical studies on public health's use of social media were limited from 2000 to 2010, but there was an uptake from 2011 to 2018 due to the advancements of the internet (Zhang et al., 2020). Some of the most common applications of social media in public health or health education research concern identifying and recruiting subjects, communicating with subjects, observing social behaviors, reporting subject behaviors/experiences, collecting, sharing, and storing data, and finally, disseminating research results. Public health professionals have utilized these applications of social media to advance the spread of health information. This has led to the promotion of research on cancer, HIV, diabetes, obesity, and alcohol use through social media. Over the years, social media has transformed from only being a reference used to share information to a platform for program participation and a data source (Zhang et al., 2020). Public health interventions have been significantly impacted by the uptake of social media in intervention methodology, measurement tools, and evaluation plans. In interventions, social media has the role of being an interactive tool to target individual risky health behaviors, peer, community, and environmental factors; as an information-distributing or measurement tool, a source of health information, real-time and big data collection, and machine learning. It also changes the evaluation design/approach to test of the effectiveness of using social media platforms.

Currently, social media has a huge reach on the general population as a tool to spread health information. Over 2.4 million users of varying ages on Facebook are accessible to public health-sponsored messages (Parackal et al., 2020; Statista, 2022a). Additionally, the current COVID-19 pandemic introduced Twitter as a key player in relaying public health information to its users. Since 2020, public health-related keywords dominated the platform and made it a top location to receive public health-related information or debate over such information (Fuentes & Peterson, 2021). With the extensive list of Facebook groups and popular Twitter accounts with millions of followers, it is possible for social media to be a tool for public health professionals to target vulnerable groups, such as low-income communities, LGBTQ + organizations, and even college students in university groups (Parackal et al., 2020; Statista, 2022b). Public health professionals and healthcare providers can find value in social media to contact these hard-to-reach groups since social media can allow one-on-one communication which is not usually possible outside of in-person interventions to focus groups (Parackal et al., 2020). There is the possibility to target difficult behaviors such as alcohol overconsumption, overeating, and drug usage, in these hard-to-reach communities that might not have had access to in-person interventions or healthcare professionals (Parackal et al., 2020).

Additionally, social media has allowed public health professionals or healthcare practitioners to communicate with the public in a highly interactive manner that allows health information to be spread quickly and directly to those seeking specific health information. A literature review showed that healthcare where social media has provided a platform for public health and healthcare professionals to interact with the population gathers information from the population itself, and create interventions that best suit the needs of the people (Tengstedt et al., 2018). Individuals that gather health information spread through social media or social networks can increase their awareness of health issues and empower their decisions regarding healthcare (Giustini et al., 2018). Since the beginning of the interaction between social media and public health, researchers have sought to determine the best method to use social media in healthcare and what specific role social media can play in promoting public health information.

A systematic review from 2013 listed seven different ways that social media is used in healthcare: to provide information on a range of issues, provide answers to medical questions, facilitate a dialogue among patients and healthcare professionals, collect data and information on patient experiences and opinions, conduct health interventions for health promotion and health education, reduce disease and stigma, and provide a mechanism for online consultations with patients and their providers (Giustini et al., 2018). Social media can be a major factor in healthcare management and disease control, such as aiding in the prevention and control of infectious or chronic diseases as well as promoting participation in studies on various populations. Social media platforms can also serve as a communication tool for healthcare providers to reach their patients. Public health professionals conducting health interventions can use social media platforms, such as Facebook, Twitter, and Mobil apps, to launch interventions or increase awareness of relevant health information. As technology and internet speed and availability are continually improved, it is easier now than ever to manage public health issues, such as disease surveillance, assessment, and control (Zhang et al., 2020). The recent influx of health information dissemination efforts of public health organizations during the COVID-19 pandemic is an example of how social media can be used to aid in disease control and assessment.

Even in this new age of health information sharing during the COVID-19 pandemic, there is still limited information on just how social media can be integrated overall into public health (Zhang et al., 2020). There is a call for more research on the benefits and downsides of using social media in a healthcare setting. Social media is a broad concept that covers different platforms that are individually designed to reach different audiences in diverse ways. A goal for future public health professionals would be to identify the strengths and weaknesses of the various platforms of social media to target specific populations and aid in the spread of factual health information to those most in need. Fig. 1.1 shows the trend in the prevalence of social media and public health peer-reviewed article publications increased dramatically from 2000 to 2021. The PubMed, CINHAL, and Embase databases were searched for article publications relating to social media and public health. After 2019 there is a clear jump in article publications related to social media and public health. Embase had the largest jump in peer-reviewed articles compared to

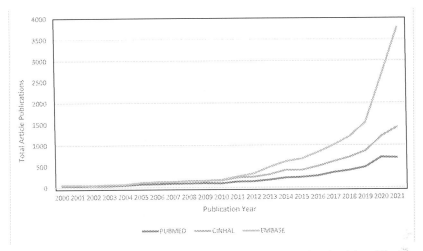

FIGURE 1.1 Prevalence of social media and public health peer-reviewed article publications from 2000 to 2021. *Database searches in PubMed, CINHAL, and Embase. Key terms "public health" and "social media."

CINHAL and PubMed. This jump could be explained by the increase in social media-integrated public health research with the advancement in social media technology and/or public usage of various social media networks, or even with the introduction of the COVID-19 pandemic.

COVID-19 and social media publications

There was an increase in article publications from 2019 to 2021 in all databases searched. One possible explanation is the events of COVID-19 led to an increase in social media-based publications due to lockdowns and an increase in individuals working from home. Social media-based public health research could have transformed their interventions and studies to include social media aspects to increase sample sizes, reach people working from home, and disseminate health information. An additional search was conducted in the previously used databases and included the search terms "public health" "social media" and "COVID-19." PubMed reported 394 articles that included topics such as mental health, vaccine hesitancy, positive and negative impacts of social media, social media influence, and psychological impacts of COVID-19. For Fig. 1.2, CINAHL reported 535 articles with results including challenges to public health during COVID-19, vaccine attitudes, antivaxxers, health measure scales designed for remote measurements, health literacy disparities, telepractice, mental health, and social media role in spreading misinformation and conspiracy theories. Embase had the highest number of COVID-19-related documents with 1335 results. Common topics included vaccine attitudes, social media efficacy, mental health, telehealth, and using social media to market positive messages like "we are in this together."

8 SECTION | I Introduction

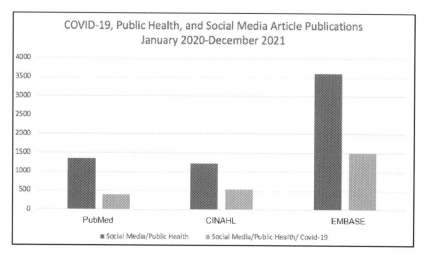

FIGURE 1.2 COVID-19, public health, and social media articles from January 2020 to December 2021.

The results show that COVID-19 did have a possible influence on the frequency of social media-related public health articles using social media as a variable in their studies. In searching for articles related to social media, public health, and COVID-19 in 2019, there were no articles produced in the databases. The jump from 2019 to 2021 in Fig. 1.2 highlights the need for social media-based public health studies due to the reduction of in-person public health research. Public health professionals can learn from the skills of social media managers to create their own public health research studies facilitated through social media. Social media is essential in starting conversations among the community and targeting a specific audience, just as someone would like to sell a product (Tufts University Relations, 2022). Public health professionals need to fill the gap in the literature and create studies that can compel users to engage in healthy behaviors and have open conversations with leaders in healthcare. It is important now more than ever, especially in the age of misinformation the COVID-19 pandemic has led us to.

Social media and the dangers of misinformation

With the introduction of the 21st century's first global pandemic, we have ushered in an age of misinformation through avenues like social media. Platforms such as Facebook, Instagram, and TikTok were first used to bolster the community and provide a source of information during isolation. However, as the pandemic progressed, the influx of misinformation began to flow across several social media platforms. At this point in 2022, we can reflect on the 950,000 total deaths, months of high hospitalization rates, and millions of cases (Vera et al., 2022). The pandemic began with public support and acceptance of healthcare, public health, and healthcare providers. There was also a push for a vaccine to help life get back to normal, but even after the development of the vaccine, public acceptance began to decline with people citing incorrect sources found on social media (Vera et al., 2022).

COVID-19 is unprecedented for current society, and there was plenty of confusion about whether or not to wear a mask, when to stay home, and how long to isolate if sick. Social media created a dangerous outlet for misinformation to spread and led to trends such as taking ivermectin or bleach as a treatment for COVID-19 (Vera et al., 2022). Many key players voiced their concerns on social media and influenced the public to take dangerous means to protect themselves from COVID-19 or to lower their perceived risk and susceptibility to contracting the virus. One major takeaway from this pandemic is how powerful a tool social media can be to spread misinformation, change social norms, and influence the population's behavior, and public health professionals must understand that social media can also be a tool to dispel incorrect information regarding this virus and future pandemic-like responses.

Public health professionals and healthcare workers alike must utilize social media in the future to spread correct information and spread correct health-related information across multiple platforms. Targeting ads or informational posts toward different age groups according to social media platforms can be an effective way to improve attitudes and behaviors regarding public health information. Many social media platforms operate on an algorithm, which means when someone interacts with a certain type of content, they are more likely to see similar content in the future (Vera et al., 2022). Public health professionals are faced with the difficult task of amplifying the correct information and phrasing messages to reach the public, but these are hard barriers to overcome (Vera et al., 2022). The population must have a trusted source to find their correct health information. There is such an influx of information regarding COVID-19 that it can be overwhelming for the average person. Public health professionals now have the job of undoing misinformation caused by political conflict and relaying health information to the public (Vera et al., 2022).

Types of social media

Social media or web-based interventions are powerful and efficient tools that can be used in tandem with health promotion and nutrition interventions to increase awareness of the health impact. These interventions have been applied in several areas, such as self-management of long-term conditions (e.g., diabetes, heart disease, arthritis, and asthma), health promotion (e.g., smoking cessation, alcohol reduction, sexual health, diet, and exercise), and mental health (e.g., depression and anxiety) (Barak et al., 2009; Murray, 2012). Several types of social media can include wikis, social networking sites, media-sharing sites, blogs and microblogs, immersive worlds, and 3-D virtual globes (Giustini et al., 2018). Social media sites will vary in their features depending on who is the desired audience (Lee Ventola, 2014). Social networking sites include Facebook, Myspace, Google Plus, and Twitter. In 2012, Facebook had over one billion users worldwide, and Twitter reported 65 million tweets with 100 million active users. Some findings about social media use were reported below (Lee Ventola, 2014).

- More professional networking is done on sites such as LinkedIn.
- YouTube and Flickr are primary sources for different forms of media sharing.
- Wikipedia is a common site for quick facts, knowledge, and information gathering.

These various platforms are the tools that healthcare professionals can use to engage with the public, promote healthy behaviors, and interact with their patients and colleagues. A survey done on 4000 physicians found that over 90% of physicians use social media in some form for personal reasons and only 65% of the surveyed physicians use it for professional matters (Lee Ventola, 2014). In this study, physician use of social media is increasing. Most are attracted to participating in online communities, listening to experts, and communicating with colleagues regarding patient issues. Interestingly, health organizations are also increasing their use of social media to communicate with the community, engage patients, market products, fund-raise, and provide customer support. In general, roughly 70% of US healthcare organizations use social media, especially Facebook, Twitter, YouTube, and blogs (Lee Ventola, 2014).

There are four major types of social media platforms, including social networking, discussion forums, video hosting, and blogging networks (Indeed Editorial Team, 2021). Table 1.1 shows the variations in what the different types of social media platforms have to offer for users.

TABLE 1.1 Variations in the top social media-based platforms.

Platform Features	Social networks	Discussion forums	Video hosting	Blogging
Communication/building connections	X	X	X	X
Content creation	X		X	X
Live stream/video curating	X		X	
Community outreach	X		X	X
Internet-based	X	X	X	X
App-based	X	X	X	X

- **Social networking** sites, such as Facebook, Twitter, Instagram, and LinkedIn, offer a form of communication among users that allows for idea sharing. There are many groups formed based on similar interests and hobbies. Individuals that use social networking sites can also create content with photo sharing, video streaming, and group discussions. Healthcare organizations can promote their message through social networking sites as businesses promote their brand and spread awareness or a positive presence online.
- **Discussions forums**, such as Reddit, Digg, Quora, and Clubhouse, are created to share ideas and news among users. People also post questions to receive feedback from fellow users. When questions are answered correctly and verified, that business or individual can show an increase in credibility. Public health and healthcare professionals can benefit from this form of communication by answering questions related to healthcare that users will identify as a trusted source.
- **Video hosting platforms**, such as YouTube, TikTok, Snapchat, Vimeo, and Instagram, can be used by content creators, filmmakers, journalists, and everyday people. This type of platform allows for video streaming and influencing people to make choices about their consumer habits. Many content creators spread information regarding health-related topics, such as nutrition, dieting, and reproductive health. TikTok is a popular platform among youths and is an opportunity for healthcare professionals to directly influence the younger generations.
- **Blogging networks**, such as Medium, WordPress, Facebook, and Tumblr, are social media networks to express ideas and thoughts related to professions, current events, and hobbies. Blogging allows for creativity, and public health professionals have the freedom to create curated content with accurate health information to disseminate to the public and quell any misinformation circulating on the internet.

Social media and public health interventions

In recent years, technology has been used as a platform to not only disseminate health information but also bring public health intervention to people in need (Félix et al., 2022; Graafland, 2018; Radovic & Badawy, 2020). Using technology to deliver interventions has also become a cost-effective public health method, and populations like adolescents are top targets for public health-related interventions (Félix et al., 2022). Adolescents have become so accustomed to using technology and social media with about 95% of all adolescents having access to a smartphone and access to social media sites like YouTube, Instagram, and Snapchat (Anderson & Jiang, 2018; Radovic & Badawy, 2020). Public health professionals can take advantage of this high period of engagement with social media and technology to deliver health interventions related to prominent public health topics, such as obesity, substance use, drinking and driving, and sexually risky behaviors (Radovic & Badawy, 2020). There is research already conducted that supports using technology to deliver interventions. For example, YouTube has shown to decrease feelings of isolation, improve support, teach coping strategies, and provide an additional resource than only visits to pediatricians or family doctors (Forgeron et al., 2019; Naslund et al., 2014; Radovic & Badawy, 2020). Additional public health interventions can utilize social media to target adolescents as well as other populations to help reduce poor health outcomes and increase public health-related knowledge.

Besides adolescents, older adults can also be targeted for health interventions. There has been an observed increase in the popularity of adults over 65 using technology, which is hopeful for public health professionals to deliver health promotion interventions. Social media is an increasingly popular source for delivering health interventions, but it has also been proven to increase user engagement, especially in programs geared toward weight loss (An et al., 2017).

Other health behaviors similar to weight loss, such as physical activity, diet quality, breastfeeding, calorie intake, health screening, tobacco use, and medication adherence, have been examined in interactive social media for public health (Petkovic et al., 2021). There are findings that suggest interactive social media is more beneficial compared to noninteractive social media (Petkovic et al., 2021). Interactive social media is defined as the activities, practices, or behaviors among communities of people who come together online to interactively share information, opinions, and knowledge. Interventions based on interactive social media have the potential to adjust health behaviors and improve well-being, such as mental health (Petkovic et al., 2021). Among adults, interactive social media interventions had a slight improvement in health behaviors, including increasing step counts but minimal to none in diet quality (Petkovic et al., 2021). So far, interactive social media is best used for small improvements, but there is a possibility for public health professionals to improve the use of social media and interactive social media to improve health outcomes. One possible method to enhance interactive social media interventions is to refer to predictable behavioral theories, like the *social cognitive theory*.

Behavioral theories like the *social cognitive theory* have helped form the framework behind social media-based interventions (An et al., 2017). The *social cognitive theory* relies on interpersonal relationships, social support, and increasing self-efficacy to instill successful behavior change in various populations (Glanz et al., 2015). Social media platforms can help reinforce positive health behaviors through increasing self-efficacy, and public health interventionists can provide guidance on weight management, physical activity, and dietary behaviors with virtual examples to help improve participants' ability to perform these behaviors successfully (An et al., 2017). With the novelty of social media in public health, we have the opportunity to adapt several behavioral health theories to achieve successful behavior change. One theory with potential to be adapted for social media-based interventions is the *social ecological model*.

Social media and social ecological model

The *social ecological model* can provide an ecological perspective to social media use and public health. Table 1.2 describes the application of social media at each level of the *social ecological model*. This model can provide an outline for public health interventionists to possibly target various levels on social media platforms. At the individual level, individual knowledge, beliefs, and attitudes can be targeted with social media interventions. This can include targeting changes in health behaviors, such as smoking status, eating habits, and physical activity levels. Social media platforms like TikTok and YouTube are examples of opportunities to create curated videos to deliver health information to the public. The interpersonal level focuses on relationships with family and peers. Social media is highly based on building connections with friends, family, and new people. Public health interventions geared toward improving social connections, isolation, and mental health outcomes such as depression and loneliness can benefit from using social media as the platform to connect participants. Platforms such as Facebook are ideal for hosting group discussions and facilitating conversations regarding topics of health with professionals and fellow group members.

A plethora of organizations uses social media to spread information and promote positivity regarding their brand, image, and products. Healthcare organizations can provide access to educational programs and support systems between healthcare providers and patients through social media, either by

TABLE 1.2 Social media applications and social ecological model.

Social ecological model	Social media application
Policy level	Lobby, disseminating information on new policies
Community level	Community awareness, capacity building, and bridge gaps between resources and community members
Organization level	Deliver educational programs, services, and support systems about available or communication services
Interpersonal level	Build social connections, support, and network with friends, family, and new people
Individual level	Individual knowledge, beliefs, attitudes, and behaviors change

spreading information about available services or performing online communication services. At the community level, government organizations, healthcare facilities, and individual practitioners can promote healthy habits through platforms such as Instagram, Facebook, and Twitter. These social media networks can also increase community awareness, and capacity building and bridge gaps between resources and community members through discussion boards, live streams, and Facebook groups. Finally, the policy level is vital in lobbying and providing more funding/resources to the government, community, and organizations using social media platforms. Lobbying and policy change may be influential by providing additional funding to health professionals can be an incentive for health interventions to incorporate social media into their programs. Platforms such as TikTok are useful in disseminating information on new policies related to healthcare since political figures and health professionals can create accounts geared toward increasing the knowledge and awareness of the public.

Social media and youth

Children and adolescents are popular users of various social media platforms, and it has led to them being one of the first generations to grow up entirely with intense usage of social media (Oxtoby, 2022). They have a plethora of information at their fingertips unlike their parents and earlier generations. A recent study in the United Kingdom found that upwards of 60% of children (3—17 years) are connecting through at least one social media platform or app, which is a significant concern for parents being able to control what is accessible through social media (Oxtoby, 2022). However, social media is only growing and becoming more of a relevant part of our culture, especially generation Z culture. So, not only do parents need to adapt and understand the unique language spoken by our most technologically savvy generation but also public health professionals hoping to improve health outcomes for these children and adolescents.

The first steps should include finding the right words to engage with youth as to what they want out of social media and why they use it. These conversations can lead to positive social media-related experiences for youth as well as give insight to parents and public health professionals on which sites, content, and influencers this generation is listening to (Oxtoby, 2022). Public health professionals can use this information to design interventions with a higher chance of success in reaching the young population. For instance, YouTube, Instagram, and TikTok are some of the most popular social media platforms used by generation Z (Oxtoby, 2022). Public health professionals can begin to understand how messages successfully reach users on these platforms and develop interventions to educate children and adolescents on various public health-related issues. Young people enjoy using social media, and it has become a major part of their lives. However, they also need tools to safely navigate social media since social media can be used to spread information seeking to harm younger individuals. That leads us to discuss the ethical issues related to using social media and the possible harms that come with it.

Ethical issues in social media

As the technology of social media continues to evolve, ethical principles can be used as a guide when navigating the uncertain waters of social media. Ethics in healthcare entails that there is a moral duty of acting health and public health professionals to uphold a certain moral standard when acting on disease prevention, life elongation, and psychological/physical well-being. When social media is used as a source of health information, data sharing, and communication with patients, public health and healthcare professionals need to understand that ethics must be taken into consideration. Populations that must be focused on when maintaining ethical conduct include children and adolescents and the elderly (Denecke et al., 2015).

Most children and adolescents grew up as generations with the most engagement with social media. They have a high social media presence and use various platforms to express ideas, communicate, and search for answers with feedback from professionals. Since Facebook and Twitter are two of the most used platforms for youths, ethical issues that can be presented by engaging youth through social media include informed consent, as well as confidentiality and privacy (Denecke et al., 2015). Social media sites can be used to gain online consent, but it is more difficult to obtain consent from those under the age of 18. Privacy can become a concern for youth interacting through social media in health interventions because some might not want health information to be accidentally shared online or even mention their involvement in health interventions.

The elderly is also targeted online for social media health interventions. Most are geared toward cognitive training and ways to improve their memory or mental skills. Similarly, to children and adolescents, the elderly must also have a full understanding of what their consent means when participating in interventions delivered through social media. Other concerns include privacy, consent, autonomy and choice, justice, inclusion, security, and dignity when interacting with health interventions and health professionals (Denecke et al., 2015). Each of these issues must be addressed by public health and healthcare professionals wishing to reach certain populations' healthcare needs.

Just as privacy is a concern with the youth and elderly, patient and physician communication through social media must consider the ethical guidelines of communicating on a social media platform. In typical patient-physician meetings, health information is shared, including the history of medication, illness, family history, and personal contact information. There is a complex process to ensure the confidentiality and privacy of said information in a traditional healthcare setting. Patients seeking treatment or communication through social media must be made aware of the risk of data breaches that can compromise confidential information. Healthcare professionals must take steps to ensure that patients' data are protected and they are aware of the risks of using social media to communicate private information (Denecke et al., 2015). All of these issues should be noted when health professionals and healthcare practitioners develop a research project or services using social media.

Challenges and opportunities of social media

Given that social media platforms will continue to change rapidly recently, it is important to further the understanding of how social media can affect those individuals actively utilizing various platforms. Social media has been observed to increase risky behaviors and diminish the sense of well-being in certain scenarios with active users as follows (Giustini et al., 2018).

- Social networking platforms can easily influence negative emotions and promote problematic behaviors.
- Challenge in using social media to promote health information is dispelling false information from spreading to the masses.
- Public health and healthcare professionals need to dispel false medical advice as well as false claims and misinformation.
- Teenagers especially are at an increased risk for the harmful effects of social media, such as isolation, depression, and cyberbullying.
- This can lead to addictive behaviors and decreased self-efficacy.

There is no overwhelming consensus on whether health-related content being communicated on social media sites can promote or jeopardize health, and the COVID-19 pandemic introduced a unique dilemma in spreading correct medical facts. The COVID-19 pandemic increased the amount of healthcare information spread over social media and has even been deemed an "infodemic" (Schillinger et al., 2020). This sparked a debate on whether public health efforts can be undermined or facilitated by social media, especially since misinformation can easily "go viral" and influence the attitudes, beliefs, norms, and behaviors of the public. The spread of misinformation can have deleterious effects, such as popularizing ineffective, unsafe, costly, or inappropriate protective measures (Schillinger et al., 2020). Since the start of the pandemic, public health professionals and healthcare professionals have needed to correct misinformation and provide true medical information that is spread throughout the US population. Even with the increase in challenges related to using social media to spread public health information, there are some opportunities as well to use social media to aid in the population's health.

Even though misinformation can spread easily through social media, it is just as feasible to correct misinformation by using social media as well. Public health professionals and organizations can prevent the spread of misinformation and increase public awareness on how to differentiate between correct and false sources of information. For example, organizations like the WHO (World Health Organization, 2020) are working closely with social media and technology companies to curb the spread of misinformation by ensuring that messages from the WHO or other official sources are appearing when people search for information related to COVID-19. Social media presents other opportunities, such as real-time surveillance of diseases, monitoring contamination exposures, tracking behavioral changes, and detecting or predicting the outbreak of disease (Schillinger et al., 2020). It is easier to contact hard-to-reach groups with social media messages as well as reach the public with quick information regarding the outbreak of disease. Additionally, populations at risk for the negative side of social media can benefit as well, such as adolescents using social media to increase their self-esteem and social capital and safe identify experimentation (Giustini et al., 2018).

One of the main benefits from social media is the efficient way to instantly spread information to billions of people with the press of a button. Not only can individuals participating on social media stay in contact with their friends, families, and leaders but also they can engage directly with willing professionals, such as public health leaders or healthcare professionals (University of South Florida, 2022). If someone has an interest in becoming healthier, such as with increased physical activity, there is a plethora of resources available to help engage this individual with the right resources. Also, once individuals engage in certain content, they will be more likely to come across similar ads or posts later on. Social media has been successfully used to connect industry leaders with potential employees, so it is only reasonable that public health professionals can also reach potential users in need of health-related information.

The literature on social media's influence has mixed results with negatives and positives attached to social media usage (Abroms, 2019). There is concern about the negative effects on an individual's well-being; however, there is a lot of optimism attached to social media when conducting public health surveillance and disseminating public health information (Abroms, 2019). Social media can give healthcare professionals a platform to interact with hard-to-reach populations as well as dispel misinformation. With the development of the internet and social media, the ability to simply access different information from various social media approaches has improved. Social media has been proven to be a feasible and effective platform for public health interventions, but there needs to be a strategic plan among public health professionals and public health agencies to understand the best practice for reaching the public via social media platforms (Thackeray et al., 2012). There is still more to be explored on the effectiveness of using social media for the good of the public, and there is a call for more involvement with social media approaches/tools with better ethical standards to ensure correct health information is spread and individuals interacting with health professionals are protected. There is endless potential in using social media as a means to disseminate health information and public health interventions. Future studies should focus on comparing the best platforms for various public health studies, the influence of social media on mental health, especially in children and adolescents, the role of social media in misinformation, and how public health professionals can make social media a safer place from the spread of dangerous information to people in need (Cabucana, 2022).

Chapter summary

Social media and public health will be intertwined in the future of health interventions. The several social media platforms are designed to engage users and encourage the spread of ideas, knowledge, and information. Public health interventions have the capability to reach previously hard-to-reach populations with technology supported by interactive social media. Adverse health outcomes like obesity, tobacco use, alcohol overconsumption, and poor mental health can be mitigated with social media-based interventions that allow public health professionals to spread accurate information to populations in need to improve public health and health-related behaviors. There are pros and cons to using social media, especially for younger generations, but the prevalence of social media around the world is so vast that public health professionals have the task to discover the best practices to use social media as an intervention platform. Due to issues like the rapid spread of misinformation in an era of mistrust in the medical field, it is important now more than ever to engage with the public with correct information as well as sources to use in the future to engage with the right social media contacts. With the rise of social media usage, public health interventions now have the opportunity to improve population health on a broad scale and make the internet a safe space to access accurate health information.

Questions for discussion

1. Expand on the negative and positive effects of using social media to interact with the general population.
2. There is no universal guideline in protecting patient information through social media interactions. What guidelines are necessary to be included in future protocols to protect patients from data leaks, discrimination, harassment, and other issues related to social media use in healthcare?
3. What are the strengths and weaknesses of social media in healthcare and public health interventions?
4. How has COVID-19 impacted public health's ability to disseminate health information?

Websites to explore

Facebook: A social networking site used globally to connect and share information
   https://www.facebook.com/
   Access Facebook through your personal account, if you have one. Once you log in, create your profile, and explore the social networking platform, search for public health-related Facebook groups. See if you can find groups, such as support groups, for different health ailments. Additionally, see if you can find groups with public health leaders as organizers. What did you find? Are there any public health-related groups? How do you think public health professionals can be a part of these groups?

   TikTok: A popular website and app for all generations to use, especially for youth
   TikTok is based on an algorithm to determine your video preferences. If you do not already have an account, sign up and begin building your "For You" page. The app will prompt you to pick subjects you like. Do this but also choose any health-related topics you see and then start scrolling through your For You feed. What do you see? Do you see any videos related to health or public health? This can *include certified dieticians, medical doctors, and nurses. How well do you think they convey their health messages in their videos? How do you think public health interventions or studies can utilize TikTok videos to gain program participants?*

Key terms—definitions

COVID-19—coronavirus of 2020 that introduced the 21st century's biggest global pandemic and resulted in the deaths of millions of people.
   Ethics—code of moral principles that govern individual actions.
   Health education—any combination of learning experiences designed to facilitate voluntary adaptations of behavior conducive to health (Green & Iverson, 1982).
   Intervention—approaches that aim to change certain behaviors or health outcomes.
   Misinformation—information spread with ill intent that incorrectly reports information to be factual without justification.
   Public health—the field of studying population health in communities to promote healthy lifestyles, behaviors, and attitudes.
   Social media—platforms to communicate, share, discover, and learn new information with people in countless friends, family, communities, and social groups.
   Social networking—the process of communicating across social media platforms to share interests, ideas, and activities with those of similar interests.

References

Abroms, L. C. (2019). Public health in the era of social media. *American Journal of Public Health, 109*, S130−S131. https://doi.org/10.2105/AJPH.2018.304947

Anderson, M., & Jiang, J. (2018). *Teens, social media & technology 2018*. Pew Research Center. https://www.pewresearch.org/internet/2018/05/31/teens-social-media-technology-2018/.

An, R., Ji, M., & Zhang, S. (2017). Effectiveness of social media-based interventions on weight-related behaviors and body weight status: Review and meta-analysis. *American Journal of Health Behavior, 41*(6), 670−682. https://doi.org/10.5993/AJHB.41.6.1

Apuke, O. D. (2016). Social and traditional mainstream media of communication: Synergy and variance perspective. *Online Journal of Communication and Media Technologies, 7*(4). https://doi.org/10.29333/ojcmt/2614

Barak, A., Klein, B., & Proudfoot, J. G. (2009). Defining internet-supported therapeutic interventions. *Annals of Behavioral Medicine, 38*(1), 4−17. https://doi.org/10.1007/s12160-009-9130-7

Cabucana, C. (2022). *The top 10 most interesting social media research topics*. Career Karma. https://careerkarma.com/blog/social-media-research-topics/#.

Denecke, K., Bamidis, P., Bond, C., Gabarron, E., Househ, M., Lau, A. Y. S., Mayer, M. A., Merolli, M., & Hansen, M. (2015). Ethical issues of social media usage in healthcare. *Yearbook of Medical Informatics, 10*(1), 137−147. https://doi.org/10.15265/IY-2015-001

Félix, S., Ramalho, S., Ribeiro, E., Pinheiro, J., de Lourdes, M., Gonçalves, S., & Conceição, E. (2022). Experiences of parent−adolescent dyads regarding a facebook-based intervention to improve overweight/obesity treatment in adolescents: A qualitative study. *Applied Psychology: Health and Well-Being, 14*(1), 122−139. https://doi.org/10.1111/aphw.12294

Forgeron, P. A., McKenzie, E., O'Reilly, J., Rudnicki, E., & Caes, L. (2019). Support for my video is support for me. *Clinical Journal of Pain, 35*(5), 443−450. https://doi.org/10.1097/AJP.0000000000000693

Fuentes, A., & Peterson, J. V. (2021). Social media and public perception as core aspect of public health: The cautionary case of @realdonaldtrump and COVID-19. *PLoS ONE, 16*(5 May), 1−10. https://doi.org/10.1371/journal.pone.0251179

Gatewood, J., Monks, S. L., Singletary, C. R., Vidrascu, E., & Moore, J. B. (2020). Social media in public health: Strategies to distill, package, and disseminate public health research. *Journal of Public Health Management and Practice, 26*(5), 489−492. https://doi.org/10.1097/PHH.0000000000001096

Giustini, D. M., Ali, S. M., Fraser, M., & Boulos, M. N. K. (2018). Effective uses of social media in public health and medicine: A systematic review of systematic reviews. *Online Journal of Public Health Informatics, 10*(2), e215. https://doi.org/10.5210/ojphi.v10i2.8270

Glanz, K., Rimer, B., & Viswanath, K. (2015). In *Health behavior: Theory, research, and practice* (5th ed.). John Wiley and Sons https://books.google.com/books/about/Health_Behavior.html?id=PhUWCgAAQBAJ.

Graafland, J. H. (2018). New technologies and 21st-century children: Recent trends and outcomes. *OECD Education Working Papers, 179*. https://doi.org/10.1787/E071A505-EN

Green, L., & Iverson, D. (1982). School Health Education. *Annual Review Public Health, 3*, 321−338. https://www.annualreviews.org/doi/pdf/10.1146/annurev.pu.03.050182.001541.

Indeed Editorial Team. (2021). *10 types of social media to promote your brand*. https://www.indeed.com/career-advice/career-development/types-of-social-media.

Kietzmann, J. H., Hermkens, K., McCarthy, I. P., & Silvestre, B. S. (2011). Social media? Get serious! Understanding the functional building blocks of social media. *Business Horizons, 54*(3), 241−251. https://doi.org/10.1016/j.bushor.2011.01.005

Lee Ventola, C. (2014). Social media and health care professionals: Benefits, risks, and best practices. *P and T, 39*(7), 491–500.

Mazer, J. P., Murphy, R. E., & Simonds, C. J. (2007). I'll see you on "Facebook": The effects of computer-mediated teacher self-disclosure on student motivation, affective learning, and classroom climate. *Communication Education, 56*(1). https://doi.org/10.1080/03634520601009710

Merriam-Webster. (2022). *Merriam-Webster: Since 1828.* https://www.merriam-webster.com/dictionary/social%20media.

Murray, E. (2012). Web-based interventions for behavior change and self-management: Potential, pitfalls, and progress. *Medicine 2.0, 1*(2), e3. https://doi.org/10.2196/med20.1741

Naslund, J. A., Grande, S. W., Aschbrenner, K. A., & Elwyn, G. (2014). Naturally occurring peer support through social media: The experiences of individuals with severe mental illness using YouTube. *PLOS ONE, 9*(10), e110171. https://doi.org/10.1371/JOURNAL.PONE.0110171

Oxtoby, K. (2022). Social media: A force for good? *Community Practitioner, 95*(4), 28–32.

Parackal, M., Parackal, S., Mather, D., & Eusebius, S. (2020). Dynamic transactional model: A framework for communicating public health messages via social media. *Perspectives in Public Health, 141*(5), 279–286. https://doi.org/10.1177/1757913920935910

Petkovic, J., Duench, S., Trawin, J., Dewidar, O., Pardo Pardo, J., Simeon, R., DesMeules, M., Gagnon, D., Hatcher Roberts, J., Hossain, A., Pottie, K., Rader, T., Tugwell, P., Yoganathan, M., Presseau, J., & Welch, V. (2021). Behavioural interventions delivered through interactive social media for health behaviour change, health outcomes, and health equity in the adult population. *Cochrane Database of Systematic Reviews, 2021*(5). https://doi.org/10.1002/14651858.CD012932.PUB2

Pew Research Center. (2021). *Social media fact sheet.* https://www.pewresearch.org/internet/fact-sheet/social-media/.

Pew Research Center Internet and Technology. (2018). *Social media fact sheet.* http://www.pewinternet.org/fact-sheet/social-media/.

Radovic, A., & Badawy, S. M. (2020). Technology use for adolescent health and wellness. *Pediatrics, 145*(Suppl_2), S186–S194. https://doi.org/10.1542/PEDS.2019-2056G

Schillinger, D., Chittamuru, D., & Susana Ramírez, A. (2020). From "infodemics" to health promotion: A novel framework for the role of social media in public health. *American Journal of Public Health, 110*(9), 1393–1396. https://doi.org/10.2105/AJPH.2020.305746

Statista. (2022a). *Facebook MAU worldwide 2022 | Statista.* https://www.statista.com/statistics/264810/number-of-monthly-active-facebook-users-worldwide/.

Statista. (2022b). *Global Facebook user age and gender distribution 2022 | Statista.* https://www.statista.com/statistics/376128/facebook-global-user-age-distribution/.

Tengstedt, M.Å., Fagerstrøm, A., & Mobekk, H. (2018). Health interventions and validity on social media: A literature review. In *Procedia computer science* (Vol. 138, pp. 169–176). https://doi.org/10.1016/j.procs.2018.10.024

Thackeray, R., Neiger, B. L., Smith, A. K., & Van Wagenen, S. B. (2012). Adoption and use of social media among public health departments. *BMC Public Health, 12*(1), 1–6. https://doi.org/10.1186/1471-2458-12-242/TABLES/5

Tufts University Relations. (2022). *Social media overview—communications.* https://communications.tufts.edu/marketing-and-branding/social-media-overview/.

University of South Florida. (2022). *Introduction to social media.* University Communications and Marketing. https://www.usf.edu/ucm/marketing/intro-social-media.aspx.

Vera, M. A., El-Khoury, J. M., Thorp, H., Tofel, R. J., Ross, J. S., Mandavilli, A., & Topol, E. (2022). Public misinformation and science communication in times of public health crises. *Clinical Chemistry, 1014.* https://doi.org/10.1093/clinchem/hvac088

World Health Organization. (2020). *Immunizing the public against misinformation.* https://www.who.int/news-room/feature-stories/detail/immunizing-the-public-against-misinformation.

Zhang, Y., Cao, B., Wang, Y., Peng, T. Q., & Wang, X. (2020). When public health research meets social media: Knowledge mapping from 2000 to 2018. *Journal of Medical Internet Research, 22*(8), 1–15. https://doi.org/10.2196/17582

Chapter 2

Gender and age-specific use of social media

Jody L. Vogelzang
Grand Valley State University, Allendale, MI, United States

Learning Objectives

1. To describe how age impacts social media access and use.
2. To discuss the use of stereotypes in social media research.
3. To evaluate the differences in social media use based on gender.

Introduction

The Internet became a mainstay for work and social interaction with the launch of Web 2.0 in 2004. Today, it is hard to imagine working, taking classes, or staying in touch with family and friends without using the web (Merolli et al., 2013). This chapter will describe how the differences in age and gender impact the consumption of web-based social media.

Social media growth and access

The first printing press by Johannes Gutenberg in 1450 was celebrated as the "world-changing milestone" because it resulted in the revolutionary and unprecedented distribution of ideas and information widely to individuals with no access to reading the information themselves. That same transformative impact can be attributed to the Internet and the proliferation of social media riding the Internet today. Indeed, social media is ubiquitous today—providing a platform for global communication to share ideas and build and support relationships. There is no need to wait for information printed on paper; in fact, most Gen Z's have no memories of a newspaper arriving on the stoop in the early morning hours.

The miracle of creating and distributing information electronically makes it accessible almost instantaneously. The "live" or near "same-time" nature of electronic communication has expanded the dimensions of human relationships and changed the geographic limitations of the past. From Facebook to Instagram, LinkedIn, interest, Snapchat and Twitter, the web host sites that appeal to gender, age, and interests. "For reference, in 1995, only 14% of the U.S. adult population had access to the Internet and by the year 2000, only 53% owned a mobile phone and 46% had access to the Internet, compared to 92% and 87%, respectively, in 2015" (George et,al., 2018, p. 79). But is access to social media available to all groups of people? Is the electronic information revolution impacting everyone to the same extent or in a similar manner? Do all genders experience online information sharing in a similar manner? Are older adults web adverse? As this chapter explains, electronic media usage does appear to decrease among the above platforms with age and gender appears to moderate online information sharing (Hruska et al., 2020). However, establishing hard and fast assumptions dependent on the fluidity of gender and wide aging experiences is cautioned. This chapter unpacks research but also cautions on stereotyping based on perceived gender and age.

Age-specific use of social media

Children

Parental influence

Many parents struggle with this question: When is my child ready to consume social media safely and independently? The answer to this question may depend on child's age and maturity and the parents' oversightents. While children may gain access to electronic devices at the earlier ages, with almost 100% of those age eight and above having access to smartphones or tablets, site accessibility and dosage may still fall under the partial control of parents or guardians (Wartella et al., 2018).

Just because almost all children have access to electronic media, does it follow that they will all be influenced to the same degree? As indicated by social learning theory, children's use of social media is indeed influenced by their parents' Internet behavior (Bandura, 1971). The adult's consumption of social media constructs an environment which impacts the child (Bandura, 1993). To fully understand a child's use of social media, one must also examine the home environment and how a parent provides guidance and support. For example, a family is out for dinner, yet instead of table conversation, all are isolated and engrossed in their portable devices, including the youngest member of the family entertained by YouTube videos. In another meal scenario, a child is talking to a parent who is actively involved in texting throughout the entire dinner. This behavior can become normative and lead children to seek affirmation and interaction through child-oriented social media.

Recently, a child version of the adult Pelatron exercise bike has been made by Little Tykes. The Little Tykes Pelican was specially made with a stand for a tablet or other portable device so children could watch social media while pedaling—just like their parents. According to Bandura (1993) "People are partly a product of their environment. Therefore, beliefs of personal efficacy can shape the course lives take by influencing choice of activities and environments" (P. 135).

But learning to use electronic media is not limited to parental behavior or home environment. Children use portable devices as early as preschool, where learning modules are assigned to augment classroom learning. Walking into a United States classroom today, it is not unusual to find the entire class wearing headsets and playing math or literacy games. These applications can keep children focused, allow students to progress at their own speed, and track their progress (Georgiev et al., 2020). Dore et al. in 2019 noted in their research of four-year-olds "children can learn new vocabulary from a single bout of playing a mobile game. Further, the stringent nature of our learning measures suggests that children were able to generalize beyond the game context" (P. 8). However, children's inclusion and perhaps dependency on social media may concern parents, who believe such media access may take away from reading paper books, while some may fear that children may go down the slippery slope of screen fixation (Connell et al., 2015). This is not an unrealistic concern. Screen time does not appear to level off as children age. According to Rideout et al. (2022), US children as young as four years of age spend an average of an hour per day on an interactive screen device. Screen time of emerging adults is approximately six hours a day without the time used for homework (Vannucci et al., 2019).

Applications for the elementary-aged children: The use of electronic media by older children is also driven by the proliferation of wide-ranging entertainment programming applications or "apps" that are available. These media applications are built to engage, amuse, and relate to children. Children can develop virtual relationships with the adventures of their favorite characters and explore special interests. Media applications related to elementary age children are widely available, and many are free. Children can augment their reading and language skills, practice math problems, do art, and learn about history using their tablets or other handheld devices (Radesky et al., 2016). The perceived educational nature of these applications generates additional attraction, if they can truly impact learning and provide educational value to young learners. But is this perception correct? Several recent studies have shown that "most apps in the academic category for Android and iOS operating system devices (Apple App Store and Google Play) have no educational value and are based on rote learning and memorization" (Papadakis et al., 2022). YouTube, a highly used app by children, has low-educational quality and is rife with advertising (Radesky et al., 2022). Furthermore, most self-proclaimed educational apps "lack clear evidence of efficacy and are not scientifically established, having received no feedback from developmental specialists during their development" (Papadakis et al., 2022).

Nevertheless, educational apps' unfounded, perceived educational value stokes electronic media use and screen time for young children. Pressure on children to use electronic media is not limited to "educational" applications. Other applications of this media are available to children that are purely entertaining. These applications can be even more attractive to young children. Moreover, these free applications often contain advertising inserted so smoothly that it appears to be a part of the game or video. Thus, the influence of this media is insidious, manipulating children to interact with products masquerading as entertainment. Most children do not discern the game from marketing and require additional adult oversight (Radesky et al., 2022).

Other free applications allow interaction with other children in real time by playing games like Dominoes and UNO. These familiar family games add an altogether new dimension to child access to electronic media. While parents may view the games as a pleasant and safe pastime, it raises questions as who are the other children in the game? Are they children? What are they communicating to each other through the games? Most public rooms used for game playing are not moderated and require adult oversight (Mommy Poppins).

The electronic media revolution does provide some help to those who desire insight into the applications that children can access. For example, Common Sense Media rates health-related games and information for children as young as three. Topics like toothbrushing, food facts, and bath time engage children and provide a positive health message. Children are intrigued, entertained, and taught using social media applications. They can also be indoctrinated and manipulated if screen time is not monitored by a responsible adult. Establishing parameters and guidelines during childhood will begin e-literacy and smooth the transition to adolescence.

Adolescence

Viewing habits: As a child matures, new challenges arise in setting limits and guidelines. Usage of social media, which, as discussed above, begins at an early age, typically increases as children become older (Anderson et al., 2018). Adolescents spend longer periods of time in independent activities largely unmonitored by an adult. Screen time connects them to others, alleviates boredom, and entertains. Surveys indicate that adolescents spend 5–7 h per day on social media. The amount of time spent online, and the media viewed, may be impacted by family income, and a predetermined "at risk" status of the adolescent (George et al., 2018).

The most heavily used social media sites for adolescents is Snapchat, followed by TikTok and Instagram. Teen users of Snapchat may check in 30 or more times a day. When YouTube is added to the mix, it is rated as teens' favorite site (Rideout et al., 2022). Thus, the highly accessed sites may aggressively recruit and draw young people to return again and again. It is not just that they find applications that attract them for periods of time; they access sites that become integral to their lives. This drives them to return repeatedly, and each viewing period reinforces their screen fixation. Moreover, specific sites draw young people to return because of the nature of the application's subject. Online influencers are savvy in drawing youth with their contemporary content and may exert a powerful influence on behavior. As adolescents try to find themselves, social media may set up unrealistic aspirations and an unattainable worldview (Table 2.1).

Adolescence and self-esteem

Researchers state that the length of time spent interacting on social media opens an additional risk to self-esteem resulting in depression, anxiety, and overall lower mental health (Ali et al., 2021; Schodt et al., 2021). While these are not necessarily the result of other users' actions, increased time on social media can increase vulnerability to intimidation and harassment. The number of hours spent on social media can increase exposure to targeted bullying and senseless trolling activities, another risk to healthy self-esteem (Santos et al., 2022). However, on the positive side, research by George et al. (2018) noted "that more time online and greater online communication was associated with fewer same-day internalizing symptoms" in at-risk children, perhaps related to increased socialization, especially through texting, which reduced anxiety symptoms. Yet, long and daily exposure to social media is associated with poorer personal relationships and involvement in risky behavior. "Past research found that the number of social media websites used by those ages 13–24 years was positively correlated with depression, anxiety, and loneliness" (Barry et al., 2017).

TABLE 2.1 Average number of minutes spent on social media by age group.

Age	Minutes/Day	Source
Children	180+ not including school	https://www.statista.com/statistics/1284561/daily-screen-time-children-use-united-states/.
Tweens	333	https://www.nytimes.com/2022/03/24/well/family/child-social-media-use.html.
Teens	519	https://www.nytimes.com/2022/03/24/well/family/child-social-media-use.html.
Adult	147	https://www.oberlo.com/statistics/how-much-time-does-the-average-person-spend-on-social-media#:~:text=your%20free%20trial-,Average%20time%20spent%20on%20social%20media, also%20the%20highest%20ever%20recorded.
Older adults (55–64)	113	https://review42.com/resources/how-much-time-do-people-spend-on-social-media/#:~:text=55%2D64%2Dyear%2Dolds,minutes%20of%20social%20media%20time.

Adolescence and health information

Adolescents may use social media for health information especially on sensitive or potentially embarrassing topics that are difficult to discuss with adults (Gulec et al., 2022; Mitchell et al., 2014). Questions relating to eating disorders, depression, sexually transmitted disease, contraception, pregnancy, and sexual orientation are a few of the topics that youth prefer to research in private without having to explain their curiosity to an adult. "Forty four percent of 15–24-year-old's who have looked for health information online sought out information about sexual health, second only to information about diseases like cancer and diabetes" (Mitchell et al., 2014). Understanding that adolescents seek health information online without interpretation by a health professional makes accuracy, reading level, and safety imperative when creating websites or blogs to ensure information online is accurate. Huo et al. (2022) found that adolescents had a low level of personal application to web-accessed health information coupled difficulty in searching and acquiring information. Difficulty is accessing health information calls adolescent E-health literacy into question. Adolescents not E-health literate may have difficulty sifting through massive amounts of varying and contradicting health information. Wartella et al. (2018) state that at least one-third of those adolescents who find and read health information act on it. Mitchell et al. (2014) put that number even higher and look at behavioral impact of online health information: "41% of adolescents indicate having changed their behavior, and 14% have sought healthcare services because of information they found online."

Teaching E-health literacy as early as middle school could increase the likelihood that young adults will access such information, act upon it, and experience improved quality health outcomes. Online access to health information can provide benefits that could not be experienced in traditional ways. It provides information 24/7, is private, stigma-free, and can benefit young people's health when found and implemented.

Adults

Viewing habits

Is it true that growing up with electronic media access predisposes adults to even higher levels of screen time than seen in adolescence? Not necessarily. Adults aged 30 years and up have grown up with access to Web 2.0, and it is estimated that they use social media for over two hours each day, which *decreases* as age increases. Although research indicates that adults are not necessarily spending an ever-increasing amount of time accessing electronic media, this does not exempt or protect them from some of the mental health implications noted for younger social media consumers. With a fully formed psyche and presumably less time to spend on social media, one might believe adult social media consumers are informed and savvy users of the Web, one might even expect that with a lifetime of experience on the web, and work and family "adult" experiences, they would be less likely to be trolled, or bullied and know what is real and what is not. Unfortunately, this too is not the case.

Adults are not immune to the siren call of social media. If honest, they too would admit to feelings of envy or discontent while scrolling LinkedIn and Facebook. For example, seeing an announcement on LinkedIn that a friend or coworker just received a big promotion or a prestigious award can generate feelings of envy and inadequacy. Seeing a picture of a cousin's 40-foot yacht with him waving from the teak deck could do the same emotional damage. Perhaps, self-doubt and discontent are triggered by a photo of a friend's family, all smiles, at a tropical location as the social media consumer sits alone in their room. The journey to adulthood does not ensure protection for a fragile self-image and anxiety. Adults interacting with social media can have the same reactions younger people have to similar situations. Some vulnerable adults, like young people, can also become heavy users of electronic media.

Vulnerability of adults in social media

Vulnerable adults who are subject to mental and personality disorders may find solace, or further angst, on social media. Their access to the electronic communication "main street" can bring comradery or panic with their existing conditions exacerbated or reinforced. For example, adults displaying narcissistic traits meet their psychological cravings by being admired on social media. For them, social media simply offers a mechanism that exacerbates their condition. The number of likes and comments on their posts is a drug for a narcissist. The more likes and compliments they see, the better they feel. This can also impact others who interact with such a person. A consumer caught up in another's narcissistic web may experience amplification of their existing personal doubts and vulnerabilities. This is especially true in the posting of personal health and wellness images. A toned, well-muscled body, beautifully clothed is particularly impactful in creating dissonance, discontent, and discomfort. These feelings are not limited to any specific age group as downward negativity, and the feeling of not measuring up occurs in adults and adolescents. The shared and universal experience of this unworthiness has attracted the curiosity of social media researchers.

While it is true that feelings of inadequacy and not measuring up to others can be exacerbated by social media, they are not necessarily due only to social media, as these negative feelings can occur in various social situations. However, researchers have used these discordant feelings to explore subclinical personalities and how they interact with social media.

A systematic review by Moor and Anderson (2019) found 26 studies that examined the many dark sides of social media, including "trolling, cyber-aggression, cyberloafing, sending unsolicited explicit images, the non-consensual dissemination of 'sexts,' cyberbullying, problematic social media usage, problematic online gaming, problematic Internet use, internet-use disorder, social media addiction, intimate partner cyberstalking, technology facilitated sexual violence and technology facilitated infidelity." These behaviors can range from mild annoyances to genuinely frighting, and they stem from a triad of conduct referred to as the "Dark Triad": narcissism, psychopathy, and Machiavellianism (Jablonska et al., 2020).

Narcissists feed on the attention that social media engagement generates and incessantly post self-photos to show hair, makeup, clothing, and activities. Their boasting and self-aggrandizement help to numb feelings of inadequacy; in short, narcissists are heavy consumers of social media. They "check-in" at food or entertainment events signaling social status and heavily track pictures and postings to compare themselves to others (Vazire et al., 2008). This behavior is associated with deliberate and persistent online engagement and in extreme cases results in cyberstalking (Kircaburun et al., 2018). Both males and females can have narcissistic tendencies.

Media users with other personality disorders are also drawn to the Internet. They misuse the anonymity of social media and tend to victimize those perceived as "weak" without empathy or self-restraint. Psychopaths, for example, use aggression and intimidation, which can lead to social media bullying and trolling (Moor et al., 2019). Machiavellians are another example of a personality disorder found in social media; these people are master manipulators. On the surface they appear outgoing and friendly, yet, they too, experience low self-esteem and may not socialize well face-to-face. Machiavellians are not "selfie"-oriented like narcissists, rather they remain in the background sowing seeds of doubt from the side lines using bullying tactics sprinkled with seduction (Towler, 2020). While these behaviors are described in the adult section of this chapter, it is wise to remember that older adults, adolescents, and even children can become victims of Dark Triad personalities.

Older adults

Social media as a means of communication: Social media use declines with age, yet that does not mean that older adults do not use social media. Some do not, but many do. The belief that older adults are not computer savvy replicates ageist stereotypes (Guest et al., 2022). However, older adults use fewer platforms than adolescents, and adults and their communication needs may be typically different. For example, common messaging apps among social media users aged 60 and above include Facebook Messenger, Zoom, Facetime, and WhatsApp (Fender, 2021). These apps provide a more private space for real-time conversation, yet don't create the same social solid ties as face-to-face interaction (Papadakis et al., 2022).

There are also material benefits to older adults from using electronic media that cannot be ignored. The goal of social media is to develop social networks and relationships (Nam, 2021). Using a handheld device, whether a tablet or smartphone, for human contact can mitigate many socialization barriers often seen in older adults. Difficulty in mobility, lack of transportation, and long-distance separations can be navigated through social media platforms and increase the quality of life. This use of social media is genuinely beneficial to older adults to connect with other people and positively impact their view of life (Cotten, 2022). Nam (2021) noted that more frequent social media users appear to feel more supported, indicating dosage may play a role in sustaining relationships. However, the types of connections fostered by older adults may be different from younger users. While maintaining relationships is vital to older adults, using social media to make new friendships is not appealing to them (Newman et al., 2021). This should be considered by those assisting and intervening with older adults. Health professionals should increase their attention on providing technology-driven access to family and friends to help mitigate isolation, depression, and lack of internalized value to society. Messaging apps are an effective tool in providing face time with family members especially when travel is difficult. This real-time interaction, on a regular basis, could lessen loneliness and increase socialization.

Schlomann et al. (2019) looked at three psychological parameters frequently associated with aging: loneliness, anomie (isolation), and autonomy, using a sample of those 80 years and older. They noted that about one in four, or about 25%, used a device with an Internet connection indicating that age does play a role in the interest and use of web-connected devices. The conclusion of the research study was that the use of web-connected Information and Communication Technology (ICT) was a "relevant person—technology interaction among the oldest-old" and its use was "positively associated with (the) different domains of subjective well-being" (*P.* 8). This finding was enlightening as caregivers struggled to connect older adults with their families during the COVID-19 pandemic.

Older adult and social media during COVID

The COVID-19 pandemic demonstrated the need for online communications skills. Older adults who relied on in-person social settings to receive and share information but could not interact in person lost that opportunity to receive information during the pandemic. The senior centers, coffee shops, and community senior meal sites closed during the pandemic, creating a social and communication void. Older adults with handheld devices were technologically able to find virtual visits invigorating. For those quarantined in assisted living, long-term care or acute care, it was not uncommon to see friends and children in lawn chairs outside of the building face timing with loved ones inside. "Virtual visits and meetings are now a part of everyday life, from virtual medical visits to gatherings with friends. It's essential that we can provide this technology to our patients and their families to give them the best possible experience amid this new normal." (University of Michigan Medicine, 2020).

Ensuring that older adults are included in social media planning and not stereotyped as "too old to learn" or labeled with the derogatory title of "luddite" is essential in preparing for the next outbreak or natural disaster that restricts face-to-face communication (Guest et al., 2022).

It is necessary to understand what causes older adults to use electronic media. Shang and Zuo (2020) relied on the Health Belief Model to further understand the motivation behind older adults who utilize social media for health information. They noted: "Older adults possess stronger self-efficacy around learning health knowledge if they perceive themselves as having higher e-health literacy" (P 354). Moreover, because older adults with high E-health literacy believe they know how to answer their health-related questions, they can benefit by applying the knowledge. Much like achieving E-health literacy in adolescents, older adults must not be forgotten and should be intentionally included in communication technology training. Social media strategies focusing on E-health literacy are important to all ages (Table 2.2).

The impact of gender on social media use

Stereotypical access and use of social media

The use and impact of social media differ not only by age group but also by gender. Several theories have been used to explain gender differences in the online sharing of information. Fishbein and Ajzen's (2011) Theory of Reasoned Behavior supports the predicted use of social media on certain characteristics. In general, if individuals perceive the use of web-based social interaction to be supportive, easy, safe, and fun, they will engage more heavily than those who find it cumbersome, unsafe, and unnecessary. However, inserting the variable of gender into this theory raises the question of the impact of gender on the perception of support, fun, and safety. Discussing the effect of gender on the use of social media is tricky. One of the challenges here is those certain assumptions about gender, and societal expectations about gender, drive specific behavior of

TABLE 2.2 Social media as intervention with older adults.

Barrier to socialization	Learning theory for intervention	Outcomes
Isolation	Reasoned behavior	Interaction
Lack of cconfidence	Social learning	E-literacy
Interpersonal ddisconnect	Reasoned behavior	Communication
Physical distance	Reasoned behavior	Face-to-face engagement

people based on their biological or anatomical features. These assumptions result in generalizations and stereotypical conclusions about the use of social media, as common norms define. But these assumptions are not inclusive, and they ignore the intersection of age, gender, and race in the consumption and interaction with social media. Also ignored are societal norms that proport women need strong social ties while men do not, thus facilitating traditional roles and stereotypes (Archer, 1996). Rubin (1985) in her seminal book looking at the role of friendship noted: "With the rise of an articulate feminist movement, a new scholarship developed that pointed quite convincingly to the ways in which these and other sex-stereotyped behaviors are shaped and maintained by the mandate of culture" (P. 80).

Common gender stereotype extends to specific social media platforms. Some of these platforms are designed and are inherently "male" and others "female." In other words, there is the assumption that has legitimately developed over time and through experience, that certain gender types of males seem to be drawn more to specific applications than others. For example, experience has shown that males tend to use gaming apps, and females are more interested in picture apps. This "reality" is assumed to be universally valid, yet completely binary in its origin. The belief shuts down the equal reality of gender fluidity and differing normative behavior. "Therefore, gender norms are seen as social rules imposed on individuals since birth and the expectations they must fulfill to be seen as integrated members of society" (Seserman, 2021, P. 41).

Traditional gender expectations include accepted norms on friendship and the desire for social interaction. "A woman wants to dissect and 'chew' to understand the underlying facts." Men are more dispassionate and not looking for "all that talk" (Rubin, 1985, p. 160—161). Following from this 20th century dichotomy, there should be no surprise that many issues are seen through a binary gendered lens. Assumptions may be made that "chatty" women would be heavy social media users and men would only use social media to help solve a problem.

This section of the chapter will describe research and draw conclusions about the differences in media consumption based on gender; however, its findings are tempered by the fact that social media consumption and interaction are also individual. The conclusions reached are reflective of differing opinions on how variables such as gender and age impact social media use (Table 2.3).

Cisgender

Social media and social support

The gender identity of individuals impact feelings about how they "fit" into society. As noted above, these assumptions are rooted in centuries of expectations. The use of social media by nonbinary individuals is a newer area of research and needs closer attention. Aspects of social media such as anonymity,

TABLE 2.3 Use of social media platforms based on gender.

Social media platform	Proportion of use among males	Proportion of use among females
Facebook	50.7%	49.3%
Youtube	56.7%	43.3%
Twitter	55.7%	44.3%
Instagram	49.5%	50.5%

Hruska, J., & Maresova, P. (2020). Use of social media platforms among adults in the United States—behavior on social media. Societies, 10(1), 27. MDPI AG. https://doi.org/10.3390/soc10010027 Hruska & Maresova, 2020.

lack of stigma, and shared norms drive the use of specific social media platforms for sexual and gender minorities (SGMs) (Escobar-Viera et al., 2021). Social media provides an interactive connection to encouragement and support for those feeling set aside or different. Earlier research initially showed that there was no association between Facebook use and perceived social support among a sample of lesbian, gay, bisexual, trans, queer, and questioning (LGBTQ) youth (McConnell et al., 2017). However, Gerke et al. (2020) noted that differences in social media use and social support were seen in males who have sex with males (MSM) and trans women. Their results indicated that gender identity did influence social media use and perceptions of social support.

Being classed as the "other" has been correlated with substance abuse, depression, and suicide. Specific research has found this to be particularly true in youth living in rural areas where SMG was at odds with social norms. Paceley et al. (2022) noted that "participants reported utilizing the internet and social media to access other SGM people, establish friendships and relationships, and access SGM-related supports" P.47. Further, social media platforms are also used to disseminate health information tailored to SGMs. Human immunodeficiency virus (HIV), condoms, preexposure prophylaxis (PrEP), sexually transmitted infections (STIs), and domestic violence are a few topics geared to this audience. Overall effectiveness has been noted in information uptake if the information was interactive. However, the beneficial impact of social media is not uniform. Despite the positive advantages of social media use in SGM populations, "SGM young adults experiencing processes and effects of minority stress (i.e., internalized stigma, depressive symptoms, and low emotional, social support) may also be at risk for problematic social media use" (Vogel et al., 2021, p. 2). When SGM populations have access to online resources, such as culturally applicable information, support groups, and safe gathering spaces, they are able to better cope with minority stress (Gerke et al., 2020).

The risk of being trolled or victimized increases with time spent online, and SMS adolescents and young adults spend about 45 min longer on social media each day compared with heterosexual adolescences and young adults (Kaiser et al., 2021). Using social media as a social support can put SGM users at risk for victimization, bulling, and cyberstalking because of their online disclosures. Earlier in this chapter, the Dark Triad was unpacked, and it is important to note that this same behavior may be used in spaces where SGMs are frequent, resulting in depression and alienation.

Females

Research has shown that gender roles are stereotyped and strictly enforced on social media (Loxton et al., 2022). Men and women are expected to portray themselves based on their gender roles driven by societal norms. Images are an essential part of establishing and displaying a genderized self. Most social media users incorporate images as a significant part of their online interaction. Selfies have been widely used as the source of these images. However, the types of images demonstrate how gender appears to drive them. As of August 2018, 82% of adults between the ages of 19 and 34 had posted at least one selfie on social media. According to Thompson (2020), women were 8.6 times more likely to post a selfie than males, more prone to show faces in their images (Zheng et al., 2016), and documented their belongings 5.4 times more than men (Thompson, 2020). Images put females are in jeopardy of being overexposed, not being authentic, and vulnerable (Loxton et al., 2022).

Snapchat and Instagram are social media platforms built on visual media and are popular with females (Statista, 2022). These platforms allow the posting of carefully curated pictures representing normalized ideals. Yet, it is not the zeal with which one posts but instead the influence of the post, or "feedback" received from the posts, which is measured by the comments and most "likes" received. This can impact how users feel about themselves. Indeed, this interaction with social media may be harmful to certain females with body dissatisfaction or narcissistic tendencies (Pedalino et al., 2022). It may even exacerbate or even create dissatisfaction or narcissistic tendencies.

Social media requires perfection, which is not compatible with being authentic (Duffy & Hund, 2019). And body dissatisfaction is not the only area where perfection appears unattainable. With the posting of multiple images with faces, families are also at risk of being judged by viewers. Some social influencers are attuned to crafting family images with exciting backgrounds, coordinated clothing that is clean and pressed, and sunny bright smiles on all family members. Failing to measure up as a mother in real life based on social media's depiction adds to the dissatisfaction and discontent associated with social networking sites.

Image altering is very common, resulting in elongated, thinned, and accentuated features, producing an unattainable image.

On the other hand, it is suggested that a high level of social media literacy can provide a protective factor against body dissatisfaction. Understanding the reasoning behind an influencer's post allows for the discernment of authenticity and may mitigate behaviors while viewing online images (Paxton et al., 2022). Extensive use of social media while parenting may present another dark side to heavy media use. Radesky et al., 2016 noted that the constant interruption and intrusion of social media into home life, including meals and general parenting, slowed language development in their children. Being distracted by mobile devices meant "not being present" in real time for the children. Parents, male or female, should acknowledge this intrusive aspect of social media.

Males

As this section discusses the use of social media by gender, it is essential to point out that although based on research, the information provided here is not universally generalizable. Making assumptions about certain groups and their media as males needs to consider other factors such as age, race, culture, education, and income. This chapter intended to look at only two variables but acknowledges that social media use occurs at the intersection of all individual identifying characteristics.

Males are robust social media users, with 65% of men reporting social network engagement. However, male preference for social media platforms differs from females, with males showing a higher appreciation for Twitter and YouTube. Their platform choice indicates a higher consumption behavior and less interactional behavior. Tifferet (2020) concludes that men give and receive less social support on social media than women. This lack of interaction may indicate that men are less prone to experience body dissatisfaction or discontent than females who are heavily engaged in image posting, liking, and commenting on others' posts; however, the lack of interaction can contribute to loneliness.

Younger men using social media appear less likely to feel this loneliness than middle-aged men. It seems that views of masculinity are changing, resulting in the acceptance, and nurturing of stronger male-to-male friendships and adjusting views of stolidness (Seidler, 2022).

Older men, who may experience loneliness more, use social media to stay connected than their younger counterparts. As mentioned earlier in this chapter, communication platforms help mitigate loneliness through social interaction.

Although males are less likely to become addicted to social media than females (Su et al., 2020), Wang et al. (2022) noted that depression, often connected to loneliness, could significantly impact social media use. Attempting to erase isolation through a greater dosage of social media use can lead to social media addiction in men and women. Along with this, extended and frequent use of social media increases the vulnerability of being exposed to other users with subclinical behaviors resulting in bullying and trolling. "In men, depression and anxiety predict greater cyberbullying victimization and perpetration, particularly among men with relatively higher levels of social media use" (Schodt et al., 2021, p. 1).

Males victimized on social media by females are not widely publicized, possibly due to strictly enforced gender norms that conclude that "real men" cannot be victimized. Victimization of males occurred most often when they did not uphold the normative view of masculinity, appearing weak and vulnerable (Loxton et al., 2022). This rare insight into male victimization warns against the strong applying stereotypical roles to social media users.

Chapter summary

This chapter presents a convincing review of existing literature demonstrating that age and gender influence social media use. Arguably, age influences social media use more than gender, perhaps because age is a quantifying factor with distinct boundaries, whereas gender is more fluid. Both tenderized norms and age stereotypes are rigidly enforced in social media interactions. These generalizations are anxiety-producing and possibly dangerous.

While marketing professionals may find gender and age essential in promoting products, those involved in health literacy, health promotion, and health education would do well to treat gender and age with respect while crafting critical E-health communications. This chapter also reinforces the need for E-health literacy throughout life, regardless of gender or age, to mitigate the more problematic uses and abuses of social media.

Discussion questions

1. Go back to the beginning of this chapter and look at the questions provided. How would you now answer these questions after reading the chapter?
   a. Is access to social media available to all groups of people? Explain your answer.
   b. Is the electronic information revolution impacting everyone the same? Explain your answer.
   c. Do all genders experience online information sharing in a similar manner? Why or why not?
   d. Are older adults web adverse? Find one additional reference (not used in this chapter) to support your answer.

Definition of key terms

Cyberloafing: Employees using work time to engage in social media.

Dark Triad: Personality traits that make up the Dark Triad are narcissism, Machiavellianism, and psychopathy. When individuals with these personalities are encountered online, they may manifest as volatile, arrogant, domineering, or manipulative.

E-health: Accessing health information via the Internet.

E-health literacy: The ability to locate and discern the validity of content when accessing health information.

Gender fluidity: Gender that is not fixed or binary.

Gen Z: Individuals born between 1997—and 2012 are referred to as "Generation Z" or "Gen Z."

Luddite—A demeaning term used to define someone who is resistant to technology.

Online influencer: A regular user of social media who generates a large audience and is seen as an expert in a certain area.

Selfie: An image of yourself, taken by yourself, with a handheld device.

Technology-facilitated sexual violence (TFSV): TFSV is digital facilitated occurrences of verbal sexual abuse, violence, and harassment.

Trolling: An individual who purposely provokes others with outlandish remarks or crude language with the intent to promote dissonance.

Websites to explore

1. **Best Health Apps and Games for Kids:** One way to promote healthy decisions is to provide educational and entertaining apps and games that spark kids' curiosity. *Select and view one of the health apps from the link below. In your opinion, what makes this a fun learning experience for children?*
   https://www.commonsensemedia.org/lists/best-health-apps-and-games-for-kids

2. **How to Deal with Internet Trolls:** This site provides information and healing for those who experienced trolling during their Internet use. *Have you ever been trolled? How did you handle it?*
   https://www.verywellfamily.com/how-to-deal-with-internet-trolls-4161018

3. **Social Media User Generated Data:** This site provides statistics that assist in understanding the penetration of social media into everyday lives globally. *Review the link provided. What statistic was most surprising to you and explain why?*
   https://www.statista.com/markets/424/topic/540/social-media-user-generated-content/#overview

4. **Twenty Free On-line Games:** These games make it easy for kids to play some of their favorite board and video games together even when they're not gathered around the same table or screen.
   https://mommypoppins.com/kids/9-free-online-games-for-kids-to-play-together.

References

Ali, S., Habes, M., Qamar, A., & Al Adwain, M. N. (2021). Gender discrepancies concerning social media usage and its influences on students' academic performance. *Utopiay Praxis Latinoamericana, 26*, 321−333. https://doi.org/10.5281/zenodo.455628

Anderson, M., & Jiang, J. (2018). *Teens, social media and technology 2018*. Pew Research Center. http://www.pewInternet.org/2018/05/31/teens-socialmedia-technology-2018. Accessed September 26, 2022.

Archer, J. (1996). Sex differences in social behavior: Are the social role and evolutionary explanations compatible? *The American Psychologist, v51*(n9).

Bandura, A. (1971). *Social learning theory*. General Learning Press.

Bandura, A. (1993). Perceived self-efficacy in cognitive development and functioning. *Educational Psychologist, 28*(2), 117. https://doi.org/10.1207/s15326985ep2802_3

Barry, C. T., Sidoti, C. L., Briggs, S. M., Reiter, S. R., & Lindsey, R. A. (2017). Adolescent social media use and mental health from adolescent and parent perspectives. *Journal of Adolescence, 61*, 1−11. https://doi.org/10.1016/j.adolescence.2017.08.005

Connell, S. L., Lauricella, A. R., & Wartella, E. (2015). Parental co-use of media technology with their young children in the USA. *Journal of Children and Media, 9*(1), 5−21. https://doi.org/10.1080/17482798.2015.997440

Cotten, S. R., Schuster, A. M., & Seifert, A. (2022). Social media use and well-being among older adults. *Current Opinion in Psychology, 45*. https://doi.org/10.1016/j.copsyc.2021.12.005

Duffy, B. E., & Hund, E. (2019). Gendered visibility on social media: Navigating Instagram's authenticity bind. *International Journal of Communication, 13*, 4983−5002.

Escobar-Viera, C. G., Melcher, E. M., Miller, R. S., Whitfield, D. L., Jacobson-López, D., Gordon, J. D., Ballard, A. J., Rollman, B. L., & Pagoto, S. (2021). A systematic review of the engagement with social media−delivered interventions for improving health outcomes among sexual and gender minorities. *Internet Interventions, 25*. https://doi.org/10.1016/j.invent.2021.100428

Fender, J. (2021). *5 Apps to keep the elderly connected through messaging*. https://aznha.org/5-apps-to-keep-the-elderly-connected-through-messaging/ Accessed May 31, 20225.

Fishbein, M., & Ajzen, I. (2011). *Predicting and changing behavior: The reasoned actionapproach*. New York, NY: Psychology Press.

George, M. J., Russell, M. A., Piontak, J. R., & Odgers, C. L. (2018). Concurrent and subsequent associations between daily digital technology use and high-risk adolescents' mental health symptoms. *Child Development, 89*(1), 78−88. https://doi.org/10.1111/cdev.12819

Georgiev, V., & Nikolova, A. (2020). Tools for creating and presenting online learning resources for preschool kids. *TEM Journal, 9*(4), 1692−1696. https://doi.org/10.18421/tem94-49 o

Gerke, D. R., Step, M. M., Rünger, D. D., Fletcher, J. B., Brooks, R. A., Davis, N., Kisler, K. A., Reback, C. J., & and of National Significance Social Media Initiative Study Group, S. P. (2020). Associations between social support and social media use among young adult cisgender MSM and transgender women living with HIV. *Health Promotion Practice, 21*(5), 705. https://doi.org/10.1177/1524839920936248

Guest, A. M., & Peckham, A. (2022). Identifying better communication practices for older adults during the next pandemic: Recommendations from the COVID-19 experience. *Journal of Communication in Healthcare, 15*(1), 11−14. https://doi.org/10.1080/17538068.2022.2029104

Gulec, H., Kvardova, N., & Smahel, D. (2022). Adolescents' disease- and fitness-related online health information seeking behaviors: The roles of perceived trust in online health information, eHealth literacy, and parental factors. *Computers in Human Behavior, 134*. https://doi.org/10.1016/j.chb.2022.107318

Hruska, J., & Maresova, P. (2020). Use of social media platforms among adults in the United States—behavior on social media. *Societies, 10*(1), 27. https://doi.org/10.3390/soc10010027. MDPI AG.

Huo, M., Gu, L., & Zhu, L. (2022). Analysis on the influence of adolescent health informational literacy using big data analysis technology under social network environment. *Journal of Environmental and Public Health*, 1−10. https://doi.org/10.1155/2022/4126217

Jablonska, M. R., & Zajdel, R. (2020). The Dark Triad traits and problematic internet use: Their structure and relations. *Polish Sociological Review, 212*, 477−495. https://doi.org/10.26412/psr212.06

Kaiser, S., Klare, D., Gomez, M., Ceballos, N., Dailey, S., & Howard, K. (2021). A comparison of social media behaviors between sexual minorities and heterosexual individuals. *Computers in Human Behavior, 116*. https://doi.org/10.1016/j.chb.2020.106638

Kircaburun, K., Jonason, P. K., & Griffiths, M. D. (2018). The Dark Tetrad traits and problematic social media use: The mediating role of cyberbullying and cyberstalking. *Personality and Individual Differences, 135*, 264−269.

Loxton, A., & Groves, A. (2022). Adult male victims of female-perpetrated sexual violence: Australian social media responses, myths and flipped expectations. *International Review of Victimology, 28*(2), 191−214. https://doi.org/10.1177/02697580211048552

McConnell E, A., Clifford, A., Korpak, A. K., Phillips, G., II, & Birkett, M. (2017). Identity, victimization, and support: Facebook experiences and mental health among LGBTQ youth. *Computers in Human Behavior, 76*, 237−244.

Merolli, M., Gray, K., & Martin-Sanchez, F. (2013). Health outcomes and related effects of using social media in chronic disease management: A literature review and analysis of affordances. *Journal of Biomedical Informatics, 46*(6), 957−969.

Mitchell, K. J., Ybarra, M. L., Korchmaros, J. D., & Kosciw, J. G. (2014). Accessing sexual health information online: Use, motivations and consequences for youth with different sexual orientations. *Health Education Research, 29*(1), 147−157. https://doi.org/10.1093/her/cyt071

Moor, L., & Anderson, J. R. (2019). A systematic literature review of the relationship between dark personality traits and antisocial online behaviours. *Personality and Individual Differences, 144*, 40−55. https://doi.org/10.1016/j.paid.2019.02.027

Nam, S.-J. (2021). Mediating effect of social support on the relationship between older adults' use of social media and their quality-of-life. *Current Psychology, 40*(9), 4590−4598. https://doi.org/10.1007/s12144-019-00399-3

Newman, L., Stoner, C., & Spector, A. (2021). Social networking sites and the experience of older adult users: A systematic review. *Ageing and Society, 41*(2), 377−402. https://doi.org/10.1017/S0144686X19001144

Paceley, M. S., Goffnett, J., Sanders, L., & Gadd-Nelson, J. (2022). Sometimes You Get Married on Facebook": The use of social media among nonmetropolitan sexual and gender minority youth. *Journal of Homosexuality, 69*(1), 41−60. https://doi.org/10.1080/00918369.2020.1813508

Papadakis, S, Alexandraki, F., & Zaranis, N. (2022). Mobile device use among preschool-aged children in Greece. *Education & Information Technologies, 27*(2), 2717−2750.

Paxton, S. J., McLean, S. A., & Rodgers, R. F. (2022). My critical filter buffers your app filter": Social media literacy as a protective factor for body image. *Body Image, 40*, 158−164. https://doi.org/10.1016/j.bodyim.2021.12.009

Pedalino, F., & Camerini, A. L. (2022). Instagram use and body dissatisfaction: The mediating role of upward social comparison with peers and influencers among young females. *International Journal of Environmental Research and Public Health, 19*(3), 1543. https://doi.org/10.3390/ijerph19031543

Radesky, J. S., Kistin, C., Eisenberg, S., Gross, J., Block, G., Zuckerman, B., & Silverstein, M. (2016). Parent perspectives on their mobile technology use: The excitement and exhaustion of parenting while connected. *Journal of Developmental and Behavioral Pediatrics, 37*(9), 694−701. https://doi.org/10.1097/DBP.0000000000000357

Radesky, J. S., Seyfried, J. L., Weeks, H. M., Kaciroti, N., & Miller, A. L. (2022). Video-sharing platform viewing among preschool-aged children: Differences by child characteristics and contextual factors. *CyberPsychology, Behavior and Social Networking, 25*(4), 230−236. https://doi.org/10.1089/cyber.2021.0235

Rideout, V., Peebles, A., Mann, S., & Robb, M. B. (2022). *The common sense census: Media use by tweens and teens.* https://www.commonsensemedia.org/research/the-common-sense-census-media-use-by-tweens-and-teens-2021. Accessed May 25, 2022.

Rubin, L. (1985). *Just.friends: The role of friendship in our lives.* New York: Harper and Row.

Santos, I. L. S., Pimentel, C. E., & Mariano, T. E. (2022). Online trolling: The impact of antisocial online content, social media use, and gender. *Psychological reports, 332941211055705.* Advance online publication. https://doi.org/10.1177/00332941211055705

Schlomann, A., Zank, S., Woopen, C., & Rietz, C. (2020). Use of information and communication technology (ICT) devices among the oldest-old: Loneliness, anomie, and autonomy. *Innovation in Aging, 42*(2).

Schodt, K. B., Quiroz, S. I., Wheeler, B., Hall, D. L., & Silva, Y. N. (2021). Cyberbullying and mental health in adults: The moderating role of social media use and gender. *Frontiers in Psychiatry, 12*, 674298. https://doi.org/10.3389/fpsyt.2021.674298

Seidler, Z. E., Wilson, M. J., Rice, S. M., Kealy, D., Oliffe, J. L., & Ogrodniczuk, J. S. (2022). Virtual connection, real support? A study of loneliness, time on social media and psychological distress among men. *International Journal of Social Psychiatry, 68*(2), 288−293. https://doi.org/10.1177/0020764020983836

Seserman, C. M. (2021). Assessing the way gender norms affect teenagers' behaviour in the context of social media. *Social Research Reports, 13*(1), 38−48. https://doi.org/10.33788/srr13.1.4

Shang, L., & Zuo, M. (2020). Investigating older adults' intention to learn health knowledge on social media. *Educational Gerontology, 46*(6), 350−363. https://doi.org/10.1080/03601277.2020.1759188

Statista. (2022). *Our research and content philosophy.* Available at: https://www.statista.com/aboutus/our-research-commitment.

Su, W., Han, X., Yu, H., Wu, Y., & Potenza, M. N. (2020). Do men become addicted to internet gaming and women to social media? A meta-analysis examining gender-related differences in specific internet addiction. *Computers in Human Behavior, 113.* https://doi.org/10.1016/j.chb.2020.106480

Thompson, T. J. (2020). *I studied 5,000 phone images: Objects were more popular than people, but women took way more selfies.* The Conversation. https://theconversation.com/i-studied-5-000-phone-images-objects-were-more-popular-than-people-but-women-took-way-more-selfies-150080 Accessed May 26, 2022.

Tifferet, S. (2020). Gender differences in social support on social network sites: A meta-analysis. *Cyberpsychology, Behavior and Social Networking, 23*(4), 199−209.

Towler, A. (2020). *Machiavellianism: What it is, how to recognize and cope with Machiavellians.* www.ckju.net/en/dossier/machiavellianism-what-it-how-recognize-and-cope-machiavellians#:~:text=Machiavelli%20encourages%20leaders%20to%20lie,boost%20their%20own%20self%2Dworth. (Accessed 21 September 2022).

University of Michigan Medicine. (2020). Amid restricted visitor policies, Zoom and Facetime are a lifetime. Available at https://hits.medicine.umich.edu/news/amid-restricted-visitor-policies-zoom-facetime-are-lifeline.

Vannucci, A., Ohannessian, C. M., & Gagnon, S. (2019). Use of multiple social media platforms in relation to psychological functioning in emerging adults. *Emerging Adulthood, 7*, 501–506.

Vazire, S., Naumann, L. P., Rentfrow, P. J., & Gosling, S. D. (2008). Portrait of a narcissist: Manifestations of narcissism in physical appearance. *Journal of Research in Personality, 42*(6), 1439–1447. https://doi.org/10.1016/j.jrp.2008.06.007

Vogel, E. A., Ramo, D. E., Prochaska, J. J., Meacham, M. C., Layton, J. F., & Humfleet, G. L. (2021). Problematic social media use in sexual and gender minority young adults: Observational study. *JMIR Mental Health, 8*(5). https://doi.org/10.2196/23688

Wang, Z., Yang, H., & Elhai, J. D. (2022). Are there gender differences in comorbidity symptoms networks of problematic social media use, anxiety and depression symptoms? Evidence from network analysis. *Personality and Individual Differences, 195*.

Wartella, E. A., Lovato, S. B., Pila, S., Lauricella, A. R., Echevarria, R., Evans, J., & Hightower, B. (2018). Digital media use by young children: Learning, effects, and health outcomes. *Child and Adolescent Psychiatry and the Media*, 173–186.

Zheng, W., Yuan, C. H., Chang, W. H., & Wu, Y. C. J. (2016). Profile pictures on social media: Gender and regional differences. *Computers in Human Behavior, 63*, 891–898.

Section II

Social media use and mental health outcomes in diverse populations

Chapter 3

Social media and young adults

Bronwyn MacFarlane[1] and Jason Kushner[2]

[1]*Department of Educational Leadership Studies, Curriculum, and Special Education, College of Education and Behavioral Science, Arkansas State University, Jonesboro, AR, United States;* [2]*University of Arkansas at Little Rock, College of Business, Health and Human Services, School of Counseling, Human Performance, and Rehabilitation, Little Rock, AR, United States*

Learning objectives

- Describe the usefulness of social media for young adults.
- Identify the prevalence of social media use among young adults.
- Discuss the challenges young adults face associated with social media use.
- Discuss ways to promote healthy habits when using social media for recreation or professional purposes.
- Identify recommendations for practitioners working with young adults to achieve mental health goals by using social media.
- Discuss questions for discussion among educators and mental health professionals.

Social media and young adults

Throughout the 20th and 21st centuries, some children (and their parents) who enjoy athletics and the arts may have aspired breaking into the big time in the professional sports or entertainment fields. Some may make it into the professional teams and artistic venues and land endorsement deals, but many choose another profession and continue with sports and the arts activities in clubs, groups, and as youth coaches and mentors. After social media gained wide use, the appearance of social media "influencers" became known as people and organizations with a purported expert level of knowledge or social influence in their field or interest area. Like professional athletes with endorsement deals, "influencers" on social media may be someone or something with the influence to encourage buying habits or other actions by uploading content to social media platforms like Instagram, YouTube, TikTok, Snapchat, Facebook, etc. Established consumer brands have enrolled and contracted with influencers who have developed a social media audience as representatives for companies in providing testimonial advertising and endorsements. While some individuals using social media may achieve "influencer" status, the majority of account holders are simply participants on social media navigating and consuming posted information. In addition to learning more about peers, news, and entertainment, special learning opportunities such as educational and professional experiences can be shared to a larger audience on social media for more young adults to learn about and access information to apply and participate. Some social media account holders may develop unhealthy habits or addiction to social media use. Teaching young people about the benefits and concerns of sharing a social media presence on the Internet should be a priority among those who work with and educate young people and their parents. The use of social media by young adults has enormous potential for both positive youth development with social and professional connections, as well as concerns for mental health and focus.

Current research-based findings about mental health among young adults as related to social media use are presented in this chapter to provide an overview of current understandings about the impact of social media use on the mental health among older adolescents and young adults. This chapter also presents educational information related to promoting healthy habits when using social media for recreational or professional purposes.

Social media usage among young adults

Social media statistics among different demographic users have been accruing exponentially by big data organizations, and specific use by young adults has shown large patterns of usage over time. Young adults were the earliest social media adopters and have continued to use social media at high levels, especially YouTube, Facebook, Instagram, Pinterest, LinkedIn, Snapchat, Twitter, WhatsApp, TikTok, Reddit, and Nextdoor (Auxier & Anderson, 2021). With a wide variety of publications about social media use, researchers have documented the use of social media by young adults (Abi-Jaoude et al., 2020).

Empirical research concerns about social media use among young adults

In a Swedish study that followed a specific group of students in 101 schools for 2 years to determine the connection between social media usage and mental health, researchers concluded that there was no longitudinal association between social media use and mental health problems, but that it could be reasonable to suggest that social media usage might be a symptom of mental health problems (Beeres et al., 2021).

In the United States and elsewhere, social media trends always change with the times, new technologies, and trends among teens and adolescents. Pew research from 2010, for example (Lenhart et al., 2010), found teens were using blogs less, while older people were using blogs more; today similar trends continue with teens using Instagram, Snapchat, and TikTok, while older adults tend to use Facebook and other legacy social media platforms. True to the characteristics of the developmental phase of adolescence, they often want what is new, the latest, or the trendiest.

In a study focused on whether consumers' habit formation may be influenced by their user satisfaction with social media, a survey concerning social media usage, level of satisfaction from social media, and habits that formed because of social media revealed that social benefit does not affect satisfaction, but dependency does affect satisfaction, and satisfaction has a significant effect on habit formation. In essence, people use social media because they feel like they need to and that use of social media leads to the formation of new habits in a variety of categories (Pluhar et al., 2019; Santoso & Sutedjo, 2019). A particular challenge of establishing a basis for comparison is that for most young people in 2022, social media has been a fairly ubiquitous presence in their lives for as long as they have been using computers, tablets, smartphones, and other communication devices (Centers for Disease Control and Prevention, 2022; Maheux et al., 2022). Moreover, elements of their identity have also been related in part to their presence in social media platforms because their socialization in real time is mixed partly with their social experience online. Orben (2020) noted that some 69% of 12–15-year-olds in the United Kingdom had a social media profile and that it is difficult to know whether screen time in and of itself leads to negative outcomes as hypothesized by other researchers (Frith, 2017). Orben suggested that though conventional wisdom may suggest less screen time is healthier, there is not yet a compendium of longitudinal evidence as such that high use of screens, in general, correlates to negative mental health outcomes. Moreover there is not consensus in the literature, some studies find social media decreases authentic relationships and contributes to loneliness and isolation, while other studies find just the opposite; and what cannot be overstated is the individual nature of the effect of social media on youth who regularly incorporate it as an integral part of their lives. Overall, Orben (2020) concluded whatever negative association there is between digital technology and mental health, the effect may be minuscule. A significant challenge of measurement is controlling for extraneous variables, and measuring social media use in native environments would yield better results. But, doing so would require cooperation and data from social media companies in order to compare perceptions of the user to actual data stored in social media accounts and by the companies who aggregate user behavior and trends.

A similar challenge for researchers and practitioners who work with teens is the myriad ways that social support and its associated activities are defined by the research community. Parents, teachers, and, importantly, teenagers benefit from having social support gained through activities and relationships they experience in their objective lives, offline, in addition to those who receive or augment social support from social media platforms, applications, and relationships of one kind or another. "Social support is generally regarded as an important protective factor for mental health at all ages" and for teenagers is associated with decreased anxiety, suicidal behavior, risky behavior, and adjustment problems (Bauer et al., 2021). Generally, social support networks are available to all teens in one form or another, but for teens in lower socioeconomic status (SES) homes, access to social media platforms is reduced, and access is often associated with higher SES or the ability to fund smartphones, computers, and apps which require Internet connections (Bohleber et al., 2016).

Cyberbullying can affect large numbers of adolescents, in a variety of settings, and while the psychological impact is not clearly understood, anecdotes about suicides and other tragedies of cyberbullied youth are profound (Guilbault & MacFarlane, 2020). Cyberbullying research has focused primarily among children and teens as students. In a New Zealand national sample, researchers found that young adults (18–25 years) experienced the highest levels of cyberbullying both in the past month and overall lifetime (Wang, Yogeeswaran, et al., 2019). Reports of cyberbullying varied slightly between women and men, with women overall reporting slightly greater levels of having ever experienced cyberbullying than men; however, this significant difference did not carry into reports of cyberbullying over the past month. On average, participants identifying as European reported lower levels of cyberbullying than Māori and Pacific Nations participants during both time frames, with Asian participants falling in the middle (Wang et al., 2019).

To explore the potential negative effects of social media on mental health, in particular, the increased opportunities on social media to make unhelpful social comparisons, Warrender and Milne (2020) concluded that social media encourages unhelpful social comparisons as users often present idealistic versions of themselves. Further, social comparison on social media can reduce self-esteem, and some patients may benefit from education about social media's impact on self-esteem. Commenting on research related to concerns for mental health among children, the UK chief medical officer advised parents and caregivers to take a "precautionary" approach and strike a "healthy balance" between the potential benefits of screen time on child development, and children's need for other essential health-promoting activities, such as sleep, exercise, and face-to-face social interaction, which screen activity should not supersede (Davies et al., 2019).

Key terms defined

1. Social media is defined by Meikle (2016) as Internet-based and networked communication platforms that allow both personal and public communication.

2. Cyberbullying is defined as "willful and repeated harm inflicted through the use of computers, cell phones, and other electronic devices" (Hinduja & Patchin, 2019, p. 334).
3. Bullying is defined as "aggressive behavior with potential to cause physical or psychological harm to the recipient" (Peterson & Ray, 2006, p. 148).
4. Social comparison refers to behavior where individuals assess themselves in comparison to certain aspects of others, such as individual behaviors, opinions, status, success, etc (Buunk & Gibbons, 2007).
5. Mental health includes emotional, psychological, and social well-being. It influences how an individual may think, feel, and act. Many factors contribute to mental health including biological factors, life experiences, and family history (mentalhealth.gov).

Mental health and counseling trends and issues

Globally and in the United States, mental health disorders among adolescents and young adults have been on the increase every successive decade since statistics have been collected. In the past decade, 14% of adolescents and young adults experienced a mental disorder of one kind or another, the most common of which are depression, anxiety, and behavioral disorders (World Health Organization, 2021). In the United States, depression, suicide, attention-deficit hyperactivity disorder (ADHD), and substance abuse are added concerns beyond common global indicators of mental disorders (Centers for Disease Control and Prevention, 2022). In the 2018–19 CDC report on adolescent mental health, 15% had a major depressive episode, 37% had persistent feelings of sadness or hopelessness, and 19% had considered attempting suicide (Berryman et al., 2018). All of these mental health indicators are exacerbated by the experience since 2020 of isolation in response to COVID-19 protocols (Marsh et al., 2022).

Abi-Jaoude et al. (2020) identified a variety of issues for young adults related to social media usage including but not limited to: the impact of social media on the adolescent sense of self, social media addiction, smartphone impact on social skills, sleep loss because of social media and mental health, and the use of evidence by physicians.

Some researchers have studied whether social media platforms may be used to survey mental health among social media users. Machine learning has been suggested to identify, study, and judge whether individual users on social media platforms may have mental health issues, may be at risk of depression or suicidal ideation, or be exposed in various data collection surveillance (Skaik & Inkpen, 2021). Certainly, during the height of the COVID-19 pandemic, a global phenomenon, researchers discovered a variety of ill effects from the inability to socialize and engage with the outside world in ways to which they were accustomed prior to the pandemic. One of the corollaries to social and emotional isolation is the "fear of missing out," colloquially referenced as "FOMO" on social media. A study by Amran and Jamaluddin (2022) demonstrated that screen time drastically increased during the period from 2020, when isolation protection measures were put in place, to later with more hybridized forms of isolation. The amount of screen time nearly tripled during the COVID-19 pandemic for a group of adolescents in Malaysia in one example. The effect of adolescents spending so much time on social media resulted in an increase in their perspectives of their own insecurities as they constantly compared their own lives to those of others whom they saw and read about online. Increased screen time led to an increase in FOMO, a reality that extended beyond social limitations imposed during lockdown periods.

Social media platforms are global phenomena, and the codification and severity of mental illness vary in different parts of the world, where, in the more developed parts of the world, and in cities, in particular, incidence rates of mental illness among teens are reportedly higher. There are many reasons for a higher number of mental health episodes, and, as is always the case, it is difficult if not impossible to find causation in the kinds of studies that report data on mental illness with respect to environmental factors such as whether teens live in urban or rural settings. An Indian study, for example, found prevalence rates of teen mental illness at 7.3% from a sample in urban and rural India, important not least of which because the population of India is nearly a billion people. Of particular interest are outcomes related to family income, parental occupation, and urban versus rural habitation. In particular, adolescents from urban communities were at higher risk for mental health problems, while students in rural areas were more likely to report problems with conduct in peers (Gunasekaran et al., 2022). In a qualitative analysis, these findings were corroborated by teachers and parents who reported associations between adolescent behavior and academic performance. Teachers observed that family problems and to some degree, social media influence, could both instigate problematic behavior. The authors propose life skill training as a force to decrease observed mental and behavioral problems.

Other international studies find similar complications from adolescents needing to stay home and away from social gatherings typically found in schools. In the country of Ghana, in sub-Saharan Africa, school closures during the COVID-19 pandemic created anxiety and stress heretofore not seen combined with the added stress of living in one of the poorest regions of the world where only 2.8% of adolescents receive needed mental health care. Even where the isolation of school closure may be in some way alleviated by online schooling and social media engagements as is common in the West, many adolescents even in cities have unreliable Internet and limited access to the technology to use it. The problem is particularly acute for students nearing the conclusion of their formal academic studies as they prepare for exams or move on to postsecondary schools, universities, and the workforce (Oppong-Asante et al., 2021). Bohleber, Crameri, Eich-Stierli, Telesko, and von Wyl recommended a companion app to fill the gap where there are deficits for teens who need prosocial and health-promoting activities. From a sample of adolescents in Switzerland, the authors noted that for adolescents who used the app, results were promising and useful, but that the app was used less than other social media approaches that for them achieved similar ends. For social media applications to be useful for adolescents, they have to be something that teens find easy to use, are useful, used by many other adolescents, and appear not to require something extra from them. Chief among the issues here is that social connectedness requires large-scale use of an app in question; even if the peer group is adolescents generally, connection necessitates using an app that is widely used in the teens' peer group. In addition, training in the use of social supports can be complicated by who is responsible for doing it; teens without the native facility for doing so may not also even ask for help they may need to engage socially and online. Teens who are more inclined for self-sufficiency are better able to help themselves use social supports for depression, anxiety, loneliness, and isolation.

All around the world, the developmental expectations of adolescents remain largely universal where researchers, parents, teachers, and humans, in general, can reasonably expect similar experiences and developmental milestones related to that period of life. Given some of the biological universals in adolescence, we would expect at times unpredictable behavior, risk-taking, seeking of acceptance from peers, and deepening personal relationships, along with a focus on the self in the context of their social environment to be relatively the same. There are different experiences apart from the biological imperatives of adolescence which also govern expectations about what life is like for adolescents and young adults based on where in the world they happen to live and cultural differences. According to the Pew Research Center, over 95% of adolescents aged 13–17 living in the United States have access to a smartphone, 44% report going online at least several times per day, and 45% of teens say they are online "nearly constantly" (Anderson & Jiang, 2018) with expectations that the proportion is even higher given the relative ease of access presently combined with new uses of tablets and smartphones as a reaction to the COVID-19 pandemic. Concurrent with the reality of social media use during the pandemic, teens' lives are becoming less distinct entities between their online and offline presence. At the same time, some adolescents and young adults feel the need to create a fantasy life online with amplification of the number of friends they have, social events attended, and experiences accomplished to augment or take the place of their actual offline lives. Explanations from developmental and social psychology provide clues as to the motivation of young adults. Erikson (1994) notes identity formation as a crucial milestone of adolescence and the theory of social comparison, which has explained behavior for several generations, provides a fitting framework for understanding why teens and young adults would seek to create or promote an extravagant life online to bolster their hopes of achieving or creating the perception of a similar status offline.

Much of the literature around teen and young adult use of social media with respect to mental health concerns is related to reports of depression, anxiety, loneliness, isolation, FOMO, social comparison, and connectedness. In nearly every case, however, social media does not create these conditions; rather, social media can be a contributor to negative psychosocial effects, and even psychiatric disorders, but it is also the case that it can very often be a way to connect people with conditions that they would have had anyway, and, it can provide the very solution to what for some other teens and adolescents are perceived by them and others to create some of the mental health concerns. Steele et al. (2020) noted four common themes in the framework of digital stress experienced by adolescents and young adults: availability stress, approval anxiety, FOMO, and communication overload. "The term (digital stress) is also used to identify the cognitive, affective, and physiological arousal that accompanies notifications from or actual use of social media (Thomee et al., 2010)." Digital stress is necessarily managed in accordance with the facility that teenagers and young adults have for coping with other forms of stress. For example, the constant barrage of messages, updates, notifications, chimes, and the need to check for messages taxes the information processing system of all recipients of these messages. Some will cope and manage that stress better than other people, and it is here that the extent to which digital stress in a social media context as a problem is defined in many ways by individual users. *Availability demands*, for example, a type of stressor whereby social media users feel that they need to constantly respond to notifications and messages from others in a timely or immediate fashion. *Approval anxiety*, another type of stressor, pertains to users' concern about how others react to, perceive, and respond to messages they post, while *FOMO*, another type of stressor in social media use, is fairly self-explanatory; that is the fear that somebody else is engaged in some rewarding activity where one is absent; finally, *connection overload* is the stressor related to the excessive amount and variety of received information through social media. These are summarized in Table 3.1 below.

Social media by itself is neither harmful nor helpful; rather, what is important are the ways in which adolescents and young adults use social media for some other end. All the problems with which this age group contends are not created by social media; social media can be used as a way to help and connect young adults with resources that could reduce stress and anxiety. As a common feature of contemporary life, social media remains a necessary force to develop and maintain connections so important to this developmental phase of life, and for younger members of the current generation, it is not sufficient nor practical to compare life without social media because for those in the developed world, it always has been a part of their lives (Steele et al., 2020).

Other research has found that anxiety and stress in general are features of adolescence that appeared to increase in contemporary times in conjunction with the advent, development, and wider use of social media platforms among

TABLE 3.1 Types of stressors related to social media.

Social media-related stressors	Definition
Approval anxiety	A type of stressor in the use of social media that pertains to users' concern about how others react to, perceive, and respond to messages they post
Availability demands	A type of stressor in social media whereby users feel that they need to constantly respond to notifications and messages from others in a timely or immediate fashion
Connection overload	A type of stressor in the use of social media that refers to the excessive amount and variety of received information through social media
Fear of missing out (FOMO)	A type of stressor in the use of social media that entails fear that somebody else is engaged in some rewarding activity where one is absent

teenagers and young adults. In a study by Mundy et al. (2020), the authors focus on the degree of usage, finding that teenagers with the highest levels of use of social media platforms report the highest degree of depressive and anxiety symptoms when compared with teenagers who have low to normal usage, with girls in particular reporting higher levels of depressive symptoms with high social media usage. Boys, by contrast, did not report higher levels of anxiety associated with social media usage. Though the study focuses on nonclinical levels of symptomatology, for practitioners assessing mental health concerns with adolescents and young adults, the degree of social media usage is an important consideration because of their associations with reported cases of depression and anxiety symptoms, particularly for girls. Marsh et al. (2022) also found the degree of engagement and emotional investment to be salient predictors of social media's influence on mental health. Because ADHD is also a common mental health condition reported by today's adolescents, its relationship to social media deserves investigation in large part due to its high representation of teens and adolescents with related conditions such as anxiety. In the Marsh study, for example, teens with ADHD reported higher incidence rates of cyber victimization compared to those without ADHD. The higher the emotional investment in social media, the higher the anxiety and depression associated with cyber victimization on social media. This is also a particular concern for practitioners and parents who work with teenagers largely because social media can prove to be both a respite from some of the victimization that they may experience in the offline world, but, for others, particularly those who spend a high portion of their spare time using social media, cyber victimization contributes to what is objectively often reported conditions associated with high social media use, in particular, anxiety and depression.

Taken to extremes, anxiety about appearance in young adults can lead to a condition known as body dysmorphic disorder, and, although incidence rates of that are low, curiously, anxiety about perceived levels of attractiveness carries over from the online and offline worlds inhabited by young adults. A study by Zimmer-Gembeck et al. (2021), framing anxiety from a cognitive behavioral model of body dysmorphic disorder identified victimization as a risk factor for appearance anxiety. A sensible question around appearance anxiety is to what degree is it associated with online and offline experiences of young adults. "Findings showed that adolescents and young adults who report more appearance victimization (face-to-face and cyber) have concurrently higher levels of offline appearance anxiety and online appearance preoccupation, and this was found for both males and females." Appearance anxiety remains stable for young women and men, but the focus of that anxiety varied by gender with girls and young women reporting only slightly higher levels of appearance anxiety or concerns, compared with boys, with a particular focus on thinness for girls and a muscular physique for boys (Karazsia et al., 2017). Nevertheless, the relationship between offline appearance anxiety and online appearance preoccupation remains stable (Zimmer-Gembeck et al., 2021).

Much of the attention related to social media and mental health focuses specifically on anxiety because it is most closely associated with fear, appearance, popularity, and missing out, and all of those concerns are equally worthy of attention as depression. There are times when depression and anxiety are linked to mental problems that can be associated with excessive social media use, but they are just as likely to be exclusively experienced for different reasons. In a scoping review of studies about depression and social media use, Vidal et al. (2020) noted that in 42 studies that examined the relationship between depression and social media use, there was a generally positive relationship between social media use and depression. Moreover, reported depression rates for young adults and adolescents generally skyrocketed in the past decade (Ettman et al., 2020), even though there is no causal relationship between the advent of social media and depression. There are a myriad reasons for increases in the stress faced by today's young adults, but the association between depression and high social media use is illustrated by a number of studies on the subject of young adults and adolescents who report depression. One difficulty noted by Vidal et al. (2020) is the proliferation of social media sites in the various categories of the ways that people interact with them makes it challenging to discern which site or type of site is most closely related to depression and suicidality. Even though no causal relationship can be found between social media use and suicide, social media as a marketplace of ideas becomes a congregation place for adolescents seeking information about support for symptoms related to depression and suicide. Even though many correlational studies find a positive relationship between depressive symptoms and high social media use, it is also true that social media can be a safe harbor for people who have trouble in their offline social relationships. Whether depression or anxiety, social media is a gathering place for adolescents with symptoms and social media platforms can magnify but not cause the symptoms as illustrated in a study of depression and appearance-related social media consciousness (Maheux et al., 2022). Social media as a gathering place like so many other gathering places extend the reach of the social experience; noted by Watson et al. (2022), "Mattering was found to significantly correlate with adolescent distress, online activity, problematic social media usage, and school connectedness, evidencing its importance in addressing adolescent mental health."

One of the hallmarks of the adolescent years is social comparisons with others on a variety of complex hierarchies. Social media like other parts of the media ecosphere is very much a mechanism to advertise products and create an image of what people think that they should look like or aspire to be. Indeed, social media influencers are a new category of pitch persons for a variety of products aimed at promoting a standard of perceived beauty, importance, and popularity. As is the case with print and televised media, those standards are often impossible to reach for the overwhelming majority of people who consume them. Because social media requires a higher degree of engagement from its users, the net effect of the content is to create a heightened level of importance of appearance and belonging. Lee et al. (2020) found in a study about positive feedback in the form of "likes" on social media that those who receive negative feedback in the form of very few likes were at greater risk for depressive symptoms. One of the confounding factors related to the immediacy of online profiles is that the degree of the teenager's popularity is immediately visible to every other viewer available to their profiles. As is the case with other mental health concerns, social media adds a spotlight to symptoms that may be present in advance. Teenagers are keenly aware that the number of followers, likes, and comments that they receive related to social media posts and pictures are a proxy for their overall popularity, and in the study participants who received the fewest likes did experience a significant amount of emotional distress and rejection due to an insufficient amount of validation in the form of positive interaction from social media. Even outside the confines of a controlled study like the Lee et al. (2020) investigation, readers can deduce adolescents are keenly aware in real time of their popularity based on the amount and degree of positive feedback they get from their social media platforms. During the teenage years, social media experience and exposure can play an important role in identity development, particularly for girls, as they toggle between the online and offline world developing social relationships and perceptions of self (Schrayer, 2022). Social comparison is a universal feature of the adolescent experience, but, for racial, cultural, ethnic, and sexual minorities, the amount and effect of these comparisons are exacerbated by the added discrimination faced online and offline (Malloy, 2022).

While many studies illustrate the challenges posed by a high degree of social media usage including body image concerns, depression, anxiety, cyber victimization, bullying in various forms, and the absence of social relationships offline, an important consideration is that social media platforms of various kinds are simply other places where adolescents and young adults congregate, and perhaps what is different and more concerning about social media as hang-out spaces is the fact that what adolescents experience offline is also in many cases experienced online and exacerbated by the much larger reach of an online profile. Social media platforms certainly have the potential to create a welcoming space for minority adolescents who belong to a marginalized group including but not limited to ethnic and racial minorities, LGBTQIA + individuals, religious minorities, and any group of people who have historic individual and group discrimination. In a study comparing sexual minority and nonminority youth, a content analysis found associations between social media and mental wellness are not well understood. Generally, LGBTQIA + individuals have positive and negative social associations, but that social capital was a positive effect of social media. Concerns for the LGBTQIA + population mostly centered around profile management and cyber victimization, which is to say that for this minority population of people, having a place to connect and make friends is vitally important, and social media provides ways to gain connections online when those connections are not possible in their own communities, very often in rural areas or in homes where being "out" could pose serious threats to safety and security. Compared with adolescents from the majority categories, LGBTQIA + adolescents require particular concern over who has access to their profile and management of their online friends and associates (Escobar-Viera et al., 2020).

Social media platforms and dating applications are increasingly a meeting space where adolescents and young adults connect with potential romantic partners. Personal safety and precautions must be understood to avoid dangerous situations. In a study by Ma et al., 2021, gender-diverse adolescents with greater experience in dating were more familiar with some of the specific concerns related to meeting potential romantic partners through social media and dating applications. Above all other concerns, Ma and colleagues, reported that safety is the number one element for gender-diverse adolescents. Reported second to safety was openness in their identity. Because gender-diverse and transgender adolescents are at greater risk for dating violence, they have to be particularly aware of whom they are meeting and the pragmatic protocols for doing so. Transgender and gender-diverse adolescents are more likely than heterosexual adolescents to meet and seek romantic partners online, so understanding the necessary precautions is in many ways built into their experiences of dating generally (Ma et al., 2021). Similarly, ethnic and cultural minorities face similar positive and negative effects based on the offline community in which they reside. Among all groups of people, individuals who experience negative thoughts and actions with social media may have associated mental problems independent of social media use.

Social media offers an important gateway to the world of similar communities for people with any number of exceptionalities, and at its best, social media platforms provide a vehicle for connection, sharing, and relationship building both online and offline. Adolescents and young adults with a variety of disability categories including autism often find a home in social media communities that would be difficult if not impossible if they live outside of an area that has opportunities to build these relationships natively. In particular, for groups of people who are marginalized in the offline communities in which they live, the springboard to a social life that social media facilities is a vital component of this developmental phase of life when establishing independent social relationships, developing a sense of personal identity, and connecting with peers is a benchmark of a healthy and full life. Deducing problematic social media behavior from other parts of behavior is difficult, and many studies have focused specifically on screen time. Authors have noted that screen time is problematic as a measurement metric because it is what adolescents and young adults are doing on the screens that is more important than a quantification of hours looking at a screen. Some researchers sought to identify and even codify problematic interactive media use (PIMU) generally around the amount of time spent using interactive media, defined in a number of ways including social media, Internet gaming, viewing inappropriate photos, and binge-watching videos and surfing websites to the degree that it involves functional impairment in other parts of life (Pluhar et al., 2019). From a mental health treatment standpoint, PIMU is categorized like other behavioral addiction disorders. PIMU is not, in and of itself, a discrete diagnosis, however.

For adolescents and young adults with a variety of mental health conditions, social media usage can both augment the problems and be a source of hope for those same problems. Mental health practitioners find that there are a variety of strategies used for other kinds of behavioral activities that can in extreme cases reveal themselves to be addictive. Still, mental health practitioners treat the specific mental disorder more so than social media usage broadly not only because of the difficulty of uncoupling social media usage from any other daily activity such as using the telephone, watching television, listening to music, playing video games, and any number of other behavioral activities that would typify how adolescents and young adults spend their time in activities not related to schooling or working. As of this writing, social media addiction is not officially a mental health diagnosis listed in the *Diagnostic and Statistical Manual of Mental Disorders* published by the American Psychiatric Association; however, social media addiction can be measured using objective instrumentation, the most common of which is the Bergen Social Media Addiction Scale (BSMAS), which employs Griffith's six core addiction elements: salience, mood modification, tolerance, withdrawal, conflict, and relapse (Schou-Andreason et al., 2016). Nevertheless, educators and mental health practitioners working with adolescents and young adults will find that because of the ubiquity of social media usage with adolescents and young adults, in particular, the specific mental conditions for which treatment is most indicated, for example, depression, anxiety, eating disorders, and adjustment, should be addressed more immediately.

As we move farther into the future and even where we stand today, it is nearly unheard of in the developed world for any adolescent or young adult to not have experience with social media of various categories, and from a treatment perspective, it is not practical or necessarily even possible to make any comparison to a world where social media is not part of young adults' lives in one way or the other. What's important for practitioners, parents, clinicians, and other helpers is to be able to understand the role, function, and individual experience that adolescents and young adults have with social media and what purpose it serves in their lives. Supported with information about social media experience including the purpose and function it serves in adolescents' lives, family and individual history of mental disorders, along with other risk factors, practical approaches can be devised to provide evidence-based, effective treatment for a variety of mental conditions, with the knowledge that social media can be both an impediment to and a contributor of outcomes that promote mental wellness and quality of life.

Summary

This chapter has provided a discussion about the use of social media by young adults that has enormous potential for both positive developments for social and professional connections, as well as concerns for mental health and focus. While new research should continue to be conducted to understand the impact of social media on young adults, existing research provides important understandings regarding mental health, social comparison, and bullying for professionals to reference in working with young adults. Technologies are ever-changing and advancing to connect people in new ways and benefit society with new tools for communication, and how people interact in both kind and unkind ways has also advanced. As technology has advanced for positive human interaction, so too has the platform availability for negative interaction (Guilbault & MacFarlane, 2020). Understanding the possibilities and the risks associated with social media usage and its impact on young people is an important responsibility for educators, caregivers, mental health service providers, and social media account users. Review the following questions and discuss with colleagues to consider the benefits and concerns for young adults in using social media.

Questions for discussion

1. List some of the possible impacts of social media use on young adults. What are some concerns about these impacts when internalized individually? What are external concerns?
2. What can professionals do to mitigate the impact of social media concerns among this population of young adults?
3. What specific mental health problems may be exacerbated by social media use?
4. What techniques can professionals use to help young adults who may exhibit unhealthy habits with social media?
5. Discuss practical approaches that are evidence-based, effective treatments for a variety of mental conditions that promote mental wellness and quality of life.

Website to explore

Pew Research Center

https://www.pewresearch.org/internet/2010/02/03/social-media-and-young-adults/

Pew Research Center is a nonpartisan fact tank that informs the public about the issues, attitudes, and trends shaping the world. They conduct public opinion polling, demographic research, content analysis, and other data-driven social science research. They do not take policy positions. Review this website. *Critically review their survey of social media use among young adults. Identify the strengths and weaknesses of the survey.*

Activity for the reader

For a practitioner working with young adults, ask a young adult to keep a log of the number of hours spent using social media accounts. On the log, record the type of social media activity, for example, reading, corresponding, creating content, etc. Ask the individual to reflect on their personal emotional response while using social media. Does the time produce positive or negative feelings toward self and toward others? How do they feel about the amount of time used for social media? What influences the amount of time spent on social media? What goals are impacted by their social media use?

Key terms used in the chapter

Approval anxiety: A type of stressor in the use of social media that pertains to users' concern about how others react to, perceive, and respond to messages they post.

Availability demands: A type of stressor in social media whereby users feel that they need to constantly respond to notifications and messages from others in a timely or immediate fashion.

Bullying: Belligerent behavior with the potential to cause physical, emotional, or psychological harm to anyone.

Connection overload: A type of stressor in the use of social media that refers to the excessive amount and variety of received information through social media.

Cyberbullying: Deliberate and repetitive harm perpetrated through the use of computers, cell phones, and other electronic devices.

Digital stress: To identify the cognitive, affective, and physiological arousal that accompanies notifications from or actual use of social media.

Fear of missing out (FOMO): A type of stressor in the use of social media that entails fear that somebody else is engaged in some rewarding activity where one is absent.

Influencers: In social media, people and organizations with a purported level of social knowledge or social influence in their field or interest area.

Mental health: Emotional, psychological, and social well-being. It influences how an individual may think, feel, and act. Many factors contribute to mental health including biological factors, life experiences, and family history.

Social comparison: Behavior that refers to individuals assessing themselves in comparison to certain aspects of others, such as individual behaviors, opinions, status, success, etc.

Social media: Internet-based and networked communication platforms that allow both personal and public communication.

References

Abi-Jaoude, E., Naylor, K. T., & Pignatiello, A. (2020, February 10). Smartphones, social media use and youth mental health. *CMAJ, 192*(6), E136−E141. https://doi.org/10.1503/cmaj.190434. PMID: 32041697; PMCID: PMC7012622.

Amran, M. S., & Jamaluddin, K. A. (2022). Adolescent screen time associated with risk factor of fear of missing out during pandemic COVID-19. *Cyberpsychology, Behavior and Social Networking, 25*(6), 398−403. https://doi.org/10.1089/cyber.2021.0308

Anderson, M., & Jiang, J. (2018). *Teens, social media and technology 2018*. Pew Research Center. https://www.pewresearch.org/internet/wp-content/uploads/sites/9/2018/11/PI_2018.11.28_teens-social-media_FINAL4.pdf.

Auxier, B., & Anderson, M. (2021). *Social media use in 2021: A majority of Americans say they use YouTube and Facebook, while use of Instagram, Snapchat, and TikTok is especially common among adults under 30*. Pew Research Center. https://www.pewresearch.org/internet/2021/04/07/social-media-use-in-2021/.

Bauer, A., Stevens, M., Purtscheller, D., Knapp, M., Fonagy, P., Evans-Lacko, S., & Paul, J. (2021). Mobilising social support to improve mental health for children and adolescents: A systematic review using principles of realist synthesis. *PLoS One, 16*(5), 1−23. https://doi.org/10.1371/journal.pone.0251750

Beeres, D. T., Andersson, F., Vossen, H. G., & Galanti, M. R. (2021). Social media and mental health among early adolescents in Sweden: A longitudinal study with 2-year follow-up (KUPOL study). *Journal of Adolescent Health, 68*(5), 953−960. https://doi.org/10.1016/j.jadohealth.2020.07.042. Epub 2020 Sep 14. PMID: 32943289.

Berryman, C., Ferguson, C., & Negy, C. (2018). Social media use and mental health among young adults. *Psychiatric Quarterly, 89*(2), 307−314. https://doi.org/10.1007/s11126-017-9535-6

Bohleber, L., Crameri, A., Eich-Stierli, B., Telesko, R., & von Wyl, A. (2016). Can we foster a culture of peer support and promote mental health in adolescence using a web-based app? A control group study. *JMIR Mental Health, 3*(3), e45. https://doi.org/10.2196/mental.5597

Buunk, A. P., & Gibbons, F. X. (2007). Social comparison: The end of a theory and the emergence of a field. *Organizational Behavior and Human Decision Processes, 102*(1), 3−21.

Centers for Disease Control and Prevention. (2022, June). *Data and statistics on children's mental health*. U.S. Centers for Disease Control and Prevention. https://www.cdc.gov/childrensmentalhealth/data.html.

Davies, S., Atherton, F., Calderwood, C., & McBride, M. (2019). *United Kingdom chief medical officers' commentary on 'screen-based activities and children and young people's mental health and psychosocial wellbeing: A systematic map of reviews*. UK: Department of Health and Social Care.

Erikson, E. H. (1994). *Identity and the life cycle*. Norton.

Escobar-Viera, C., Shensa, A., Hamm, M., Melcher, E. M., Rzewnicki, D. I., Egan, J. E., Sidani, J. E., & Primack, B. A. (2020). I don't feel like the odd one: Utilizing content analysis to compare the effects of social media use on well-being among sexual minority and nonminority US young adults. *American Journal of Health Promotion, 34*(3), 285−293. https://doi.org/10.1177/0890117119885517

Ettman, C. K., Abdalla, S. M., Cohen, G. H., Sampson, L., Vivier, P. M., & Galea, S. (2020). Prevalence of depression symptoms in US adults before and during the COVID-19 pandemic. *JAMA Network Open, 3*(9), e2019686. https://doi.org/10.1001/jamanetworkopen.2020.19686

Frith, E. (2017). *Social media and children's mental health: A review of the evidence*. Education Policy Institute. https://epi.org.uk/wp-content/uploads/2018/01/Social-Media%5fMental-Health%5fEPI-Report.pdf Accessed 8 January 2020.

Guilbault, K., & MacFarlane, B. (2020). Bullying. In C. Callahan, & J. Plucker (Eds.), *Critical issues and practices in gifted education: What the research says* (3rd ed., pp. 75−87). Prufrock Press.

Gunasekaran, K., Vasudevan, K., & Srimadhi, M. (2022). Assessment of mental health status among adolescents in Puducherry, India—A mixed method study. *Journal of Family Medicine and Primary Care, 11*(6), 3089−3094. https://doi.org/10.4103/jfmpc.jfmpc_2420_21

Hinduja, S., & Patchin, J. W. (2019). Connecting adolescent suicide to the severity of bullying and cyberbullying. *Journal of School Violence, 18*(3), 333−346.

Karazsia, B. T., Murnen, S. K., & Tylka, T. L. (2017). Is body dissatisfaction changing across time? A cross-temporal meta-analysis. *Psychological Bulletin, 143*(3), 293. https://doi.org/10.1037/bul0000081

Lee, H. Y., Jamieson, J. P., Reis, H. T., Beevers, C. G., Josephs, R. A., Mullarkey, M. C., O'Brien, J. M., & Yeager, D. S. (2020). Getting fewer "likes" than others on social media elicits emotional distress among victimized adolescents. *Child Development, 91*(6), 2141−2159. https://doi.org/10.1111/cdev.13422

Lenhart, A., Purcell, K., Smith, A., Zickuhr, K., & Pew Internet and American Life Project. (2010). *Social media & mobile internet use among teens and young adults. Millennials*. Pew Internet & American Life Project.

Ma, J., Korpak, A. K., Choukas-Bradley, S., & Macapagal, K. (2021). Patterns of online relationship seeking among transgender and gender diverse adolescents: Advice for others and common inquiries. *Psychology of Sexual Orientation and Gender Diversity, 9*(3), 287−299. https://doi.org/10.1037/sgd0000482

Maheux, A. J., Roberts, S. R., Nesi, J., Widman, L., Choukas, B. S., & Choukas-Bradley, S. (2022). Longitudinal associations between appearance-related social media consciousness and adolescents' depressive symptoms. *Journal of Adolescence, 94*(2), 264−269. https://doi.org/10.1002/jad.12009

Malloy, D. L. (2022). The effects of negative social media in the depressive symptomology of African American adolescents [ProQuest Information & Learning]. In *Dissertation abstracts international: Section B: The sciences and engineering* (Vol. 83, Issue 5−B).

Marsh, N., Fogleman, N., Langberg, J., & Becker, S. (2022). Too connected to being connected? Adolescents' social media emotional investment moderates the association between cyber-victimization and internalizing symptoms. *Research on Child Adolescent Psychopathology, 50*(3), 363−374. https://doi.org/10.1007/s10802-021-00867-0

Meikle, G. (2016). *Social media: Communication, sharing, and visibility*. Routledge: Taylor & Francis Group.

Mundy, L. K., Canterford, L., Moreno-Betancur, M., Hoq, M., Sawyer, S. M., Allen, N. B., & Patton, G. C. (2020). Social networking and symptoms of depression and anxiety in early adolescence. *Depression and Anxiety, 38*(5), 563−570. https://doi.org/10.1002/da.23117

Oppong-Asante, K., Quarshie, E. N.-B., & Andoh-Arthur, J. (2021). COVID-19 school closure and adolescent mental health in sub-Saharan Africa. *International Journal of Social Psychiatry, 67*(7), 958−960. https://doi-org.library.capella.edu/10.1177/0020764020973684.

Orben, A. (2020). Teenagers, screens and social media: A narrative review of reviews and key studies. *Social Psychiatry, 55*, 407−414. https://doi.org/10.1007/s00127-019-01825-4

Peterson, J. S., & Ray, K. E. (2006). Bullying and the gifted: Victims, perpetrators, prevalence, and effects. *Gifted Child Quarterly, 50*(2), 148−168.

Pluhar, E., Kavanaugh, J. R., Levinson, J. A., & Rich, M. (2019). Problematic interactive media use in teens: Comorbidities, assessment, and treatment. *Psychology Research & Behavior Management, 12*, 447−455. https://doi.org/10.2147/PRBM.S208968

Santoso, S., & Sutedjo, B. (2019). Structural relationship between social benefit, dependency, satisfaction, and habit formation on the use of social media. *Binus Business Review, 10*(1), 51. https://doi.org/10.21512/bbr.v10i1.5407

Schou-Andreassen, C., Billieux, J., Griffiths, M. D., Kuss, D. J., Demetrovics, Z., Mazzoni, E., & Pallesen, S. (2016). The relationship between addictive use of social media and video games and symptoms of psychiatric disorders: A large-scale cross-sectional study. Psychology of addictive behaviors. *Journal of the Society of Psychologists in Addictive Behaviors, 30*(2), 252−262. https://doi.org/10.1037/adb0000160

Schrayer, A. (2022). The not-so imaginary audience: The impact of social media on identity development in adolescent girls [ProQuest Information & Learning]. In *Dissertation abstracts international: Section B: The sciences and engineering* (Vol. 83, Issue 5−B).

Skaik, R., & Inkpen, D. (2021). Using social media for mental health surveillance: A review. *ACM Computing Surveys, 53*(6), 1−31.

Steele, R. G., Hall, J. A., & Christofferson, J. L. (2020). Conceptualizing digital stress in adolescents and young adults: Toward the development of an empirically based model. *Clinical Child and Family Psychology Review, 23*(1), 15−26. https://doi-org.library.capella.edu/10.1007/s10567-019-00300-5.

Thomee, S., Dellve, L., Harenstam, A., & Hagberg, M. (2010). Perceived connections between information and communication technology use and mental symptoms among young adults: A qualitative study. *BMC Public Health, 10*(66). https://doi.org/10.1186/1471-2458-10-662836296

Vidal, C., Lhaksampa, T., Miller, L., & Platt, R. (2020). Social media use and depression in adolescents: A scoping review. *International Review of Psychiatry, 32*(3), 235−253. https://doi-org.library.capella.edu/10.1080/09540261.2020.1720623.

Wang, M., Yogeeswaran, K., Andrews, N., Hawi, D., & Sibley, C. (2019, November). How common is cyberbullying among adults? Exploring gender, ethnic, and age differences in the prevalence of cyberbullying. *Cyberpsychology, Behavior, and Social Networking*, 736−741. https://doi.org/10.1089/cyber.2019.0146

Warrender, D., & Milne, W. (2020). How use of social media and social comparison affect mental health. *Nursing Times [online], 116*(3), 56−59.

Watson, J. C., Prosek, E. A., & Giordano, A. L. (2022). Distress among adolescents: An exploration of mattering, social media addiction, and school connectedness. *Journal of Psychoeducational Assessment, 40*(1), 95−107.

World Health Organization. (2021, November). *Adolescent mental health*. World Health Organization. https://www.who.int/news-room/fact-sheets/detail/adolescent-mental-health.

Zimmer-Gembeck, M. J., Rudolph, J., Webb, H. J., Henderson, L., & Hawes, T. (2021). Face-to-face and cyber-victimization: A longitudinal study of offline appearance anxiety and online appearance preoccupation. *Journal of Youth & Adolescence, 50*(12), 2311−2323. https://doi.org/10.1007/s10964-020-01367-y

Further reading

Mental Health.gov. (2022). *What is mental health?*. https://www.mentalhealth.gov/basics/what-is-mental-health.

Pew Research Center. (2018). *How teens and parents navigate screen time and device distractions*. August.

Chapter 4

Social media: utility versus addiction

Ram Lakhan[1], Bidhu Sharma[2] and Manoj Sharma[3]
[1]*Department of Health and Human Performance, Berea College, Berea, KY, United States;* [2]*School of Arts and Science, University of Louisville, Louisville, KY, United States;* [3]*Department of Social and Behavioral Health, School of Public Health, University of Nevada, Las Vegas, NV, United States*

Objectives

✔ Describe the utility and addiction of social media.

✔ Explain how social media meets the addiction definition.

✔ Discuss the theoretical basis of addictions to social media.

✔ Identify the prevalence of social media usage worldwide.

✔ Illustrate the role of the COVID-19 pandemic in increasing social media usage.

✔ Assess mental, physical, and social consequences of excess social media use.

Introduction

What is social media in our life? To answer this question, let us imagine you are about 40 years old, then try to go back to 20 years ago in time imagining you are sitting in a classroom with 20 other students. Your instructor has not arrived yet. What was everyone doing? Maybe working on assignments, reading books, standing in the hallway, and chatting with friends about academics, sports, or some other topics of their interest. Now bring yourself back to today's class. What do you see? Most of the students are on their phones.

They are not necessarily on social media, but can you say all of them are doing what used to happen about 20 years ago in such situations? Which act is more beneficial, being on the phone in modern classes or with friends and books way back when we lacked advanced and affordable technology? Let us imagine another scenario. You invited relatives, friends, and families to celebrate an accomplishment, wedding anniversary, birthday, or something else. About seven to eight children aged range from 4 to 10 years came along with their parents to this party. All of the adults are sitting and chatting. What are these kids doing? Can you think about a time when we did not have mobile phones? What would their behaviors have been? They would have been running around and doing all sorts of behaviors ranging from screaming, fighting with other kids, asking the attention of their parents, not allowing them to talk or do anything, and so on. In today's situation, you might see all these kids sitting quietly with their electronic devices in their hands completely disconnected from their surroundings. No wonder, they might have headphones on their ears, so no one feels disturbed by the noise. People at the party might want to talk with these kids, but they must find ways to get their attention even for seconds. Decades ago, these kids use to talk nonstop, and now their talking got limited to yes and no words in such settings. The behavior of adults no matter young, middle age, older adults, male, female, poor, rich, rural, urban, educated, or uneducated has gotten strongly associated with social media.

The question then arises, why has social media become such a strong component of life in all societies today? The simple answer is that humans inherently need social networking. If we look at the operating mechanism of any structure in society including family, relatives, friends, fraternities, social or interest groups, parties, or even larger communities and populations, the single most important component that connects individuals in and outside of these units is considered communication. Communication is an innate need of humans. To fulfill communication needs, humans have developed a variety of tools and techniques since the inception of civilization (Dissanayake, 2003). Language, sound symbols, and scripts are the major inventions in human history that have been helping people to express their needs and keep social structures functioning. Letter writing and sending it to people far might be the first step in human history for keeping people connected at distance. Later, the invention of the phone made communication even easier, faster, and real. A recent invention of the Internet has advanced all forms of communication. Now with the help of the Internet, a written message can be sent in seconds and people can talk seeing each other on-screen in real time, irrespective of distance. The need and availability of resources to fulfill is a perfect regime of behavior expression. The need for communication and connectedness and the availability of electronic devices and the Internet complement each other behaviorally and lead to social media usage.

Social media provides innumerable benefits including psychological, social, learning, cognitive development, social capital, entertainment, and even in business and employment (Andreassen et al., 2016; Gómez-Galán et al., 2020; Picazo-Vela et al., 2012). The Internet is replacing many established structures that were used as a source of information previously. For example, how many people carry paper maps in their vehicles? How many people write letters nowadays? Earlier, people use to store print photos in albums, and now these pictures are stored digitally. People can easily find out answers to most of their questions. All of these conveniences increase the usage of social media. It is not an exaggeration that we have started operating in a virtual world. Traditional societies have started turning into virtual societies (Dey et al., 2019). Excess usage of social media is also creating multiple challenges. Psychological problems, lack of communication skills, reduced listening skills, emotional withdrawal, anxiety, depression, sleep problems, eating habits, self-esteem, relationship difficulties, and self-decentralization are the most challenging negative consequences of social media (Yayman & Bilgin, 2020; Gómez-Galán et al., 2020; Magnuson & Dundes, 2008; Molavi et al., 2018). The excessive use of social networking sites has been found to have negative effects on individuals and their relationships (Zheng & Lee, 2016).

Prevalence of social media use

It is estimated that out of the current world population of 7.96 billion people, five billion people (equal to 62.81% of the total world population) use the Internet. About 4.6 billion, which is 61.55% of the total world population, use social media. On average, social media users spent about 147 min per day on social media (Statista, 2022). Presently, Asia leads with the highest number of social media users. Presently, about 26% of total social media users in the world come only from Eastern Asia followed by Southeast (18.44%), Southern Asia (11%), and Western Asia (3.57%). North America stands in the second position with 9.47% of users, while 6.68% of users come from South America. About 5.7% of users come from Central & Western Europe, 4% from Eastern Europe, and 3.99% from Southern Europe. About 3.36% of users come from Northern Africa, and less than 2% from Western and Eastern Africa each. A survey conducted in 2021 looked at the purpose of social media usage and found that close to half (47.6%) of the user population use social media for staying in touch with family and friends. The second leading purpose of social media usage was found to fill spare time (36.3%), read the news (35.1%), find needed content (31.6%), making new contacts (23.9%), posting about their own life (21.4%), and many other causes including watching live streams, following celebrities, what is being talked about, sharing, and discussing opinions of other people (24%) (Dixon, 2022a). In the year 2020, about 3.6 billion people in the world had used social networks. It is projected that this number will cross 4.4 billion users by 2025. The projection of the top 10 social media-using nations is provided in Fig. 4.1 (Dixon, 2022b). On average, females spend about 63 and males 49.5 min per day on social media

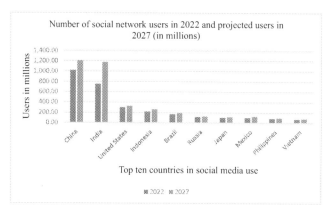

FIGURE 4.1 Number of social network users in 2022 and projected users in 2027 (in millions). Note: The information was adapted from the Statista website available at this link (https://www.statista.com/statistics/278341/number-of-social-network-users-in-selected-countries/) (Dixon, 2022b).

(Ceci, 2022a). Age-wise, people under 24 years spend under 49.8 min, in the age group 30 to 49 spend between 58 and 59.5 min, and those 50 and above spend 56.4 min per day (Ceci, 2022b). According to the Pew Research Center (2022), there were only 5% of adults in America used social media in 2005. By February 2021, about 72% of adults have used some type of social media (Pew Research Center, 2022).

Social media during the COVID-19 pandemic

Social media use has been on an increasing trend right from its inception because of its associated benefits. The COVID-19 pandemic appeared as a blessing for social media to expand even faster. The pandemic created one of the worst situations in human history. A worldwide lockdown, social distancing, and several other restrictions placed on people to contain the virus from spreading forced everyone to stay away from each other. Work from home, school from home, and even recreational activities from home were encouraged and practiced (Sobaih et al., 2020). A lack of socialization was one of the major issues felt by most people. The rate of depression, anxiety, stress, sleep problems, interpersonal relationship issues, boredom, loneliness, physical inactivity, and suicidal ideation were some of the issues which started increasing drastically around the world (Misirlis et al., 2020). Kids were not able to go to school and interact with friends, while adults were not able to see and interact with their colleagues the way they were used to before. The need of retaining socialization to prevent psychological health issues during a challenging time of increasing mental health issues became a major concern (Boursier et al., 2020). Online interaction was promoted. The crisis created opportunities for companies to develop more user-friendly online websites for various purposes including education, training, meetings, business, health, and for socialization. From every side, online products and behavior were promoted. People staying at home had more time than before. Even people who never thought of having social media accounts became lenient to it and started using them. Young kids who spend a majority of their time either at school or daycare centers needed something to do after completing their online schoolwork at home. Social media was found to be one of the best tools in that situation, and that was opted by parents to engage their kids and they also used the same platform for interaction with their extended family members, relatives, friends, and colleagues (Mirbabaie et al., 2020). Never in history did social media receive such a favorable response in any community as it got during the COVID-19 crisis. Presently, the world appears to be out of the pandemic, but the learned behavior related to social media is not easy to undo. Two years of need-based use of social media have become a part of many people's lives now. Indirectly, it can be inferred that the COVID-19 pandemic established social media strongly in people's lives and it would only be expanding in the future (Cheng & Lau, 2022; Hamilton et al., 2022).

The epigenetic effect is a future concern

Social media use was increasing in the population at an exponential rate, but its justified expansion during COVID-19 made a larger population adapt to the new behavior of using these technical resources, for a variety of reasons. A new virtual environment was developed during the pandemic phase (Zhao & Zhou, 2021). An environment can lead to epigenetic influences, and in the case of COVID-19, it is happening in the population which cannot be denied (Sen et al., 2021). Epigenetics is explained as the environment having the capability of changing phenotype in progeny without altering the genotype. The change could be very favorable to the environment and setting, but that is hard to predict. Young people who planned their babies during this pandemic had strong environmental effects and a good chance of having an epigenetic effect on them which might have been expressed as a technology-favoring attitude among the babies who were conceived during that time. If there is any such epigenetic influence that happened, then most likely a large percentage of children would seek a virtual world to operate in the future. A longitudinal study carried out during the pandemic phase has found an epigenetic effect of passing stress-related traits among babies who were born from parents who were stressed out because of COVID-19 (Provenzi et al., 2021b). Another study also suggested similar epigenetic effects on babies born from parents who experienced anxiety and stress during COVID-19 (Provenzi et al., 2021a). Studies conducted by Provenzi et al. (2021b) and Provenzi et al. (2021a) can support the epigenetic effect on babies for technology.

Addiction to social media

People are becoming cognizant of the negative effects of social media. The behavior of social media usage has gotten extremely strong and become a core activity of one's daily routine. Despite their best efforts, people find it difficult to stop it. Several studies have found the overuse of social media associated with low work performance, less healthy social relationships, low life satisfaction, sleep problems, anxiety, and depression (Sun & Zhang, 2021). Poor self-regulation and compulsive use of social media are being termed social media addiction. It is often used interchangeably with problematic use of social media terms in the literature. Studies show how difficult it is to regulate social media use. It appears like following an addictive pattern, but should we call it social media addiction (Bányai et al., 2017)? Addiction as a term has historical connections in defining substance use disorders. Researchers believe that pathologizing social media usage may undermine the severity of psychiatric disorders (Carbonell & Panova, 2017).

The addiction definition which applies to substance-related issues does fit partially for habitual social media use. Experts believe that addiction can be categorized into two broad categories, chemical and behavioral addiction. The set four criteria for chemical addiction are very strong (McNeely & Adam, 2020) and do not meet optimally for social media use. Probably, a psychological view of social media use in relation to a psychiatric view of addiction can help to understand it as a behavioral addiction (Alavi et al., 2012).

The true concept of addiction applies to substance use, and it does not fit fully in explaining social media addiction. However, consistent substance addiction and excessive social media use lead to social impairment, functional impairment in work, social relationships, and social situations (Andreassen & Pallesen, 2014). The point-to-point comparison provided in Table 4.1 also indicates several full and partial similarities with addiction criteria set for chemical addiction. Following Peele's model of addiction, several researchers developed the idea that addiction does not necessarily have to involve the abuse of substances (Alavi et al., 2012; Griffiths, 1999; Young, 2004). According to Peele's model, addiction is a dependency on a set of experiences. Substance addiction is only one of the most common examples (Alavi et al., 2012; Peele, 1977).

The theoretical basis of social media addiction

Excess use of social media whether it meets the criteria of being called addiction or not can certainly be considered a behavioral addiction or at least problematic social media. Chemical dependency originates through the physiological responses of the body, while behavioral dependency emerges through behavioral responses on an individual that he receives in the social-ecological frame which includes the individual self, people in network,

TABLE 4.1 Social media as a behavioral addiction.

Addiction criteria	Indicators	Social media	Behavioral addiction
Impaired control	Consuming the substance in larger amounts and for a longer amount of time than intended. Persistent desire to cut down use. Unsuccessful attempts to stop in past. Spending a great deal of time obtaining, using Persistent desire to use the substance.	There is evidence that many people spent too much time on social media. They attempt to regulate that behavior but find it unsuccessful. People who are not using social media find that they are in persistent need of using it. The criteria of spending a great deal of time accessing social media are less relevant here.	Behaviorally, easy access and affordable access, and social acceptance of social media use have played a great role in promoting all other indicators for impaired control. Out of four set indicators, social media use meets at least three indicators partially and one indicator acts as a setting reinforcer for strengthening the behavior.
Social impairment	Impairs ability to fulfill major obligations. Continued use even significantly affects social or interpersonal life. Reduces recreational, social, and occupational activities.	For social media, all these indicators may be true to a great extent. Heavy users of social media find it difficult to meet the major obligations of home, school, and work. Several studies have found that social media impacts people's social and interpersonal life. Social media is itself a strong recreational activity, so this indicator needs to be looked at critically.	Comparing social media use with addiction definition, most of the indicators were found to be strongly correlated. However, in substance use addiction, individual increases disconnection with recreational activities, while in the case of social media, it is one of the strongest setting

Continued

TABLE 4.1 Social media as a behavioral addiction.—cont'd

Addiction criteria	Indicators	Social media	Behavioral addiction
			reinforcers that lead to a reduction of an individual's participation in other activities. This factor could be seen as one of the strongest setting reinforcers that leads to overuse of social media. Often the objective of being connected with people triggers social media use. The question arises, should a person need to follow a variety of recreational activities, or it should be seen that social media use decreases all other types of recreational activities including physical, sports, and cognitive activities carried out face to face for pleasure and intellectual stimulation?
Risky use	Frequent use in physically unsafe environments. Persistent	Substance use leads to physical and psychological effects on the body. Users and abusers often understand the ill effects of	The social media can be riskier in case a person due to low

TABLE 4.1 Social media as a behavioral addiction.—cont'd

Addiction criteria	Indicators	Social media	Behavioral addiction
	substance uses even with the knowledge that it leads to physical and psychological poor health.	substances. The use of substances in physically unsafe places is common despite knowing the related risk. In relation to social media, the physically unsafe place is less relevant. However, there is a risk of exploring the risker website and getting a victim of scams. Also, certain content could be more attractive to the people depending on their interests as well as personality attributes which may lead to excess exploration of such content and drift from the major objective of life. The physical effect of social media is not direct, but spending more time on social media reduces the time for physical activities and physical inactivity leads to poor health.	understanding and awareness or by choice might use predatory platform. Physical consequences are indirect, while poor psychological health has been associated with the overuse of social media. In relation to it is safe to say that social media partially meets this criterion of addiction definition. However, the connection is not as strong as with substances.
Pharmacologic	Individual increases their dose to achieve a level of satisfaction (tolerance). Individuals experience psychological uneasiness and emptiness. The physical discomfort arises in the absence of used substances, and physiological response to it leads to altered	The use of social media is a behavioral act; therefore, its comparison with substances in relation to pharmacological effects is quite difficult. Behaviorally, it is known if an activity that leads to pleasure or any type of gratification, individuals tend to increase time on the activity to prolong gratification. Studies have found an increasing trend in terms of time spend on it among beginners. More time spends on social media leads to less time for other activities. Therefore, it can	Behaviorally, excess social media use can have partial tolerance and withdrawal effect. However, this connection cannot be seen like substance addiction, but the underlying mechanism and indirect outcome match with these criteria.

Continued

TABLE 4.1 Social media as a behavioral addiction.—cont'd

Addiction criteria	Indicators	Social media	Behavioral addiction
	functions and physician discomfort (withdrawal).	be said that tolerability increases against the other important activities. It is difficult for an individual to enjoy comfort if routine pleasure is missing. In other words, people who use social media on a daily basis and if not can use or stop, find it difficult to manage their temptation. These feelings could be like a withdrawal effect. Neurobiological theories explain how the feeling of happiness while using social media triggers a release of neurotransmitters. Neurotransmitters are chemical agents which act on physiology.	

institutions, policies, communities, and populations. Some theoretical understanding of how the behavior of social media usage is achieving behavioral addiction in the population is needed. Currently, social media addiction is an important research topic in public health. Intervention can be only planned effectively if there is a good theoretical basis available that explains the addiction or habitual and problematic social media usage (Yayman & Bilgin, 2020). However, there is not yet any exclusive theory available that explains addictive social media use. Currently, several preexisting theories of various disciplines ranging from psychological, social, and neurobiological are being explored. Recently, Sun and Zhang (2021) have presented and categorized several theories found applicable at some level in explaining social media addiction in their review of 55 empirical studies. A total of 25 theories/models were found to be researched in recent literature which Sun and Zhang (2021) have presented in eight categories on a basis of broader discipline and the origin of theories. All eight categories of all 25 theories/models presented by Sun and Zhang (2021) are briefly described here for the readers (Table 4.2).

The theoretical explanation of social media addiction indicates that social and psychological reasons play a pivotal role. The social and emotional relatedness, attachment, connection and ties with family, relatives, and friends,

TABLE 4.2 Common theories/models applicable in explaining social media addiction.

Categories with their perspective	Theory/Model	Relevance to addiction
Dispositional difference	Attachment theory Time perspective theory Social cognitive perspective theory	Argues that people with certain dispositions are more likely to develop overuse of social media. In this group, attachment theory is used the most. According to that, people see relationships and connections in the way they experienced their childhood with parents and caregivers. People often operate in attachment anxiety and attachment avoidance. Attachment anxiety promotes closeness with the people who approve of the person, and attachment avoidance is something where a person avoids contact and practices self-reliance. Social media is a comfortable and easy platform that fulfills both side objectives for the person.
Motivational	Uses and gratification theory Self-determination theory Flow theory Belongingness theory Self-escape theory	Theories basically operate in the realm of motivation. Social gratification and sociopsychological needs act as driving forces for people to use social media. Receiving social support, maintaining relationships, and interacting with others are some basic objectives fulfilled for social gratification. A feeling of belongingness and relatedness is also considered motivational needs which fulfill an individual's psychosocial needs. People can easily fulfill many of the above-listed needs through social media by making presentations and posting about their achievements.
Neurobiological	Incentive-sensitization theory Dual system theory	These theories explain how pleasant stimulation activates the reward center in the brain. The brain gradually starts seeking stimulation for activating the reward system over and over. This cycle gets physiologically strong as a person

Continued

TABLE 4.2 Common theories/models applicable in explaining social media addiction.—cont'd

Categories with their perspective	Theory/Model	Relevance to addiction
		indulges more and more in social media use.
Decision-making	Rational addiction theory Theory of planned behavior	These theories basically apply in the sense that a person makes a subjective perception that social media is beneficial to them. The value these people may see from the overuse of social media may not be true, but that is how they rationalize and end up promoting more social media use.
Learning	Classical conditioning Operant conditioning Stimulus-response reinforcement framework Social learning theory Social cognitive theory	Learning a new skill, no matter how strongly it is associated with reward or how it originated provides some sense of happiness, accomplishment, and overall impacts on self-efficacy. The confidence of a person is encouraged in the process. Social media is something where people can see their success instantly. For example, someone posting a picture on Facebook may receive likes within seconds. The action and reward, especially too quick in connection to social media play an important role in increasing the likelihood of its use.
Technology-focused	Theory of technology frame Technology acceptance model Needs affordances-features	Focuses on the ease of using and accessing a variety of quality information. The ability to customize and present in multiple ways. Technology can be also a stressor, but in the same way, it can be a distractor, or in another world, it can play role in destressing a person. For example, a person who watches movies when gets stressed due to anything that happens in her life. In such an event, it is hard to find time to travel and spend money to go watch the movie in the theater. Social media can provide the option of watching one's favorite movie or show instantly.

TABLE 4.2 Common theories/models applicable in explaining social media addiction.—cont'd

Categories with their perspective	Theory/Model	Relevance to addiction
Social network-focused	Social influence theory Social capital model	The social network itself sometimes comes as a cognitive and social force to use social media. Social ties and social support act as structural social capital and cognitive social capital for the person. Individuals want to maintain support which indirectly promotes social media use.
Internet-specific model	Davis's cognitive-behavioral; model of pathological Internet use Caplan's social skill model of problematic Internet use Interaction of person-affect cognition-execution model	These models in a nutshell talk about the interconnection between psychology, social cognition, personality, and psychological conditions. The predisposing factors of the four-dimension interplay and lead to maladaptive use of Internet behavior which further increases as overuse or addictive social media.

Note. Adapted from Sun, Y., & Zhang, Y. (2021). A review of theories and models applied in studies of social media addiction and implications for future research. Addictive Behaviors, 114, 106699. https://doi.org/10.1016/j.addbeh.2020.106699.

avoidance, social and cognitive gratifications, sense of social and cognitive capital, recognition and acknowledgment, self-presentation, and acquisition of all types of information quickly and in various formats are some of the fundamental social needs that people are easily able to fulfill through social media (O'reilly, 2020; Pittman, 2018; Khan et al., 2014; O'Keeffe et al., 2011). It is hard to undo any need to reduce the use of social media (Abd Rahman, 2014). A meta-analysis of 32 studies conducted by Cheng et al. (2021) showed that the prevalence of addictive social media is higher in Asia, Africa, and the Middle East compared to higher-income nations in North America and Western and Northern Europe (Cheng et al., 2021). The current rate of addictive social media is most likely to increase. On one side, social media provides numerous benefits, but its addiction leads to many harmful effects (O'reilly, 2020; Thorisdottir et al., 2019; Yayman & Bilgin, 2020). Some of the problematic use of social media and their outcomes are discussed next.

Studies presented in Table 4.3 from several countries have found that social media and social networking site addiction or overuse of social media and compulsive use of media have significant negative relationships with mental and physical health. The rate of depression, anxiety, stress, poor sleep quality,

TABLE 4.3 Studies demonstrating problematic consequences of social media addiction.

Authors	Country	Study design	Findings
Giordano et al. (2021)	USA	Cross-sectional research design. A national sample of 428 adolescents was included. Relationships between social media addiction, school connectedness, depression/anxiety, and cyberbullying were assessed.	A positive association was observed between excess social media use and depression and anxiety combinedly.
Stockdale and Coyne (2020)	USA	Longitudinal data of three years were used. The cross-sectional research design was followed. Sample size 385. The relationship between motivations for social networking site use and behavior was analyzed.	Initially social networking alleviates boredom, but as time passed, social media use became problematic and lead to anxiety.
Holmgren and Coyne (2017)	USA	Cross-sectional research design. Sample size was 442. Examined association of self-regulation with pathological use of social networking sites and rational aggression and depression.	Some level of association was found with social networking site usage and rational aggression and depression. However, rational aggression and depression do not associate with higher use of social media.
He et al. (2021)	China	Cross-sectional design. The sample included 218 females. The research investigated the relationship between family socioeconomic status, stress, impulsiveness and inhibitory control, and social media addiction.	A positive relationship was found between stress with problematic social media use.

TABLE 4.3 Studies demonstrating problematic consequences of social media addiction.—cont'd

Authors	Country	Study design	Findings
Chen et al. (2020)	China	Cross-sectional. The sample included 437 Chinese young adults. The relationship between social networking site addiction with social anxiety was assessed.	Social anxiety was found positively associated with social networking site addiction.
Kim et al. (2020)	China	Cross-sectional study. The sample included 209 Chinese high school students. Social media addiction was observed in association with fatigue and anxiety.	A positive relationship was found between social networking sites' addiction to fatigue and anxiety.
Hou et al. (2019)	China	Cross-sectional research. The sample included 641 Chinese college students. Data on perceived stress, depression/anxiety, psychological resilience, social support, and problematic SNS use were collected and analyzed.	The problematic social networking site (SNS) use was found associated with depression, anxiety, and stress.
Wang et al. (2018)	China	Cross-sectional research design. Sample population 365. A theoretical model was assessed to see the association between social networking site addiction and depression. Mediation and moderated mediated analysis were conducted.	A positive association was found between social networking sites addiction and depression.
Kircaburun et al. (2020a)	Turkey	Cross-sectional design. The sample included 344 university students. The study aimed to assess the association between problematic use of social media and depression and dissociation.	A favorable association between problematic social media use was found with depression.

Continued

TABLE 4.3 Studies demonstrating problematic consequences of social media addiction.—cont'd

Authors	Country	Study design	Findings
Kircaburun et al. (2020b)	Turkey	Cross-sectional research design. Sample 460 young adults. The association between personality traits and mental health factors including depression and loneliness was assessed.	Problematic social media was found associated with depression.
Kilincel and Muratdagi (2021)	Turkey	Cross-sectional research design. The sample included 1142 children between the ages of 12 and 18. The relationship of social media use during the pandemic period with anxiety and a sense of loneliness was assessed.	Association between social media use and anxiety was observed.
Kircaburun et al. (2019)	Turkey	Cross-sectional research. A sample of 470 was used. Emotional intelligence as a risk factor for problematic social networking sites was assessed with the presence of depression.	A positive relationship between problematic social networking sites and depression was observed.
Wong et al. (2020)	Hongkong	Cross-sectional design. Sample size 300. Association of Internet gaming and social media with sleep quality, anxiety, depression, and stress was observed.	Social media use and Internet gaming were found positively associated with poor sleep quality, depression, anxiety, and stress.
Brailovskaia et al. (2021)	Lithuania and Germany	Cross-sectional study design. The sample included college students in Lithuania 1640 and Germany 727. The association between problematic social media use was assessed with the life satisfaction factors	In both student groups, an association was observed with social media addiction and life satisfaction factors including

TABLE 4.3 Studies demonstrating problematic consequences of social media addiction.—cont'd

Authors	Country	Study design	Findings
		(depression, anxiety, and stress).	stress, anxiety, and depression.
Mitra and Rangaswamy (2019)	India	The cross-sectional research design used data from 264 participants. The study examined the relationship between overuse of social media and depression.	A significant positive correlation was found between overuse of social media and depression.
Sujarwoto et al. (2021)	Indonesia	Cross-sectional research. The sample included 709 university students. Association between social media addiction and mental health.	Association was found between social media addiction and mild depression.
Apaolaza et al. (2019)	Spain	Cross-sectional research. Sample 346 young adults. Research aimed to see the effect of mindfulness in lowering stress and social anxiety.	A positive correlation was observed between compulsive mobile social networking sites with stress.
Hökby et al. (2016)	Estonia, Hungary, Italy, Lithuania, Spain, Sweden, and the United Kingdom	Cross-sectional. A random sample of 2286 adolescents was collected. The aim of the study was to assess mental health's association with the time spent on the Internet and the type of website were visited.	Association between time spent with loss of sleep was found. Over a period, this led to mental health issues.
Thomée et al. (2012)	Sweden	Longitudinal data for one year were followed. Sample included 4163 young adults. The objective was to assess if high computer use is a prospective risk factor for developing mental health symptoms.	Results indicated the loss of sleep in males and mental health issues in females.
Kadam et al. (2016)	India	Cross-sectional. The sample included 890 college students. The relationship of	Mobile use at night was found associated with

Continued

TABLE 4.3 Studies demonstrating problematic consequences of social media addiction.—cont'd

Authors	Country	Study design	Findings
		sleep with mental health issues and other risk factors was assessed.	poor sleep quality and feelings of depression.
Brunborg et al. (2011)	Norway	Cross-sectional. The total sample was 2500. Postal letters were sent to all. 816 returned which were analyzed. The study attempted to investigate the use of social media in the bedroom and its effect on sleep habits.	Use of computers and mobile phones in the bedroom was found associated with poor sleep habit.
Rod et al. (2018)	Denmark	Cross-sectional design. High-resolution information on the timing of smartphone activity (based on >250,000 phone actions). Sample 850. The study assessed smartphone uses at night and its effect on sleep and physical and mental health among adolescents.	Smartphone users were found to have about 48 min of less sleep due to sleep disturbances.
Liu et al. (2019)	China	Cross-sectional study. The sample included 11,831 adolescents. The study examined the association between the duration of mobile phone use and depressive symptoms.	Prolong use of mobile phones found to be associated with depressive symptoms.
Brailovskaia and Margraf (2020)	Germany	Cross-sectional design. A total of 638 users of social media were recruited. The connection between social media use, physical activity, and depression was explored.	Increased social media use decreased physical activities. Depression level increased. Also, for those who had depression, social media addiction occurred.

TABLE 4.3 Studies demonstrating problematic consequences of social media addiction.—cont'd

Authors	Country	Study design	Findings
Masthi et al. (2017)	India	Cross-sectional research design. The sample included 760 high school students. Research aimed to explore the relationship between social media use and health concerns.	Over 60% of students have excessive social media use. About 19.9% person considered to be addicted. Social media-addicted students had at least one physical issue ranging from neck pain, tension, the strain on eyes, and fatigue, while psychological issues including anger, loneliness, and frustration; and behavioral changes, that is, sleep disturbance and neglect of personal hygiene were found commonly.
Azizi et al. (2019)	Iran	The study type was cross-sectional. A total of 360 students were included in the sample. The relationship between social media use and academic performance was observed.	Social networking was found higher among males than females and associated with poor academic performance.
Ndubuaku et al. (2020)	Nigeria	Cross-sectional research design. The sample included 400 students.	Social networking addiction was found associated with poor

Continued

TABLE 4.3 Studies demonstrating problematic consequences of social media addiction.—cont'd

Authors	Country	Study design	Findings
			academic performance among students.
Alotaibi et al. (2022)	Saudi Arabia	Cross-sectional research design. The sample included 545 undergraduate students. The study aimed to see the relationship between social media use with academic performance and physical health.	Addicted participants were found to have poor academic performance, be physically inactive, have poor sleep, have a mental illness, be overweight/obese, and have pain in their shoulder, eyes, and neck.

and poor physical health are the main negative consequences of social media found in cross-sectional studies (Alotaibi et al., 2022; Brailovskaia & Margraf, 2020; Kadam et al., 2016; Rod et al., 2018; Shannon et al., 2022).

Social media use, even if it is not addictive, also brings several types of risks for its users. The young and adolescents face even greater risks than grown-up adults. The social media-related risk for adolescents can be put into three main categories including peer-peer, poor understanding of privacy concerns, inappropriate content, and influence from outsiders including advertising groups (O'Keeffe et al., 2011). Cyberbullying is a very common problem that is being encountered by young children. Cyberbullying rates have been on an increase. The National Center for Education Statistics has reported that a majority of school children between the ages of 12 and 18 reported at least two incidences of cyberbullying in 2015 (Robers et al., 2015). Bullied children experience feelings of loneliness, loss of interest in pleasurable activities which they used to enjoy, depression, anxiety, self-injurious behavior, and suicidal ideation (Peker, 2020; Elgar et al., 2014). Sexting is another problem that affects young children and adolescents the most. Sexting is considered sending or receiving sexually explicit content, messages, photographs, and something of that nature. According to a study, about 20% of teens have shared or posted their nude or half-nude pictures or videos by themselves (O'Keeffe et al., 2011). A study that included 626 adolescents in a school in the southeastern USA found that 62.5% sexted last year (Maheux et al., 2020). Another study in Sweden found sexting and sending explicit sexual images are common among children (Dahlqvist & Gillander Gådin, 2020). Sexting may not appear to have dangers for the sender and receiver if they are in a mutual agreement, but this behavior can lead to serious legal trouble and other adverse effects on people (Cornelius et al., 2020; Strasburger et al., 2019). Females that receive explicit sexual content experience adverse well-being (Dahlqvist & Gillander Gådin, 2020). A lack of privacy, sharing too much information with each other, and putting false information about themselves and others are other risks of problematic use of social media. Some information can damage the image of people and can also put some at legal and other risks. Predators might steal information and/or cause emotional harm. Visiting some websites is not always safe because many websites keep a record of visitors and their activities no matter the user deleting their history to the best of their knowledge. This information is called a digital footprint. In the future, such a digital footprint may be used against the person's reputation or create risks in jobs and lead to emotional concerns (O'Keeffe et al., 2011). Many advertising websites pop up on several social media websites and influence visitors to buy a product or behavior. Young kids often get influenced from those advertisements (Lapierre et al., 2017; Folkvord et al., 2016). Phishing is a form of online identity theft. People on social media are vulnerable to phishing if they are not very careful. Phishing attempts were recorded to increase by 60% on social networks in 2018 on Facebook, and by 74% on Facebook and Instagram together in 2019. These statistics are alarming and indicate a continued risk for social media users (Parker & Flowerday, 2020). Elderly people are more susceptible to phishing attacks because many of them find it difficult to operate social media effectively, have memory issues, and have less computer and Internet awareness (Grilli et al., 2021; Alwanain, 2020). Social media got so prevalent that parents even find it difficult to get children off their electronic devices at the dinner table. It reduces the quality time that usually working families use to enjoy traditionally. Many parents, especially young children feel frustrated with this influx of use of the Internet and social media (Hiniker et al., 2016). Studies have also reported some level of frustration among young kids who can get enough attention from their parents because they often find themselves occupied with social media when they want to talk (Hiniker et al., 2016). A cross-sectional study conducted in Canada that included 9732 students aged 11 to 20 studied the quality of the parent—child interaction and relationship and found that the parent or child who spends more than two hours on social media has a negative interaction with others (Sampasa-Kanyinga et al., 2020). Exposure to alcohol and tobacco shown in traditional media such as movies and TV serials has been found associated with the likelihood of substance use. Social media is even more powerful in influencing such behaviors because it can combine both sides of media, including traditional and interactive. It is assumed that social media can even influence drinking behavior more strongly (Groth et al., 2017;

Moreno & Whitehill, 2014). Certain content in close relationships is needed to be carried out face to face in which a person's gestures, tone, facial expression, proximity, and language matter, but social media use has promoted expression in short forms, abbreviations, and lighter language that has reduced the ability to have coherent interpersonal face-to-face communication (Chasombat, 2014). Communication through social media with family members and in other interpersonal relationships was reported to be misunderstood in a study conducted in Canada that comprised a sample of 120 college-attending students (Lopez & Cuarteros, 2020).

Social media does provide several benefits to users. Its importance has even increased during and after COVID-19. However, its use is not free from harmful effects on health and social life. Experts are addressing the harmful effects of social media by guiding people on the proper and safe use of it. The growing use of social media and simultaneously increasing health concerns are major issues that must be addressed seriously. Social media is being used in every field including public health for educating and carrying out health education and health promotional work (Moore et al., 2021). However, the consequences of addictive social media on people remain unaddressed in public health. It is very important to explore public health intervention in addressing social media-related health concerns.

Summary

Social media offers innumerable benefits in several areas including social networking, communication, learning, cognitive development, business, and as a forum to express oneself to a larger community. Its benefits have increased its use tremendously worldwide since its inception. Communication is a prime need for humans. Traditional methods of communication such as face-to-face can be intimidating for certain populations and may not be feasible in all situations for most of us. Social media, contrary to traditional methods, is easy to use, fast, and reachable to distanced people. Advancements in technology, Internet reach to remote places, and affordable prices of electronic devices made social media more popular and affordable to people. The COVID-19 pandemic has created a very favorable atmosphere for social media to grow and the need for the larger population to use it. Due to restrictions placed to restrain the virus from spreading, people had to work from home. Most of these activities were carried out online. This situation was used for media companies to produce more user-friendly and need-addressing websites. On the other side, people cut off from the external world started experiencing psychological concerns including loneliness, depression, anxiety, stress, and sleep problems. Social media came in handy to connect with people. This online-favoring atmosphere promoted social media use further. At present out of the 7.9 billion total population, close to five billion people in the world use social media. Projections are suggesting that social media users would keep increasing. China and India have the highest number of social media users. A social media user spends about 147 min per day. The excess use of social media is turning out as a problem. Experts also started referring to the overuse behavior of social media as an addiction. Addiction has come out of psychiatry that mainly applies to chemical addiction. However, certain criteria of addiction can be easily seen on social media. Correlating with addiction, the social media indicators can be classified as a behavioral addiction. It does not matter whether the behavior has a positive or negative or even mixed outcome. It should be explained with theories about what behavior happens and why. There are several theories being explored in current research to understand the theoretical basis of social media use. Dispositional, motivational, and learning theories were found very relevant in explaining social media use. This information is highly important to address the negative and addictive practices of social media. Addiction leads to negative consequences despite its type and nature. Similarly, social media has been found to be associated with several negative health, social, psychological, and learning problems. Depression, anxiety, stress, sleep problems, social isolation, poor self-esteem, interpersonal issues, poor academic performance, reduced communication skills, social skills, and physical inactivity are some of the major harmful effects of social media which have been documented heavily in the literature. People with low self-esteem and social anxiety are more vulnerable to addiction. The COVID-19 pandemic is almost phased out, but it has left a parallel system of working, learning, and interacting with people. More online work is being carried out. This influx of online systems is leading even greater risk for vulnerable populations such as children and older adults. Young children are vulnerable to putting inappropriate content on social networking sites without adequate knowledge and understanding, while older adults are experiencing a higher risk of becoming the victim of scams and phishing. The likelihood of increasing harmful effects of social media with its anticipated increase can't be denied. COVID has led to situations also posing an epigenetic risk for developing a social media-favorable attitude among children who were conceived during the pandemic. It is important to plan public health interventions to address the growing risk of harmful effects of social media addiction and promote its beneficial use.

Conclusion

Social media is highly beneficial, but its growing addiction will continue to lead to harmful effects on users. People with low self-esteem and social anxiety benefit from the use of social media, but due to their psychological vulnerability, they are more susceptible to addiction. Excessive and problematic use of social media is believed to be an addiction, but it lacks a definition. Several indicators of substance addiction definition are true and make social media overuse a behavioral addiction. A clearer understanding and a definition of social media addiction are needed. Several theories coming from psychology, social science, neurobiology, and Internet-based model are being used to explain social media addiction, but a more exclusive theory is needed to explain social media addiction. Public health intervention is suggested to address harmful risks leading to poor health outcomes in the population.

Questions

✔ Social media overuse may not lead to strong physiological effects on the body like substance addiction, but still its excessive use can be labeled as a behavioral addiction. Do you think it is appropriate to term problematic excess use of social media addiction? Provide your explanation of why or why not.

✔ What methods can be applied to reduce the problematic use of social media?

✔ What public health theory/model can be used to reduce negative health outcomes of excessive social media use?

✔ People with lower self-esteem and social anxiety feel benefited from social media, but their condition makes them more vulnerable to social media addiction. How these people can be helped?

Definition of key terms

Addiction: Addiction is basically used to describe dependence on substances/chemicals in psychiatry. It is considered a disorder pertaining to the constant urge to use a certain substance, regardless of its potential negative effects.

Adolescent: A term used to describe an individual in a stage of life between childhood and adulthood. This phase ranges from age 10 to 19. During this phase of development, children experience rapid physical, cognitive, and psychological growth.

Attachment theory: It is based on evolutionary principles of psychological relationships. It emphasizes that child needs to have a good psychological relationship with at least one caregiver to have normal social and emotional development.

Behavioral addiction: Addiction has been classified based on chemical use as well with the process. Certain behaviors are kept in process/behavior addiction when they happen in excess and lead to negative consequences. Gambling, exercise, shopping, and obsession with sex are some good examples of behavior addiction. Social media is a behavioral addiction. People addicted to it have a constant urge to partake in it, regardless of its potential negative effects.

Belongingness theory: Theory emphasizes the need for belongingness to fulfill human emotional needs for interpersonal relationships and connectedness. The desire for death is caused by the feeling of failed interpersonal process.

Caplan's social skill model: Emphasizes the connection between measures of psychosocial well-being and preference for online interactions. Low well-being (depression and loneliness) may be associated with more online social interaction.

Classical conditioning: Based on neurobehavioral principles. The theory explains the likelihood of behavior when a biologically potent stimulus is paired with a neutral stimulus.

Davis's cognitive-behavioral: Predisposed psychological conditions play a major role in deciding the type of Internet use. It could be generalized in which a person explores content that is global acceptance such as information search while another one is pathological, in which an individual explores content that is considered maladaptive such as online sex content, etc.

Dual system theory: According to this theory, a decision or thought arises from two psychological processes including unconscious and conscious processes. Immature impulse control and higher reward sensitivity combinedly lead to risk-taking behavior.

Epigenetic: The term used to describe how changes in an individual's environment and behavior can affect one's phenotype without altering its genotype. For example, if parents living in extreme conditions conceive, the baby might get phenotype (characteristics of fighting adversity like parents in that situation) without having any change in their genetic material.

Flow theory: This theory emerged from positive psychology. It tells that individual drive please from engaging deeply in a process even though that process may not provide a direct or indirect reward.

Incentive-sensitization theory: This theory is based on neurobiological principles. When gets into the system, the body adapts it and feel the need when the level of certain chemical goes down. In psychological terms, an individual receives internal clues to take that substance.

Interaction of person-affect cognition-execution model: According to the latest understanding of the model, an addictive behavior develops because of an interaction between three main factors, (1) predisposing factors, (2) emotional/affective and cognitive responses, and (3) executive functions.

Needs affordances-features: Individual looks at the environment as the way how in which needs can be afforded. In simple terms, the person looks at the feasibility or affordability in relation to the environment in which he lives.

Operant conditioning: Next level of associative learning theory that emerged after classical conditioning. In it, the reinforcement or punishment applies for strengthening or weakening the behavior.

Phishing: Phishing is a type of social engineering in which people use names of established companies and send fraudulent messages to trick people to get their personal information such as passwords, account details, credit card information, etc.

Psychological health: Mainly includes health related to feeling, emotion, and behaviors. Minor mental health issues such as mild to moderate depression, anxiety, stress, and sleep disturbance are some examples of psychological health.

Rational addiction theory: It is based on an individual's rational decision. Often individuals plan for optimal consumption of a substance.

Sexting: Sexting is an exchange of sending and receiving sexually explicit content. It may include a variety of content of sexual nature including text messages, nude or seminude photos, and videos of oneself or others.

Self-determination theory: This theory involves two psychological constructs named intrinsic and extrinsic motivation. The individual is driven by both or either one to fulfill three psychological needs: autonomy, competence, and relatedness.

Self-escape theory: It is based on premise that people try to remove themselves from a physical location where they think have a negative perception of themselves.

Social capital: A term referring to the relationships among members in a particular social group, and the benefits of said relationships to the members.

Social networking: The usage of online networks such as social media to create and maintain social relationships with other people.

Social cognitive perspective theory: Explains the outcome of self-theory where individuals indulge in problem-solving, seek information, and make their own decision.

Social learning theory: New behavior originates from observing and imitating others. Personal and environmental factors play role in the development of behavior.

Social cognitive theory: Is an extension of social learning theory. A new component of self-efficacy was added that signifies the self-confidence of an individual that plays role in developing the behavior. It denotes the triadic reciprocity between personal factors, environment, and behavior.

Stimulus-response reinforcement framework: Theory is consistent with classical conditioning. Two aspects stimulus and response lead the learning.

Social influence theory: It tells that people are more likely to do the behavior that is seen as a norm.

Social capital model: Is a sum of actual and potential resources in the network of the individual. Probably to retain the social capital, an individual decides to follow behavior.

Time perspective theory: Theory basically explains that our perceptions of time are responsible for including our emotions, perceptions, and actions.

Theory of planned behavior: Three core psychological components, subjective norms, favorable attitude, and perceived behavior control lead to a behavior. A very popular theory in the field of public health, education, and psychology.

Theory of technology frame: Three factors associated with the technology named usefulness, importance, and significance combinedly influence individual behavior.

Technology acceptance model: It is information system theory. It explains how a person comes to accept technology and decides to use it.

Uses and gratification theory: This theory has a strong connection with social media. It is a mass communication theory. Basically, an individual focuses on the needs, motives, and gratification that they receive from using the media.

Website to explore

Substance Abuse and Mental Health Services Administration

https://www.samhsa.gov.

SAMHSA is a branch of the U.S. Department of Health and Human Services. Standing for Substance Abuse and Mental Health Services Administration, the website provides information and resources about mental illness and substance abuse. The website states its mission "to reduce the impact of substance abuse and mental illness on America's communities." Explore this website. *What did you find? Did you find any connection between substance use and mental illness? Does substance use addiction have any similarity to social media addiction?*

Popular social networking websites/applications

There are several social networking sites around the world that people can choose to use depending on their interests. Some of the most popular sites are facebook.com, linkedin.com, youtube.com, instagram.com, pinterest.com, seniormatch.com, skype.com, and twitter.com. People can connect with one another using these websites by posting pictures, sending messages, calling each other, etc. You might have an account in of those. *Share your experience and how you feel benefited by using your social media website. Does it have any effect on your social and emotional well-being? What kind of problems do you experience with your social media account?*

References

Abd Rahman, S. H. (2014). Can't live without my FB, LoL: The influence of social networking sites on the communication skills of TESL students. *Procedia-Social and Behavioral Sciences, 134*, 213–219. https://doi.org/10.1016/j.sbspro.2014.04.241

Alavi, S. S., Ferdosi, M., Jannatifard, F., Eslami, M., Alaghemandan, H., & Setare, M. (2012). Behavioral addiction versus substance addiction: Correspondence of psychiatric and psychological views. *International Journal of Preventive Medicine, 3*(4), 290–294. PMC3354400.

Alotaibi, M. S., Fox, M., Coman, R., Ratan, Z. A., & Hosseinzadeh, H. (2022). Smartphone Addiction Prevalence and its association on academic performance, physical health, and mental well-being among \university students in Umm Al-Qura University (UQU), Saudi Arabia. *International Journal of Environmental Research and Public Health, 19*(6), 3710. https://doi.org/10.3390/ijerph19063710

Alwanain, M. I. (2020). Phishing awareness and elderly users in social media. International Journal of. *Computer Science and Network Security, 20*(9), 114−119. https://doi.org/10.22937/IJCSNS.2020.20.09.14

Andreassen, C. S., Billieux, J., Griffiths, M. D., Kuss, D. J., Demetrovics, Z., Mazzoni, E., & Pallesen, S. (2016). The relationship between addictive use of social media and video games and symptoms of psychiatric disorders: A large-scale cross-sectional study. *Psychology of Addictive Behaviors, 30*(2), 252−262. https://doi.org/10.1037/adb0000160

Andreassen, C. S., & Pallesen, S. (2014). Social network site addiction - an overview. *Current Pharmaceutical Design, 20*(25), 4053−4061. https://doi.org/10.2174/13816128113199990616

Apaolaza, V., Hartmann, P., D'Souza, C., & Gilsanz, A. (2019). Mindfulness, compulsive mobile social media use, and derived stress: The mediating roles of self-esteem and social anxiety. *Cyberpsychology, Behavior, and Social Networking, 22*(6), 388−396. https://doi.org/10.1089/cyber.2018.0681

Azizi, S. M., Soroush, A., & Khatony, A. (2019). The relationship between social networking addiction and academic performance in Iranian students of medical sciences: A cross-sectional study. *BMC Psychology, 7*(1), 28. https://doi.org/10.1186/s40359-019-0305-0

Bányai, F., Zsila, Á., Király, O., Maraz, A., Elekes, Z., Griffiths, M. D., Andreassen, C. S., & Demetrovics, Z. (2017). Problematic social media use: Results from a large-scale nationally representative adolescent sample. *PloS One, 12*(1), e0169839. https://doi.org/10.1371/journal.pone.0169839

Boursier, V., Gioia, F., Musetti, A., & Schimmenti, A. (2020). Facing loneliness and anxiety during the COVID-19 isolation: The role of excessive social media use in a sample of Italian adults. *Frontiers in Psychiatry, 11*. https://doi.org/10.3389/fpsyt.2020.586222

Brailovskaia, J., & Margraf, J. (2020). Relationship between depression symptoms, physical activity, and addictive social media use. *Cyberpsychology, Behavior and Social Networking, 23*(12), 818−822. https://doi.org/10.1089/cyber.2020.0255

Brailovskaia, J., Truskauskaite-Kuneviciene, I., Kazlauskas, E., & Margraf, J. (2021). The patterns of problematic social media use (SMU) and their relationship with online flow, life satisfaction, depression, anxiety and stress symptoms in Lithuania and in Germany. *Current Psychology*, 1−12. https://doi.org/10.1007/s12144-021-01711-w

Brunborg, G. S., Mentzoni, R. A., Molde, H., Myrseth, H., Skouverøe, K. J., Bjorvatn, B., & Pallesen, S. (2011). The relationship between media use in the bedroom, sleep habits and symptoms of insomnia. *Journal of Sleep Research, 20*(4), 569−575. https://doi.org/10.1111/j.1365-2869.2011.00913.x

Carbonell, X., & Panova, T. (2017). A critical consideration of social networking sites' addiction potential. *Addiction Research & Theory, 25*(1), 48−57. https://doi.org/10.1080/16066359.2016.1197915

Ceci, L. (2022a). *Average daily time spent by users worldwide on social media apps from October 2020 to March 2021, by gender*. Available at: https://www.statista.com/statistics/1272876/worldwide-social-apps-time-spent-daily-gender/.

Ceci, L. (2022b). *Global users daily time on social media apps 2020-2021, by age groups*. Available at: https://www.statista.com/statistics/1272883/worldwide-social-apps-time-spent-daily-age/.

Chasombat, P. (2014). *Social networking sites impacts on interpersonal communication skills and relationships.* https://repository.nida.ac.th/handle/662723737/3112.

Cheng, C., & Lau, Y. C. (2022). Social Media Addiction during COVID-19-mandated physical distancing: Relatedness needs as motives. *International Journal of Environmental Research and Public Health, 19*(8), 4621. https://doi.org/10.3390/ijerph19084621

Cheng, C., Lau, Y. C., Chan, L., & Luk, J. W. (2021). Prevalence of social media addiction across 32 nations: meta-analysis with subgroup analysis of classification schemes and cultural values. *Addictive Behaviors, 117*, 106845. https://doi.org/10.1016/j.addbeh.2021.106845

Chen, Y., Li, R., Zhang, P., & Liu, X. (2020). The moderating role of state attachment anxiety and avoidance between social anxiety and social networking sites addiction. *Psychological Reports, 123*(3), 633−647. https://doi.org/10.1177/0033294118823178

Cornelius, T. L., Bell, K. M., Kistler, T., & Drouin, M. (2020). Consensual sexting among college students: the interplay of coercion and intimate partner aggression in perceived consequences of sexting. *International Journal of Environmental Research and Public Health, 17*(19), 7141. https://doi.org/10.3390/ijerph17197141

Dahlqvist, H., & Gillander Gådin, K. (2020). Swedish teens' comprehension of sexting and explicit sexual images and consequences for well-being. *European Journal of Public Health, 30*(Suppl. ment_5). https://doi.org/10.1093/eurpub/ckaa165.092

Dey, B. L., Sarma, M., Pandit, A., Sarpong, D., Kumari, S., & Punjaisri, K. (2019). Social media led co-creation of knowledge in developing societies: SME's roles in the adoption, use and appropriation of smartphones in South Asia. *Production Planning and Control, 30*(10−12), 1019−1031. https://doi.org/10.1080/09537287.2019.1582106

Dissanayake, W. (2003). Asian approaches to human communication: Retrospect and prospect. *Intercultural Communication Studies, 12*(4), 17−38. https://web.uri.edu/iaics/files/02-Wimal-Dissanayake.pdf.

Dixon, S. (2022b). *Number of worldwide social media users 2022, by region.* https://www.statista.com/statistics/454772/number-social-media-user-worldwide-region/.

Dixson, S. (2022a). *Leading social media usage reasons worldwide 2021.* https://www.statista.com/statistics/715449/social-media-usage-reasons-worldwide/.

Elgar, F. J., Napoletano, A., Saul, G., Dirks, M. A., Craig, W., Poteat, V. P., Holt, M., & Koenig, B. W. (2014). Cyberbullying victimization and mental health in adolescents and the moderating role of family dinners. *JAMA Pediatrics, 168*(11), 1015−1022. https://doi.org/10.1001/jamapediatrics.2014.1223

Folkvord, F., Anschütz, D. J., Boyland, E., Kelly, B., & Buijzen, M. (2016). Food advertising and eating behavior in children. *Current Opinion in Behavioral Sciences, 9*, 26−31. https://doi.org/10.1016/j.cobeha.2015.11.016

Giordano, A. L., Prosek, E. A., & Watson, J. C. (2021). Understanding adolescent cyberbullies: Exploring social media addiction and psychological factors. *Journal of Child and Adolescent Counseling, 7*(1), 42−55. https://doi.org/10.1080/23727810.2020.1835420

Gómez-Galán, J., Martínez-López, J.Á., Lázaro-Pérez, C., & Sarasola Sánchez-Serrano, J. L. (2020). Social networks consumption and addiction in college students during the COVID-19 pandemic: Educational approach to responsible use. *Sustainability, 12*(18), 7737. https://doi.org/10.3390/su12187737

Griffiths, M. D. (1999). Internet addiction: Internet fuels other addictions. *Student British Medical Journal, 7*(1), 428−429.

Grilli, M. D., McVeigh, K. S., Hakim, Z. M., Wank, A. A., Getz, S. J., Levin, B. E., Ebner, N. C., & Wilson, R. C. (2021). Is this phishing? Older age is associated with greater difficulty discriminating between safe and malicious emails. *The Journals of Gerontology. Series B,*

Psychological Sciences and Social Sciences, 76(9), 1711−1715. https://doi.org/10.1093/geronb/gbaa228

Groth, G. G., Longo, L. M., & Martin, J. L. (2017). Social media and college student risk behaviors: A mini-review. *Addictive Behaviors, 65*, 87−91. https://doi.org/10.1016/j.addbeh.2016.10.003

Hamilton, J. L., Nesi, J., & Choukas-Bradley, S. (2022). Reexamining social media and socioemotional well-being among adolescents through the lens of the COVID-19 pandemic: A theoretical review and directions for future research. *Perspectives on Psychological, 17*(3), 662−679. https://doi.org/10.1177/17456916211014189

He, Z. H., Li, M. D., Ma, X. Y., & Liu, C. J. (2021). Family socioeconomic status and social media addiction in female college students: The mediating role of impulsiveness and inhibitory control. *The Journal of Genetic Psychology, 182*(1), 60−74. https://doi.org/10.1080/00221325.2020.1853027

Hiniker, A., Schoenebeck, S. Y., & Kientz, J. A. (February 2016). Not at the dinner table: Parents' and children's perspectives on family technology rules. In *Proceedings of the 19th ACM conference on computer-supported cooperative work & social computing* (pp. 1376−1389). https://doi.org/10.1145/2818048.2819940

Hökby, S., Hadlaczky, G., Westerlund, J., Wasserman, D., Balazs, J., Germanavicius, A., Machín, N., Meszaros, G., Sarchiapone, M., Värnik, A., Varnik, P., Westerlund, M., & Carli, V. (2016). Are mental health effects of internet use attributable to the web-based content or perceived consequences of usage? A longitudinal study of European adolescents. *JMIR Mental Health, 3*(3), e31. https://doi.org/10.2196/mental.5925

Holmgren, H. G., & Coyne, S. M. (2017). Can't stop scrolling!: Pathological use of social networking sites in emerging adulthood. *Addiction Research & Theory, 25*(5), 375−382. https://doi.org/10.1080/16066359.2017.1294164

Hou, X. L., Wang, H. Z., Hu, T. Q., Gentile, D. A., Gaskin, J., & Wang, J. L. (2019). The relationship between perceived stress and problematic social networking site use among Chinese college students. *Journal of Behavioral Addictions, 8*(2), 306−317. https://doi.org/10.1556/2006.8.2019.26

Kadam, Y. R., Patil, S. R., Waghachavare, V., & Gore, A. D. (2016). Influence of various lifestyle and psychosocial factors on sleep disturbances among the college students: A cross-sectional study from an urban area of India. *Journal of Krishna Institute of Medical Sciences (JKIMSU), 5*(3), 51−60.

Khan, G. F., Swar, B., & Lee, S. K. (2014). Social media risks and benefits: A public sector perspective. *Social Science Computer Review, 32*(5), 606−627. https://doi.org/10.1177/0894439314524701

Kilincel, S., & Muratdagi, G. (2021). *Evaluation of factors affecting social media addiction in adolescents during the COVID-19 pandemic.* https://www.doi.org/10.4328/ACAM.20541.

Kim, M. R., Oh, J. W., & Huh, B. Y. (2020). Analysis of factors related to social network service addiction among Korean high school students. *Journal of Addictions Nursing, 31*(3), 203−212. https://doi.org/10.1097/JAN.0000000000000350

Kircaburun, K., Demetrovics, Z., Király, O., & Griffiths, M. D. (2020a). Childhood emotional trauma and cyberbullying perpetration among emerging adults: A multiple mediation model of the role of problematic social media use and psychopathology. *International Journal of Mental Health and Addiction, 18*(3), 548−566. https://doi.org/10.1007/s11469-018-9941-5

Kircaburun, K., Griffiths, M. D., & Billieux, J. (2019). Trait emotional intelligence and problematic online behaviors among adolescents: The mediating role of mindfulness, rumination, and depression. *Personality and Individual Differences, 139*, 208−213. https://doi.org/10.1016/j.paid.2018.11.024

Kircaburun, K., Griffiths, M. D., Sahin, F., Bahtiyar, M., Atmaca, T., & Tosuntaş, S. B. (2020b). The mediating role of self/everyday creativity and depression on the relationship between creative personality traits and problematic social media use among emerging adults. *International Journal of Mental Health and Addiction, 18*(1), 77−88. https://doi.org/10.1007/s11469-018-9938-0

Lapierre, M. A., Fleming-Milici, F., Rozendaal, E., McAlister, A. R., & Castonguay, J. (2017). The effect of advertising on children and adolescents. *Pediatrics, 140*(Suppl. 2), S152−S156. https://doi.org/10.1542/peds.2016-1758V

Liu, J., Liu, C. X., Wu, T., Liu, B. P., Jia, C. X., & Liu, X. (2019). Prolonged mobile phone use is associated with depressive symptoms in Chinese adolescents. *Journal of Affective Disorders, 259*, 128−134. https://doi.org/10.1016/j.jad.2019.08.017

Lopez, A. G., & Cuarteros, K. G. (2020). Exploring the effects of social media on interpersonal communication among family members. *Canadian Journal of Family and Youth/Le Journal Canadien de Famille et de la Jeunesse, 12*(1), 66−80. https://doi.org/10.29173/cjfy29491

Magnuson, M. J., & Dundes, L. (2008). Gender differences in "social portraits" reflected in MySpace profiles. *Cyberpsychology & Behavior: The Impact of the Internet, Multimedia and Virtual Reality on Behavior and Society, 11*(2), 239−241. https://doi.org/10.1089/cpb.2007.0089

Maheux, A. J., Evans, R., Widman, L., Nesi, J., Prinstein, M. J., & Choukas-Bradley, S. (2020). Popular peer norms and adolescent sexting behavior. *Journal of Adolescence, 78*, 62−66. https://doi.org/10.1016/j.adolescence.2019.12.002

Masthi, N. R., Pruthvi, S., & Mallekavu, P. (2017). A comparative study on social media addiction between public and private high school students of urban Bengaluru, India. *ASEAN Journal of Psychiatry, 18*(2), 1−10.

McNeely, J., & Adam, A. (2020). *Substance use screening and risk assessment in adults [Internet]*. Johns Hopkins University. https://www.ncbi.nlm.nih.gov/books/NBK565474/.

Mirbabaie, M., Bunker, D., Stieglitz, S., Marx, J., & Ehnis, C. (2020). Social media in times of crisis: learning from hurricane harvey for the coronavirus disease 2019 pandemic response. *Journal of Information Technology, 35*(3), 195−213. https://doi.org/10.1177/0268396220929258

Misirlis, N., Zwaan, M. H., & Weber, D. (2020). *International students' loneliness, depression and stress levels in COVID-19 crisis*. The role of social media and the host university. https://doi.org/10.48550/arXiv.2005.12806. arXiv preprint arXiv:2005.12806.

Mitra, R., & Rangaswamy, M. (2019). Excessive social media use and its association with depression and rumination in an Indian young adult population: A mediation model. *Journal of Psychosocial Research, 14*(1), 223−231. https://doi.org/10.32381/JPR.2019.14.01.24

Molavi, P., Mikaeili, N., Ghaseminejad, M. A., Kazemi, Z., & Pourdonya, M. (2018). Social anxiety and benign and toxic online self-disclosures: An investigation into the role of rejection sensitivity, self-regulation, and internet addiction in college students. *The Journal of Nervous and Mental Disease, 206*(8), 598−605. https://doi.org/10.1097/NMD.0000000000000855

Moore, J. B., Harris, J. K., & Hutti, E. T. (2021). Falsehood flies, and the truth comes limping after it': Social media and public health. *Current Opinion in Psychiatry, 34*(5), 485−490. https://doi.org/10.1097/YCO.0000000000000730

Moreno, M. A., & Whitehill, J. M. (2014). Influence of social media on alcohol use in adolescents and young adults. *Alcohol Research: Current Reviews, 36*(1), 91−100. PMC4432862.

Ndubuaku, V., Inim, V., Ndudi, U. C., Samuel, U., & Prince, A. I. (2020). Effect of social networking technology addiction on academic performance of university students in Nigeria. *International Journal of Recent Technology and Engineering (IJRTE)*, 173−180.

O'Keeffe, G. S., Clarke-Pearson, K., & Council on Communications and Media. (2011). The impact of social media on children, adolescents, and families. *Pediatrics, 127*(4), 800−804. https://doi.org/10.1542/peds.2011-0054

O'Reilly, M. (2020). Social media and adolescent mental health: The good, the bad and the ugly. *Journal of Mental Health (Abingdon, England), 29*(2), 200−206. https://doi.org/10.1080/09638237.2020.1714007

Parker, H. J., & Flowerday, S. V. (2020). Contributing factors to increased susceptibility to social media phishing attacks. *South African Journal of Information Management, 22*(1), 1−10. https://doi.org/10.4102/sajim.v22i1.1176

Peele, S. (1977). Redefining addiction I. Making addiction a scientifically and socially useful concept. *International Journal of Health Services: Planning, Administration, Evaluation, 7*(1), 103−124. https://doi.org/10.2190/A7JM-3YQ7-NPAK-MWTL

Peker, H. (2020). The effect of cyberbullying and traditional bullying on English language learners' national and oriented identities. *Bartın University Journal of Faculty of Education, 9*(1), 185−199. https://dergipark.org.tr/en/pub/buefad/issue/51796/664122.

Pew Research Center. (2022). *Social media fact sheet.* https://www.pewresearch.org/internet/fact-sheet/social-media/?menuItem=d102dcb7-e8a1-42cd-a04e-ee442f81505a.

Picazo-Vela, S., Gutiérrez-Martínez, I., & Luna-Reyes, L. F. (2012). Understanding risks, benefits, and strategic alternatives of social media applications in the public sector. *Government Information Quarterly, 29*(4), 504−511. https://doi.org/10.1016/j.giq.2012.07.002

Pittman, M. (2018). Happiness, loneliness, and social media: Perceived intimacy mediates the emotional benefits of platform use. *The Journal of Social Media in Society, 7*(2), 164−176.

Provenzi, L., Grumi, S., Altieri, L., Bensi, G., Bertazzoli, E., Biasucci, G., Cavallini, A., Decembrino, L., Falcone, R., Freddi, A., Gardella, B., Giacchero, R., Giorda, R., Grossi, E., Guerini, P., Magnani, M. L., Martelli, P., Motta, M., Nacinovich, R., Pantaleo, D., & MOM-COPE Study Group. (2021a). Prenatal maternal stress during the COVID-19 pandemic and infant regulatory capacity at 3 months: A longitudinal study. In *Development and psychopathology* (pp. 1−9). Advance online publication. https://doi.org/10.1017/S0954579421000766

Provenzi, L., Mambretti, F., Villa, M., Grumi, S., Citterio, A., Bertazzoli, E., Biasucci, G., Decembrino, L., Falcone, R., Gardella, B., Longo, M. R., Nacinovich, R., Pisoni, C., Prefumo, F., Orcesi, S., Scelsa, B., Giorda, R., & Borgatti, R. (2021b). Hidden pandemic: COVID-19-related stress, SLC6A4 methylation, and infants' temperament at 3 months. *Scientific Reports, 11*(1), 15658. https://doi.org/10.1038/s41598-021-95053-z

Robers, S., Zhang, A., & Morgan, R. E. (2015). *Indicators of school crime and safety: 2014.* National Center for Education Statistics, 2015 http://eric.ed.gov/?id=ED557756.

Rod, N. H., Dissing, A. S., Clark, A., Gerds, T. A., & Lund, R. (2018). Overnight smartphone use: A new public health challenge? A novel study design based on high-resolution smartphone data. *PloS One, 13*(10), e0204811. https://doi.org/10.1371/journal.pone.0204811

Sampasa-Kanyinga, H., Goldfield, G. S., Kingsbury, M., Clayborne, Z., & Colman, I. (2020). Social media use and parent-child relationship: A cross-sectional study of adolescents. *Journal of Community Psychology, 48*(3), 793−803. https://doi.org/10.1002/jcop.22293

Sen, R., Garbati, M., Bryant, K., & Lu, Y. (2021). Epigenetic mechanisms influencing COVID-19. *Genome, 64*(4), 372−385. https://doi.org/10.1139/gen-2020-0135

Shannon, H., Bush, K., Villeneuve, P. J., Hellemans, K. G., & Guimond, S. (2022). Problematic social media use in adolescents and young adults: Systematic review and meta-analysis. *JMIR Mental Health, 9*(4), e33450. https://doi.org/10.2196/33450

Sobaih, A. E. E., Hasanein, A. M., & Abu Elnasr, A. E. (2020). Responses to COVID-19 in higher education: Social media usage for sustaining formal academic communication in developing countries. *Sustainability, 12*(16), 6520. https://doi.org/10.3390/su12166520

Statista. (2022). *Global digital population as of April 2022*. Available at: https://www.statista.com/statistics/617136/digital-population-worldwide/.

Stockdale, L. A., & Coyne, S. M. (2020). Bored and online: Reasons for using social media, problematic social networking site use, and behavioral outcomes across the transition from adolescence to emerging adulthood. *Journal of Adolescence, 79*, 173−183. https://doi.org/10.1016/j.adolescence.2020.01.010

Strasburger, V. C., Zimmerman, H., Temple, J. R., & Madigan, S. (2019). Teenagers, sexting, and the law. *Pediatrics, 143*(5), e20183183. https://doi.org/10.1542/peds.2018-3183

Sujarwoto, Saputri, R., & Yumarni, T. (2021). Social media addiction and mental health among university students during the COVID-19 pandemic in Indonesia. *International Journal of Mental Health and Addiction, 21*, 96−119. https://doi.org/10.1007/s11469-021-00582-3

Sun, Y., & Zhang, Y. (2021). A review of theories and models applied in studies of social media addiction and implications for future research. *Addictive Behaviors, 114*, 106699. https://doi.org/10.1016/j.addbeh.2020.106699

Thomée, S., Härenstam, A., & Hagberg, M. (2012). Computer use and stress, sleep disturbances, and symptoms of depression among young adults–a prospective cohort study. *BMC Psychiatry, 12*, 176. https://doi.org/10.1186/1471-244X-12-176

Thorisdottir, I. E., Sigurvinsdottir, R., Asgeirsdottir, B. B., Allegrante, J. P., & Sigfusdottir, I. D. (2019). Active and passive social media use and symptoms of anxiety and depressed mood among Icelandic adolescents. *Cyberpsychology, Behavior and Social Networking, 22*(8), 535−542. https://doi.org/10.1089/cyber.2019.0079

Wang, P., Wang, X., Wu, Y., Xie, X., Wang, X., Zhao, F., Ouyang, M., & Lei, L. (2018). Social networking sites addiction and adolescent depression: A moderated mediation model of rumination and self-esteem. *Personality and Individual Differences, 127*, 162−167. https://doi.org/10.1016/j.paid.2018.02.008

Wong, H. Y., Mo, H. Y., Potenza, M. N., Chan, M., Lau, W. M., Chui, T. K., Pakpour, A. H., & Lin, C. Y. (2020). Relationships between severity of internet gaming disorder, severity of problematic social media use, sleep quality and psychological distress. *International Journal of Environmental Research and Public Health, 17*(6), 1879. https://doi.org/10.3390/ijerph17061879

Yayman, E., & Bilgin, O. (2020). Relationship between social media addiction, game addiction and family functions. *International Journal of Evaluation and Research in Education, 9*(4), 979−986. https://doi.org/10.11591/ijere.v9i4.20680

Young, K. S. (2004). Internet addiction: A new clinical phenomenon and its consequences. *American Behavioral Scientist, 48*(4), 402−415. https://doi.org/10.1177/0002764204270278

Zhao, N., & Zhou, G. (2021). COVID-19 stress and addictive social media use (SMU): Mediating role of active use and social media flow. *Frontiers in Psychiatry, 12*, 635546. https://doi.org/10.3389/fpsyt.2021.635546

Zheng, X., & Lee, M. K. (2016). Excessive use of mobile social networking sites: Negative consequences on individuals. *Computers in Human Behavior, 65*, 65−76. https://doi.org/10.1016/j.chb.2016.08.011

Chapter 5

Social media use among older adults and their challenges

Ram Lakhan[1], Bidhu Sharma[2] and Manoj Sharma[3]

[1]*Department of Health and Human Performance, Berea College, Berea, KY, United States;* [2]*School of Arts and Science, University of Louisville, Louisville, KY, United States;* [3]*Department of Social and Behavioral Health, School of Public Health, University of Nevada, Las Vegas, NV, United States*

Objectives

- Describe the utility of social media for older adults.
- Identify the prevalence of social media usage among older adults.
- Discuss the challenges faced in using social media among older adults.
- Delineate the importance of public health interventions for the safe and satisfying use of social media.

Introduction

Have you ever thought that when people take off their minds from the immediate task, what is the most common thing that people think about? Let's do an experiment here. Please close your eyes for a few seconds, take off your mind completely from the immediate task or any crisis that you are dealing with, and try to observe your thinking and feelings. What did you start thinking about? Many of us might start thinking about people in our connections. These people may be parents, spouses, siblings, relatives, friends, teachers, favorite actors and actresses, sports players, politicians, and probably a weird person that is living next door. Your thinking might revolve around feelings about experiences you have had or might have in the future. What does this tell you?

Probably, you would agree that most of the time, we think more about relationships, society, and our connections in social ecology than anything else. Most of our actions especially social actions are carried out for experiencing belongingness. Right from birth, we cultivate belongingness, and we grow our horizon of social connections. No matter how much success and material people acquire, ultimately a wealth of social connections and relationships are needed for the fulfillment of life. Poor social life due to poor social networks and social capital affects the well-being and health of the people (Abeliansky et al., 2021).

Most people work harder throughout their lives with the intention of enjoying free time in their older age with their loved ones. Is this objective really achieved by older people? The reality is mixed. Old age itself brings its own challenges and vulnerability. Health issues related to sensory organs, heart, lungs, kidneys, and other organs increase, muscle strength declines, and immune system weakens. These deteriorated physical health factors make older people more vulnerable. The need for care and support increases (Jaul & Barron, 2017). Mental health issues, mainly stress-related, anxiety, depression, and neurological disorders, are highly prevalent among older people. It is estimated about 20% of adults over 60 years of age experience some form of mental health issue (Petrova & Khvostikova, 2021). The need for social connectedness increases with the presence of a physical and mental vulnerability in older adults (Abeliansky et al., 2021).

Growing older population and its consequences

Advances in medical sciences, public health interventions, and improved living conditions have increased the longevity in past few decades and created a positive challenge in caring for increasing populations of older people around the world. Experts also believe that the trend in increasing the older population will continue for the next 20 years (Jaul & Barron, 2017). According to Rural Health Information Hub, currently, in the United States, there are 46 million older adults aged 65 years and older living. By the end of 2030, this number would be 64 million, which means each one person out of five would be classified as old. The population of older adults aged 65 years and more would expect to double to about 90 million by 2050 (RHIHub, 2022). The older population is also increasing around the globe. The worldwide population of 60 years of age and older is estimated to be 2.1 billion by 2050 (WHO, 2022a).

According to the World Health Organization, "Health is a state of complete physical, mental and social well-being and not merely the absence of disease or infirmity" (WHO, 2022b). Old age does bring challenges and previously mentioned issues, but it is not necessary that older adults have to suffer from poor health. Not being happy is equal to being ill. A study that looked at depressive symptoms among older adults found that geriatric people were overwhelmed with negative thoughts including helplessness, hopelessness, sadness, and uselessness (Wu & Chiou, 2020). These negative emotions and feelings do not convey good health. These factors could lead to more serious psychological issues among older people. About 10%−30% of people around the world experience depressive symptoms. The WHO estimates that 322 million older people who are about 4.4% of the total older population peak in depressive symptoms.

Several studies have documented a higher rate of depressive symptoms in older populations including 41.2% in Taiwan (Wu & Chiou, 2020), 20.6% in Japan (Fukunaga et al., 2012), 11.6% in China (Li et al., 2016), 30.3% in Korea (Park et al., 2016), 16.5% in Malaysia (Vanoh et al., 2016), 11.6% New Zealand (Barak et al., 2020), 15% in the USA (Gatchel et al., 2019), 13.9%−37.1% in Germany (Zebhauser et al., 2015), and up to 15.2% in developing countries, which included six low and middle-income countries namely China, Ghana, India, Mexico, the Russian Federation, and South Africa (Lotfaliany et al., 2019). Interconnection between negative emotions such as uselessness, helplessness, hopelessness, and sadness is paramount with depression. Do negative emotions lead to depression or other serious psychological issues or do the psychological issues increase negative emotions? No matter what precedes the other one, the mediating factors must be explored to help the older population affected with both and those who are on the brim of being affected.

Role of social support and social connectedness

A strong connection between negative emotions with life experiences including current living conditions, physical limitations, social support, and past experiences has been found (Wu & Chiou, 2020). Studies that attempted to look at the connection with depressive symptoms have found a strong correlation with three factors: living alone, social isolation, and loneliness (Gu et al., 2020; Vanoh et al., 2016). Several studies have found that stronger social support, positive intergenerational relationships, and stronger emotional cohesion with children alleviate depressive symptoms (Fukunaga et al., 2012; Silverstein et al., 2006; Stahl et al., 2017). Indirectly, it is assumed that various components of social support and engagement may also relieve negative thoughts and emotions and aid to the overall health of older people. To remain healthy, it is important that social connectedness is ensured for the older population, so their age and living situation do not influence their thinking negatively. What is social connectedness, how it is perceived and met, and how it impacts the brain in keeping an individual's thinking and feeling need to be understood.

The concept of social support varies for the people; however, experts have tried to define it as an exchange of resources among individuals which provides a sense of help and aid to their well-being (Khan & Husain, 2010; Mohd et al., 2019). Experts believe that social support can be explained in five categories such as material aid, behavioral assistance, intimate interaction, feedback, and positive social interaction (Barrera & Ainlay, 1983). Social support is multidimensional where each category of it fulfills certain psychological needs of an individual. Another explanation of social support categories it in three major areas: (a) structural support, (b) functional support, and (c) appraisal support (Cobb, 1976; Mohd et al., 2019; Wu & Chiou, 2020) (Table 5.1).

Probably a knowledgeable person with abundant resources might be able to fulfill several structural needs without taking any support from the social network, but what about the other two types of support needs? Those must be reciprocated by supporters. Some of the older people might have accumulated a lot of resources, but due to their age, the need for all types of support emerges for them. A review of several studies demonstrated the lowering effect of social support and large social network on depressive symptoms among older people (Mohd et al., 2019). A recent study conducted by Zhang et al. (2021) attempted to see the role of social media in meeting the need for social support for older people have conceptualized a social support framework that focuses on two elements. According to them, social support can be categorized into perceived social support and received social support. Perceived social support is an anticipated feeling of support available when needed, while received support is something that is actually received when needed. Authors found perceived social support as being more helpful in addressing feelings of loneliness (Zhang et al., 2021). A recently proposed self-determination theory also signifies the role of social connectedness in the overall well-being of the people. According to this theory, three psychological needs *relatedness, competence,* and *autonomy* should be met by the individuals. Relatedness—is a sense of belonging that shows how a person feels connected with their social network. Competence refers the individual capabilities and their ability to use competence in daily life. Autonomy refers to a sense of independence when individuals feel responsible for their choice and self-expression (Clark & Moloney, 2020; Deci & Ryan, 2008).

Loneliness is one of the major emotions experienced by older adults. Loneliness can affect physical health by altering physiological and behavioral mechanisms including neuroendocrine effects, gene effects, and immune functioning (Wang et al., 2018). Loneliness can be of two types, social and emotional. Social loneliness is felt when the person does not have a larger social network as desired. This can lead to boredom, exclusion, and a feeling of marginality. Emotional loneliness is affected when a person is missing an intimate relationship that leads to distress and apprehension (Gierveld & Tilburg, 2006; Wang et al., 2018). The neurobiological theory has demonstrated that loneliness and social isolation trigger proinflammatory and neuroendocrine stress responses (Cacioppo et al., 2014). According to cognitive theory, social isolation affects executive functioning and leads to

TABLE 5.1 Categories of social support with examples.

Categories	Subareas	Examples
Structural support	Size of the network, how often support is provided, and source of support	Person "A," a man, has a large family, several relatives, and many close friends who are always willing to help him. His two close friends, three relatives, and his own younger brother are very rich. All of them feel happy in providing financial assistance whenever it is needed by person A. They often check on person A for his and his family's well-being. One of his relatives lives within five minutes of walking distance. When person A and his spouse have to go out of town for business, their relative keeps their son at home and takes care of everything.
Functional support	It includes emotional, instrumental, and informational support	Person "B," a man, has taken a loan from a bank and is currently unable to pay the installment. The bank is coming after him. His business is also at a loss, and his son has gotten in bad company; he started missing his school and spending time with friends who are misguiding him. He is very upset with everything and is not able to cope. He does not know how to deal with so many factors which all of a sudden came into his life. His maternal uncle calls him and tells him that he has gone through with all the above. He tells him that everything will be all right. He just needs to stay calm and take wise steps. He suggested he provides his income and business situation to the bank and request their support. Further, he suggested a name of a consultant who can review the situation of his business and advise steps to improve the business. For his son, he suggested he spends more time with him. Try to understand why he is doing, what he is doing, and also try to support him.
Appraisal support	Satisfaction and social isolation	Person "C," a man, is a very successful academician. He deals with theories and conducts research. He comes from an uneducated family. None of the family members hardly understand and value his accomplishments. Mr. B feels a little unacknowledged and unsatisfied with what he does. He wants to see an appreciation of his achievements more in the eyes of his family members. One of his academician colleagues became a friend to him. He often calls Mr. B, listens to his research, and acknowledges his achievements. He often calls for a celebration of his accomplishments and keeps a small party that includes members of both families. Mr. B feels satisfied and connected.

poor social cognition, distrust in others, and poor ability in dealing with negative social stimuli (Cacioppo & Hawkley, 2009). The "Need to Belong" theory tells that people are predisposed genetically through the process of natural selection for connecting with others (Baumeister & Leary, 2017). Further adding to this theory, social contact activates the reward system of the brain similar to food and sex (Sescousse et al., 2013).

Social media

Social media is an electronic way of communicating with individuals, and groups, near and far with the help of computer-based technology. Social media can be used to share ideas, provide information, exchange photos, post videos, and so on. Internet outreach to millions in many countries, the development of social media websites, and their friendly usage have connected a significant population of worlds with people across boundaries. This development is proving very helpful in meeting the social needs of the people. Social media platforms are heavily popular among each section and age group of people. Older people can also use the digital platform to connect with people to address their social isolation by connecting with people outside of their social network. People can satiate their reward system by interacting with others on social media. When people receive a like on Facebook, Instagram, and Twitter, it activates the brain's reward system. The person feels connected and the feeling of loneliness and social isolation reduces (Meshi et al., 2013). By the end of 2019, over 2.5 billion people which is about one-third of the total population use social network sites to connect with others (Meshi et al., 2013). The restriction placed during the COVID-19 pandemic has increased digital mediums for work and connectivity. Social media use doubled in two years. Currently, about 4.8 billion which is about 58.6% world population actively use social media. Among all, China is the leading country with 999.9 million people on social media followed by India with 639.4 million, and the USA with 295.4 million. Age-wise, the use of social media is highest at about 59.2% in the age group of 30–49 years of age. Females account for 62.3% of all users (Dixson, 2022). About 47.6% use social media to stay in touch with family and friends, 36.3% for filling their free time, 35.1% read news stories, 31.6% for finding content, and 23.9% to make new connections (Dixson, 2022).

Social media has clearly taken place as a major mode of social connection. Its popularity among each age group around the world indicates that as we progress further in technology and Internet reach, the usage of social media will continue to grow as the most used medium for social connection and interactions. Although social media outlets have numerous disadvantages (Drahošová & Balco, 2017), they still have a strong potential for the older population to help them with their psychological health and overall well-being (Myhre et al., 2017). Looking at an increasing trend in the future, its usage rate would not revert no matter how much its negative impact is discussed. Currently, the literature is acknowledging several disadvantages of social media usage for individuals as well as its overall negative impact on society, but in long run, the focus of research would turn toward improving the social media experience and associated disadvantages. It can be assumed that social media is going to stay here with more benefits. Work would be done toward making it, even more, user-friendly, safe, and less harmful.

Research-based evidence of benefits of social media usage among older adults

Numerous studies have shown the benefits of social media usage in all dimensions of health among older adults. In a study conducted among 2003 Facebook users, 65 years of age and older people reported increased perceived social support (Yu et al., 2018). Another study observed results of Facebook usage and found older people improved their complex working memory and other executive functions (Myhre et al., 2017). A review study has found that social media usages have the potential of decreasing social isolation among seniors (Khosravi et al., 2016). A study conducted by Hutto et al. (2015) with a sample size of 141 participants, age range 51—91 years old, 61% and over females in the United States has found that Facebook users were more satisfied with their current social roles. Researchers in their analysis controlling cofactors age, gender, socioeconomic status including education and income also found that direct communication on social media had reduced loneliness and also increased social role satisfaction (Hutto et al., 2015). A randomized study conducted with 213 senior citizens older than 60 years of age in the United States found improved cognitive functions demonstrated as higher scores on Mini-Mental State Exam (MMSE) for those who used regular online social networking. The longer use was found to be associated with even higher scores on the test. Researchers concluded that regular use of social media may maintain cognitive functions and prevent age-related decline (Kim & Kim, 2014). A study conducted on 4194 participants with a mean age of 66.6 years in the Netherlands showed reduced social loneliness with the use of social media. Their subjective well-being also increased; however, emotional loneliness remained unchanged (van Ingen et al., 2017). A study conducted in the United States with a nationally representative sample of 1620 older Americans revealed social media increases the feeling of social well-being with nonkin relationships and perceived support and feeling of connectedness increased with friends. In a kin relationship, the level of perceived support also increased. Researchers believe social media is helpful for reducing feelings of helplessness and developing a sense of connectedness. However, what differential approaches need to imply so older adults may also receive perceived social support from adults in a kin relationship (Yu et al., 2016). A study conducted on 142 older Americans in the United States observed higher social role satisfaction and social connectedness among social media users. Researchers believe that the use of social media is highly useful for older adults in keeping ties with their families and feeling less lonely (Bell et al., 2013). The use of social media helped older people to overcome feelings of reduced physical mobility, stay connected with family, and maintain cognitive stimulation (Quinn et al., 2016). Another experimental study found that executive functions improved with the use of social media among older adults (Quinn, 2018). A study conducted in South Africa, with a sample of 59 people, 60 years and older age, demonstrated improved quality of life with the use of social media (Rylands & Belle, 2017). A recent study conducted during the COVID-19 pandemic included a sample of 3810 people from Norway, the United Kingdom, the USA, and Australia found a strong connection between social media use and lower social loneliness among older people (Bonsaksen et al., 2021). A study indicated that an online community could offer three major benefits including social support, self-empowerment, and well-being improvement and can support resilience factors such as a positive perspective on life, tenacity, and independence (Kamalpour et al., 2020). Another study found that older workers allowed to use social media have a positive impact on it. Authors speculated that such practice may have the potential effect of delaying retirement (Ma et al., 2021).

Policy approaches to increase social connectedness

Policy changes have also been used with regard to popularizing social media among older adults. An excellent example is that of European local authorities using social media to connect with elderly people in their municipalities. Older people just do not feel socially isolated, lonely, and helpless in many communities, their self-esteem and self-image also get affected. As a policy measure, local authorities in Sweden started using social media to visually show older adults as being happy, active, sociable, and physically capable. This approach brought 2-fold benefits: first, older people developed and maintained a sense of worthiness, improved self-image, and were socially appreciated; second, the community developed a positive image of the older people living in their communities. This approach addressed the negative stereotypes such as disempowered state, vulnerability, and infirmity attached to older adults (Makita et al., 2021; Xu & Larsson, 2021).

Challenges associated with social media for older adults

Several studies have found beneficial effects of social media in helping older people. In the future, its usage would increase. However, inconsistent with the above-discussed studies, social media has been also found to be associated with problematic effects on people. Research has also shown an association between social media use and social isolation among the younger population.

However, the causal relationship is not clear whether people who use social media tend to feel socially isolated or the ones who are socially isolated use more social media (Primack et al., 2017). If social media is encouraged among older people to address the need for social connectedness, what would be the effect on the younger population with whom older adults would connect? The type of cognitive process followed in social media can have a negative effect on the person. A person's social and cognitive comparison with others may increase symptoms of depression and reduce well-being (Tandoc et al., 2015; Verduyn et al., 2015). Older adults who might have mobility concerns may use social media to access. That may lead to no or less time for off-screen interaction which may even lead to greater perceived isolation (Meshi et al., 2013). It is important that benefits and harms must be evaluated carefully to make any general impressions. It is important to identify current challenges faced by older people in using social media for further research and interventions.

Looking at a large population of the world on social media, the use of social media appears easy and stress-free. However, when it comes to people of advanced age, the reality of using social media is not the same as with young people. There could be multiple challenges associated that might prevent or limit older people to use social media. These challenges could be of a variety of nature including manual dexterity to psychological. Based on the literature review, these challenges can be classified mainly in the following four areas: (a) social media literacy, (b) physical challenges, (c) financial challenges, and (d) psychological challenges.

Social media literacy-related challenges among older adults

Literacy skills are required for successful use of social media. A person requires some level of reading and writing ability. Many operations on items that are used for communication require a person to enter numbers, write text, and follow prompts. The literacy rate about 50 years ago was not the same as today, which means people above 60 years of age in many countries might have not received any formal education and may not have the ability to read, write, and follow the simple instructions on the devices for its use. Social media has generated huge enthusiasm for language learning; yet, the language barrier is prominent. Content, competencies, and interrelation between these two are the components of social media literacy. A person should have the ability to understand and analyze the content, be able to reciprocate, and produce a response (Cho et al., 2022). Schreurs et al. (2017) that many older adults lag behind in engagements with digital technology (Schreurs et al., 2017). They provided a conceptual framework of media discourse that keeps people older than 60 years of age behind in using social media (Fig. 5.1).

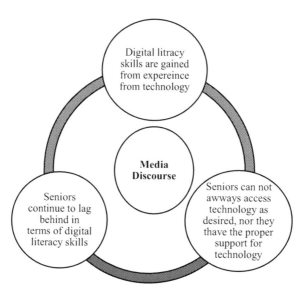

FIGURE 5.1 Problematizing the digital literacy paradox explaining media discourse.

Physical challenges related to social media use among older adults

Most commonly electronic devices, computers, tablets, iPads, and phones are used for social media. A desktop attached with a standard keyboard is among the largest devices, while all others are smaller is size and require better dexterity to operate. Older people due to their age itself often experience poor physical capacity. Eyesight is also a concern. Partial to complete blindness can create a challenge for the person to use social media even with the help of technology (Charness & Boot, 2009). Screen lights of these devices are a serious concern for many people especially those who have photosensitivity. Older people even experience greater problems dealing with such lights. Hearing ability is another barrier that plays a negative role in case person experiences (Nymberg et al., 2019; Wilson et al., 2021). Trembling of hands can pose a challenge for an older person to interact with smaller devices (Nymberg et al., 2019).

Financial challenges related to social media among older adults

For using social media, the person needs two things, a device, and Internet access. Money for buying a device and for paying for Internet service is required. People in high-income countries are better off compared to poor nations when it comes to affording electronic devices and data plans. However, still, many older people living on their social security may not have enough left each month after managing their rent and other necessary needs including paying for utility and buying groceries. A survey conducted in 2021 reported that about 24% of people living in a household with $30,000 and less annual income do not have a smartphone, 43% do not have home broadband, and 41% do not have a computer or laptop (Vogels, 2022). Internet access to Americans has increased heavily during the COVID-19 pandemic, but still only 85.5% of households have Internet access which could be very small data on phone for a member of the family (Statista, 2022).

Psychological challenges associated with social media use among older adults

These can be also addressed as functional barriers. These barriers may also fall in social literacy skills and physical barriers partially, but their interconnected outcome falls better in psychological challenges. The cognitive and perceptual abilities are very important neurological and psychological skills that a person requires to use social media successfully and meaningfully (Charness & Boot, 2009). Often, the cognitive skills of many older adults are compromised as their increase due to underlying diseases including vitamin deficiency, hormonal issues, infections, diabetes, blood pressure, and even mild cognitive impairment and psychological issues such as depression, anxiety, and stress (Arvanitakis et al., 2006). Attitude and belief are two main psychological factors that have been found responsible for using social media (Arvanitakis et al., 2006). Research has found a negative attitude as one of the psychological barriers to using social media among older adults despite having the financial ability to afford the device and Internet, social literacy skills, and physical capability of operating it (Lee et al., 2011). Lee et al. (2011) have noticed that among overall negative attitudes, interpersonal, functional, and structural attitudes play a major role in initiating and sustaining social media usage. Interpersonal barriers often found in this population pertaining to social media use are lack of help from own people and self-doubt for learning. For example, I want to learn the use of computers and phones so I can use social media to interact with others, but it requires a set of complex skills and I have no one to teach me. The other example is that I am too old for that. I have passed my age to learn all these. I am good with old methods. The structural attitude can be explained such as the cost of a smartphone is too high and then I also must get Internet, I can afford it (Lee et al., 2011). Though a person has the financial capability of affording it but interpreting it in other ways. Another functional attitude that is often encountered is that I am too old, so my memory is affected. I forgot everything, so it is difficult to remember all that (Baumeister & Scher, 1988; Lee et al., 2011). Research has found that psychological barriers such as negative attitudes are more associated with people living in poverty and having low education. Higher income and education are less associated with negative attitudes (Lee et al., 2011; Leist, 2013).

Studies conducted by Xie et al. (2012) and Lehtinen et al. (2009) to understand psychological barriers have highlighted how a negative attitude develops at the first experience or as a beginner in social media (Lehtinen et al., 2009; Leist, 2013; Xie et al., 2012). So, there are additional psychological challenges in the use of social media by older adults. Some of these are:

(a) *Social media image:* Some social media websites receive negative news in regular media about fraud such as data and private information being stolen. It is true that people get cheated and tricked on social media by scammers (Vu et al., 2019). Such views and information can make novices uncomfortable and scared, which prevents many older people to avoid using social media in the first place.

(b) *Mismatched social norms*: People operate under social norms in each society. The social norms for online community versus offline community vary. Usually, people follow informal language and communication style with close people including their family members, relatives, and friends, but a formal way of communication is used with new people. Social media uses informal language which can be an uncomfortable situation, especially for older people (Che & Ibrahim, 2018).

(c) *Self-cheer leader:* Culturally and in society, people are trained to be polite and humble in sharing their experiences and achievements. Social media follows the opposite norm. People indulge heavily in self-representation. People post pictures, achievements, and a variety of content to present themselves. While the person who does that receives some level of favorable response and feels happy, while others might see it as a purely empty way of gaining popularity. It might prevent some older adults to avoid social media because they do not perceive it as a good act (Lehtinen et al., 2009; Leist, 2013).

(d) *Lack of individuality:* In day-to-day life, people encounter new people in some known areas or professions for which they have some idea. For example, a new faculty member meets another faculty member in a different department, a worker in a supermarket meets another worker of another supermarket in a parking lot, a person meets a relative of his relative, and a public health professor meets several people at a conference, but knows that most of them are connected with public health field in some ways. While on social media, encountering random strangers is common. Many people find it very interesting, and they meet completely unknown people enthusiastically, while others feel uneasy in transiting mentally to connect with random individuals (Xie et al., 2012).

(e) *Control expertise:* Social media use requires the navigation of the website, authentication of the website, entering information, registration, making and remembering passwords, reentering it at every use, and putting a lot of personally identifiable information. There is fear that information might be compromised, and they may not have the capacity of protecting it. Several

Social media for older adults **Chapter | 5 111**

safeguards are available and used for many people, but for others, especially older adults, it could be a difficult task to have a sense of control (Charness & Boot, 2009; Charness & Boot, 2009; Leist, 2013; Pfeil et al., 2009; Xie et al., 2012).

Challenges and barriers associated with social media usage for older adults can be categorized in multiple ways. A recent scoping review conducted by Wilson et al. (2021) of 14 studies in which they attempted to discuss barriers to using e-health of older adults is used here with additional information from other studies (Leist, 2013; Wilson et al., 2021). These barriers are also highly common for social media usage for older people. These barriers can be placed under five main categories and are depicted next in Fig. 5.2 (Leist, 2013; Wilson et al., 2021). Please refer to Fig. 5.2 for these barriers.

Usage of social media has mixed results in the population, while its usage is more common among the younger population where its negative consequences are also high (Oberst et al., 2017; Richards et al., 2015). On the other hand, social media is not used as commonly among older people (Braun, 2013). Older people also experience harmful consequences of social media usage, but the older population deals with more loneliness, social isolation, restricted mobility, retirement, lack of meaningful activities, sad feelings of being in poor health, loss of spouse and relatives, loss of age group friends due to their early demise, and fear of death due to advancing age. Older people need to be engaged in society for keeping their social health and well-being better (Vogel et al., 2021). Social media is an existing and advancing field that has the ability to provide very fast momentary reach to near and far distanced own relatives, friends, and others. This mode can be helpful in meeting some of the social and psychological needs. However, as we discussed, older people experience several structural and psychological barriers to

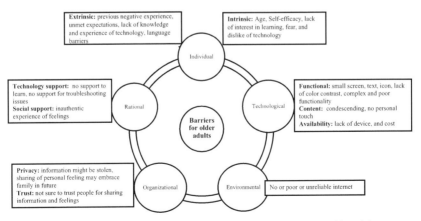

FIGURE 5.2 Factors that play a role in using social media among older adults.

using social media platforms. There is no better alternative than an in-person meeting and interacting with people, but looking at the current reality of increasing population of older people, increasing opportunity cost for families to keep and care for older family members in the homes, and increasing prevalence of depressive symptoms, loneliness, and social isolation in the older population, social should be promoted with caution of minimizing its harmful consequences and risks (Brinda et al., 2014; WHO, 2022; Wu & Chiou, 2020).

The growing number of social media among older people is a good sign of its acceptability and popularity. However, many older people do not benefit maximum from its benefit or find it difficult to use for some reasons, and many of them are still unable to start. A conducive and enabling atmosphere can encourage novices to use social media and extend more benefits to current users. The technology field appears to be more connected when it comes to making user-friendly devices and website navigation. However, the use of social media is highly dependent on behavioral factors, and its outcomes are clearly connected with overall health, well-being, happiness, and quality of life. Public health appears to be the most relevant field where experts should attempt to develop plans for its promotion among older adults (Giustini et al., 2018).

Some of the public health theories have been applied to understand its application in social networking and social media usage in various populations. *Theory of planned behavior* for social networking for college students (Cameron, 2010), *social cognitive theory* in social networking behavior for older adults (Krendl et al., 2022), and *health belief model* for understanding the intention of older people in using health information on social media (Shang et al., 2020) are a few public health theories that have been used in this area with specific purposes. It is important to fill the gap in the implication of public health theories in this area.

A recently developed fourth-generation *multi-theory model* (MTM) of health behavior change, receiving recognition in a variety of public health issues, can be considered in promoting social media usage among older people (Sharma, 2022, pp. 272–286; Sharma et al., 2021). The model has a unique ability to assess an intention and sustenance of behavior change, while most other behavior theories explain only the factors which contribute to the learning of the behavior. This model has two submodels: initiation (starting) and sustenance (maintenance). The initiation model has three constructs, participatory dialogue (comprising advantages of using social media outweighing disadvantages), behavioral confidence (surety of using social media), and changes in the physical environment (availability and accessibility to social media). The sustenance model also has three constructs that include changes in the social environment (having peer support to use social media), practice for change (ability to practice and monitor social media usage), and emotional transformation (ability to use feeling in social media usage) (please see Fig. 5.3). The application of this model can also guide appropriate intervention to promote safe and satisfying social media usage among older people. A brief plan based on MTM is proposed in Table 5.2.

Summary

A person's social connections and emotions are paramount to their general well-being. When faced with a decline in said well-being, older adults observe many examples of it. This includes a decline in physical health, such as issues with sensory organs, the heart, the kidneys, the immune system, etc. Along with this, many older people face a high prevalence of mental health issues such as anxiety, depression, and neurological disorders such as dementia. With this lowering of one's well-being, older people have a higher need for social connectedness. This issue has risen in prevalence due to a growing population of older people (due to growth caused by various advances in medical science, living conditions, etc.). It is established that social connectedness is an important factor of overall well-being. Studies have found a correlation between depressive symptoms and low social connectedness (living alone, loneliness, and social isolation). Furthermore, social connectedness falls into multiple categories including structural support, functional support, and appraisal support. While one of these categories may be partly fulfilled by oneself, such as structural support with regard to financial independence, the others necessitate input by other people. Recently, one method of obtaining such social support has grown in popularity. This method is social media. Social media is a method to electronically communicate with groups and individuals over any distance with the use of computer-based technology. This means of communication is vastly popular, with upwards of 4.8 billion people using social media sites to communicate with others. Social media has been established as a major vector for social connection, and its popularity is projected to continue increasing in the future. Due to its ease of access and the freedom to share information with anyone no matter the distance in its use,

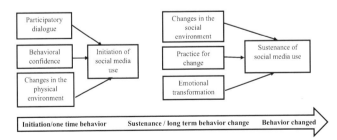

FIGURE 5.3 Framework of multi-theory model (MTM) of health behavior change for promoting the use of social media.

114 SECTION | II Social media use and mental health outcomes

TABLE 5.2 MTM-based plan to promote social media among older adults.

	Initiation model			Sustenance model		
Constructs	Situation	Suggested activities	Constructs	Situation		Suggested activities
Participatory dialogue	– Many older adults have limited information about its use and benefits – Many older adults are wary of new technology	– Help people to understand what they will achieve with its usage – Use modeling to show benefits – Demonstrate how to overcome disadvantages	Changes in the social environment	– Older adults living alone – Older adults lack social support—this is the objective of the MTM application here		– Initially, establish connections with other older adults through outside help – These people can act as catalysts and reinforce communication – This activity can also enhance the confidence of the users
Behavioral confidence	– Many older adults lack self-surety – Many older adults have	– Arrange training sessions in which skills can be acquired in small steps – Arrangement of need-based help to	Practice for change	– Older adults using social media for the first-time encounter anxiety – Older adults forget how to use social media and are		– Provide more guided opportunities for interaction

	lost the desire to learn – Many older adults think of being too old for new learning	troubleshoot potential barriers	with family and friends – Promote goal setting and self-monitoring
Changes in the physical environment	– Availability of devices may be an issue – Accessibility of the Internet – The physical and sensory issues related to older age might impact the usage of social media, for example, small icons and smaller devices	– Policy changes to make devices and the Internet accessible – Computers with a larger screen and keyboard – Technology improvement for making devices and web experiences more user-friendly—Support through the help of the individual's family and friends	
Emotional transformation	not able to sustain the usage – Many older adults despite having the ability to use it do not use it – Lack of directionality of feelings		– Whatever, good feelings they feel by using social media are highlighted – Feelings are directed toward goal setting

social media is an effective technological advancement. Social media brings with it many benefits to older populations. Studies in older populations have found that social media use leads to an increase in perceived social support; it also has the potential to decrease social isolation among seniors. Unfortunately, social media and its benefits for older adults do not come without its setbacks. Older populations face challenges with adopting social media; physical, financial, psychological, and literacy issues are the main setbacks. Older people often face issues with their eyesight or dexterity, limiting their ability to effectively use smaller devices. They may also have trouble financially affording a device and an Internet plan. Moreover, psychological challenges such as attitudes and beliefs toward social media can also prevent older people from utilizing it. Thus, promoting the usage of social media among older people and overcoming the said setbacks is an important task. This objective of promoting social media among older adults may be reached using a newer health behavior theory, the MTM of health behavior change.

Conclusion

Due to declines in various factors of well-being, older populations are vulnerable and need support to maintain a good quality of life. One crucial aspect of this well-being is that of their social connectedness and emotional health, both of which go hand in hand. The usage of social media can benefit older people with these issues. Unfortunately, there are some structural and psychological challenges associated with social media use in older populations. However, with the use of the fourth-generation, MTM of health behavior change, and a vigilant effort to make social media more accessible to the elderly, we can promote the use of social media among older people and help combat these issues of poor social connectedness and declining well-being among them.

Questions

✔ How to reduce emotional loneliness through social media usage?

✔ Is the heavy experience of informal communication on social media affecting users to follow formal communication in formal settings?

✔ What public health behavioral theory can be applied best to promote the safe and satisfying use of social media among older people?

✔ Would the growing rate of social media usage among older people be minimizing offline interaction opportunities in the long run?

Definition of key terms

Behavioral confidence: This is a construct of the initiation model. It is the confidence in one's ability to perform a given behavior and can be futuristic. The sources of this behavior can be self or external personifications like powerful others, the Almighty, a deity, etc. *Please also see the multi-theory model (MTM) of health behavior change; initiation.*

Changes in the physical environment: This is a construct in the initiation model that involves changes in availability, accessibility, preparedness, and obtainability of physical resources that facilitate the modification of the behavior. *Please also see the multi-theory model (MTM) of health behavior change; initiation.*

Changes in the social environment: This is a construct in the sustenance model. It involves having social support in the existing environment toward maintaining the target behavior change. It can come from family, friends, health professionals, teachers, social media, etc. *Please also see the multi-theory model (MTM) of health behavior change; sustenance.*

Emotional transformation: This is a construct in the sustenance model. It is an individual's ability to transform one's feelings toward a specific goal for maintaining behavior change. *Please also see the multi-theory model (MTM) of health behavior change; sustenance.*

Initiation: This is a submodel of the multi-theory model (MTM) of health behavior change. It measures the likelihood of starting a behavior. This prediction is based on the scores of three constructs of this submodel: *participatory dialogue, behavioral confidence, and changes in the physical environment. Please also see the multi-theory model (MTM) of health behavior change.*

Older adults: People who are at least 60 years of age or above.

Participatory dialogue: This is a nonverbal dialogue process within an individual or a verbal or written dialogue between individuals (the facilitator and the client or clients). The person evaluates the pros and cons of a given behavior in terms of its advantages and its disadvantages. The outcome of this process partially decides whether the person wants to initiate the behavior or not. *Please also see the multi-theory model (MTM) of health behavior change; initiation.*

Practice for change: This is a construct in the sustenance model. It involves active evaluation and reflection on the progression of behavior change. The individual actively identifies barriers that emerge during behavior change processes and addresses them. *Please also see the multi-theory model (MTM) of health behavior change; sustenance.*

Social media: An electronic way of communicating with individuals and groups, near and far, with the help of computer-based technology. *Please also see social media literacy.*

Social media literacy: The ability to read and write with the use of electronic devices such as smartphones and computers. This includes understanding and having the functional ability to navigate social media through different aspects such as creating an account, entering payment sources, making a user id and password, sending, receiving messages and attachments, etc. *Please also see social media.*

Sustenance: This is a submodel of the multi-theory model. It measures the likelihood of a behavior being continued. This prediction is based on the scores of three constructs of this submodel: *changes in the social environment, practice for change, and emotional transformation. Please also see the multi-theory model (MTM) of health behavior change.*

The multi-theory model (MTM) of health behavior change: A newly developed behavior model that emerged from several constructs of previously developed behavior change theories. This model looks at behavior from two standpoints, initiation (starting) and sustenance (maintenance). *Please also see behavioral confidence; changes in the physical environment; changes in the social environment; emotional transformation; initiation; participatory dialogue; practice for change; sustenance.*

Websites to explore

Facebook: A popular social networking site among older adults
https://www.facebook.com/

Explore this website if you have an account. If not, then create an account and attempt networking through this website. *How was your experience? Discuss the advantages and disadvantages that older adults might encounter in networking using this website.*

LinkedIn: A popular website for professionals networking
https://www.linkedin.com.

Visit this website and explore the pages for some of the contributing authors of this book, if available. *What did you find? What do you think are the pros and cons of this website for older adults in dealing with linking with professionals?*

YouTube: A popular website for watching and creating video clips
https://www.youtube.com/

Visit this website and explore videos on a topic of your choice. Watch a YouTube video on how to create a video recording. *Please create a video recording on a topic of your choice for posting on YouTube. What do you think might be the advantages or disadvantages of this social media for older adults in watching and creating video clips?*

Additional websites to explore, open an account (as needed), and discuss the pros and cons of older adults using these social media sites: Instagram https://www.instagram.com/; Pinterest https://www.pinterest.com/; Senior match https://www.seniormatch.com/; Skype https://www.skype.com/en/; and Twitter https://twitter.com.

References

Abeliansky, A. L., Erel, D., & Strulik, H. (2021). Social vulnerability and aging of elderly people in the United States. *SSM-population Health, 16,* 100924. https://doi.org/10.1016/j.ssmph.2021.100924

Arvanitakis, Z., Wilson, R. S., & Bennett, D. A. (2006). Diabetes mellitus, dementia, and cognitive function in older persons. *The Journal of Nutrition, Health and Aging, 10*(4), 287.

Barak, Y., Leitch, S., Greco, P., & Glue, P. (2020). Fatigue, sleep and depression: An exploratory interRAI study of older adults. *Psychiatry Research, 284,* 112772. https://doi.org/10.1016/j.psychres.2020.112772

Barrera, M., & Ainlay, S. L. (1983). The structure of social support: A conceptual and empirical analysis. *Journal of Community Psychology, 11*(2), 133–143. https://doi.org/10.1002/1520-6629(198304)11:2<133::aid-jcop2290110207>3.0.co;2-l.

Baumeister, R. F., & Leary, M. R. (2017). The need to belong: Desire for interpersonal attachments as a fundamental human motivation. *Interpersonal Development, 117*(3), 497–529.

Baumeister, R. F., & Scher, S. J. (1988). Self-defeating behavior patterns among normal individuals: Review and analysis of common self-destructive tendencies. *Psychological Bulletin, 104*(1), 3. https://doi.org/10.1037/0033-2909.104.1.3

Bell, C., Fausset, C., Farmer, S., Nguyen, J., Harley, L., & Fain, W. B. (May 2013). Examining social media use among older adults. In *Proceedings of the 24th ACM Conference on Hypertext and social media* (pp. 158–163). https://dl.acm.org/doi/abs/10.1145/2481492.2481509.

Bonsaksen, T., Ruffolo, M., Leung, J., Price, D., Thygesen, H., Schoultz, M., & Geirdal, A.Ø. (2021). Loneliness and its association with social media use during the COVID-19 outbreak. *Social Media+ Society,* 1–10. https://journals.sagepub.com/doi/pdf/10.1177/20563051211033821.

Braun, M. T. (2013). Obstacles to social networking website use among older adults. *Computers in Human Behavior, 29*(3), 673–680. https://doi.org/10.1016/j.chb.2012.12.004

Brinda, E. M., Rajkumar, A. P., Enemark, U., Attermann, J., & Jacob, K. S. (2014). Cost and burden of informal caregiving of dependent older people in a rural Indian community. *BMC Health Services Research, 14*(1), 1–9. https://doi.org/10.1186/1472-6963-14-207

Cacioppo, S., Capitanio, J. P., & Cacioppo, J. T. (2014). Toward a neurology of loneliness. *Psychological Bulletin, 140*(6), 1464. https://doi.org/10.1037/a0037618

Cacioppo, J. T., & Hawkley, L. C. (2009). Perceived social isolation and cognition. *Trends in Cognitive Sciences, 13*(10), 447–454. https://doi.org/10.1016/j.tics.2009.06.005

Cameron, R. (2010). *Ajzen's theory of planned behavior applied to the use of social networking by college students.* Available at: https://digital.library.txstate.edu/handle/10877/3298.

Charness, N., & Boot, W. R. (2009). Aging and information technology use: Potential and barriers. *Current Directions in Psychological Science, 18*(5), 253–258. https://doi.org/10.1111/j.1467-8721.2009.01647.x

Che, C. W. I. R. B., & Ibrahim, W. (2018). Social media tools for informal language learning: A comprehensive theoretical framework. *Asian Social Science, 14*(4), 46–50. https://pdfs.semanticscholar.org/92bd/98635fdb6d67ae6975c52cc82051f712a894.pdf

Cho, H., Cannon, J., Lopez, R., & Li, W. (2022). Social media literacy: A conceptual framework. *New Media and Society.* https://doi.org/10.1177/14614448211068530

Clark, R., & Moloney, G. (2020). Facebook and older adults: Fulfilling psychological needs? *Journal of Aging Studies, 55,* 100897. https://doi.org/10.1016/j.jaging.2020.100897

Cobb, S. (1976). Social support as a moderator of life stress. *Psychosomatic Medicine, 38*(5), 300−314. https://doi.org/10.1097/00006842-197609000-00003

Deci, E. L., & Ryan, R. M. (2008). Self-determination theory: A macro theory of human motivation, development, and health. *Canadian Psychology/Psychologie Canadienne, 49*(3), 182−185. https://doi.org/10.1037/a0012801

Dixson, S. (2022). *Statista. Social media-statistics and facts.* https://www.statista.com/topics/1164/social-networks/#topicHeader__wrapper.

Drahošová, M., & Balco, P. (2017). The analysis of advantages and disadvantages of use of social media in European Union. *Procedia Computer Science, 109*, 1005−1009. https://doi.org/10.1016/j.procs.2017.05.446

Fukunaga, R., Abe, Y., Nakagawa, Y., Koyama, A., Fujise, N., & Ikeda, M. (2012). Living alone is associated with depression among the elderly in a rural community in Japan. *Psychogeriatrics, 12*(3), 179−185. https://doi.org/10.1111/j.1479-8301.2012.00402.x

Gatchel, J. R., Rabin, J. S., Buckley, R. F., Locascio, J. J., Quiroz, Y. T., Yang, H. S., Vannini, P., Amariglio, R. E., Rentz, D. M., Properzi, M., Donovan, N. J., Blacker, D., Johnson, K. A., Sperling, R. A., Marshall, G. A., & Harvard Aging Brain Study. (2019). Longitudinal association of depression symptoms with cognition and cortical amyloid among community-dwelling older adults. *JAMA Network Open, 2*(8), e198964. https://doi.org/10.1001/jamanetworkopen.2019.8964

Gierveld, J. D. J., & Tilburg, T. V. (2006). A 6-item scale for overall, emotional, and social loneliness: Confirmatory tests on survey data. *Research on Aging, 28*(5), 582−598. https://doi.org/10.1177/0164027506289723

Giustini, D., Ali, S. M., Fraser, M., & Kamel Boulos, M. N. (2018). Effective uses of social media in public health and medicine: A systematic review of systematic reviews. *Online Journal of Public Health Informatics, 10*(2), e215. https://doi.org/10.5210/ojphi.v10i2.8270

Gu, L., Yu, M., Xu, D., Wang, Q., & Wang, W. (2020). Depression in community-dwelling older adults living alone in China: Association of social support network and functional ability. *Research in Gerontological Nursing, 13*(2), 82−90. https://doi.org/10.3928/19404921-20190930-03

Hutto, C. J., Bell, C., Farmer, S., Fausset, C., Harley, L., Nguyen, J., & Fain, B. (2015). Social media gerontology: Understanding social media usage among older adults. In , Vol 13. *Web intelligence* (pp. 69−87). IOS Press, 1.

van Ingen, E., Rains, S. A., & Wright, K. B. (2017). Does social network site use buffer against well-being loss when older adults face reduced functional ability? *Computers in Human Behavior, 70*, 168−177. https://www.sciencedirect.com/science/article/abs/pii/S0747563216308871.

Jaul, E., & Barron, J. (2017). Age-related diseases and clinical and public health implications for the 85 Years old and over population. *Frontiers in Public Health, 5*, 335. https://doi.org/10.3389/fpubh.2017.00335

Kamalpour, M., Watson, J., & Buys, L. (2020). How can online communities support resilience factors among older adults? *International Journal of Human−Computer Interaction, 36*(14), 1342−1353. https://doi.org/10.1080/10447318.2020.1749817

Khan, A., & Husain, A. (2010). Social support as a moderator of positive psychological strengths and subjective well-being. *Psychological Reports, 106*(2), 534−538. https://doi.org/10.2466/PR0.106.2.534-538

Khosravi, P., Rezvani, A., & Wiewiora, A. (2016). The impact of technology on older adults' social isolation. *Computers in Human Behavior, 63*, 594−603. https://dl.acm.org/doi/10.1016/j.chb.2016.05.092.

Kim, H., & Kim, J. (2014). The impact of senior citizens' use of online social networks on their cognitive function. *International Journal of Research Studies in Educational Technology, 3*(2), 21−30. http://consortiacademia.org/wp-content/uploads/IJRSET/IJRSET_v3i2/844-3205-1-PB.pdf.

Krendl, A. C., Kennedy, D. P., Hugenberg, K., & Perry, B. L. (2022). Social cognitive abilities predict unique aspects of older adults' personal social networks. *The Journals of Gerontology: Series B, 77*(1), 18−28. https://doi.org/10.1093/geronb/gbab048

Lee, B., Chen, Y., & Hewitt, L. (2011). Age differences in constraints encountered by seniors in their use of computers and the internet. *Computers in Human Behavior, 27*(3), 1231−1237. https://doi.org/10.1016/j.chb.2011.01.003

Lehtinen, V., Näsänen, J., & Sarvas, R. (2009). "A little silly and empty-headed:" Older adults' understandings of social networking sites. *People and Computers XXIII Celebrating People and Technology*, 45−54. https://dl.acm.org/doi/10.5555/1671011.1671017

Leist, A. K. (2013). Social media use of older adults: A mini-review. *Gerontology, 59*(4), 378−384. https://doi.org/10.1159/000346818

Li, N., Chen, G., Zeng, P., Pang, J., Gong, H., Han, Y., Zhang, Y., Zhang, E., Zhang, T., & Zheng, X. (2016). Prevalence of depression and its associated factors among Chinese elderly people: A comparison study between community-based population and hospitalized population. *Psychiatry Research, 243*, 87−91. https://doi.org/10.1016/j.psychres.2016.05.030

Lotfaliany, M., Hoare, E., Jacka, F. N., Kowal, P., Berk, M., & Mohebbi, M. (2019). Variation in the prevalence of depression and patterns of association, sociodemographic and lifestyle factors in community-dwelling older adults in six low-and middle-income countries. *Journal of Affective Disorders, 251*, 218−226. https://doi.org/10.1016/j.jad.2019.01.054

Makita, M., Mas-Bleda, A., Stuart, E., & Thelwall, M. (2021). Ageing, old age and older adults: A social media analysis of dominant topics and discourses. *Ageing and Society, 41*(2), 247−272. https://doi.org/10.1017/S0144686X19001016

Ma, Y., Liang, C., Gu, D., Zhao, S., Yang, X., & Wang, X. (2021). How social media use at work affects improvement of older people's willingness to delay retirement during transfer from demographic bonus to health bonus: Causal relationship empirical study. *Journal of Medical Internet Research, 23*(2), e18264. https://doi.org/10.2196/18264

Meshi, D., Morawetz, C., & Heekeren, H. R. (2013). Nucleus accumbens response to gains in reputation for the self relative to gains for others predicts social media use. *Frontiers in Human Neuroscience, 439*. https://www.frontiersin.org/articles/10.3389/fnhum.2013.00439/full.

Mohd, T. A. M. T., Yunus, R. M., Hairi, F., Hairi, N. N., & Choo, W. Y. (2019). Social support and depression among community dwelling older adults in Asia: A systematic review. *BMJ Open, 9*(7), e026667. https://doi.org/10.1136/bmjopen-2018-026667

Myhre, J. W., Mehl, M. R., & Glisky, E. L. (2017). Cognitive benefits of online social networking for healthy older adults. *The Journals of Gerontology: Series B, 72*(5), 752−760. https://doi.org/10.1093/geronb/gbw025

Nymberg, V. M., Bolmsjö, B. B., Wolff, M., Calling, S., Gerward, S., & Sandberg, M. (2019). 'Having to learn this so late in our lives…' Swedish elderly patients' beliefs, experiences, attitudes and expectations of e-health in primary health care. *Scandinavian Journal of Primary Health Care, 37*(1), 41−52. https://doi.org/10.1080/02813432.2019.1570612

Oberst, U., Wegmann, E., Stodt, B., Brand, M., & Chamarro, A. (2017). Negative consequences from heavy social networking in adolescents: The mediating role of fear of missing out. *Journal of Adolescence, 55*, 51−60. https://doi.org/10.1016/j.adolescence.2016.12.008

Park, J. I., Park, T. W., Yang, J. C., & Chung, S. K. (2016). Factors associated with depression among elderly Koreans: The role of chronic illness, subjective health status, and cognitive impairment. *Psychogeriatrics, 16*(1), 62−69. https://doi.org/10.1111/psyg.12160

Petrova, N. N., & Khvostikova, D. A. (2021). Prevalence, structure, and risk factors for mental disorders in older people. *Advances in Gerontology, 11*(4), 409−415.

Pfeil, U., Zaphiris, P., & Wilson, S. (2009). Older adults' perceptions and experiences of online social support. *Interacting with Computers, 21*(3), 159−172.

Primack, B. A., Shensa, A., Sidani, J. E., Whaite, E. O., Lin, L. Y., Rosen, D., Colditz, J. B., Radovic, A., & Miller, E. (2017). Social media use and perceived social isolation among young adults in the U.S. *American Journal of Preventive Medicine, 53*(1), 1−8. https://doi.org/10.1016/j.amepre.2017.01.010

Quinn, K. (2018). Cognitive effects of social media use: A case of older adults. *Social Media+ Society*, 1−9.

Quinn, K., Smith-Ray, R., & Boulter, K. (July 2016). Concepts, terms, and mental models: Everyday challenges to older adult social media adoption. In *International conference on human Aspects of IT for the aged population* (pp. 227−238). Cham: Springer.

Richards, D., Caldwell, P. H., & Go, H. (2015). Impact of social media on the health of children and young people. *Journal of Paediatrics and Cchild Health, 51*(12), 1152−1157.

Rural Health Information Hub. (2022). *Demographic changes and aging population*. https://www.ruralhealthinfo.org/toolkits/aging/1/demographics

Rylands, D., & Belle, J. P. V. (May 2017). The impact of Facebook on the quality of life of senior citizens in Cape Town. In *International conference on social Implications of Computers in developing countries* (pp. 740−752). Cham: Springer.

Schreurs, K., Quan-Haase, A., & Martin, K. (2017). Problematizing the digital literacy paradox in the context of older adults' ICT use: Aging, media discourse, and self-determination. *Canadian Journal of Communication, 42*(2), 359−377. https://doi.org/10.22230/cjc.2017v42n2a3130

Sescousse, G., Caldú, X., Segura, B., & Dreher, J. C. (2013). Processing of primary and secondary rewards: A quantitative meta-analysis and review of human functional neuroimaging studies. *Neuroscience and Biobehavioral Reviews, 37*(4), 681−696. https://doi.org/10.1016/j.neubiorev.2013.02.002

Shang, L., Zhou, J., & Zuo, M. (2020). Understanding older adults' intention to share health information on social media: The role of health belief and information processing. *Internet Research, 31*(1), 100−122. https://doi.org/10.1108/INTR-12-2019-0512

Sharma, M. (2022). *Theoretical foundations of health education and health promotion* (4th ed.). Burlington, MA: Jones and Bartlett.

Sharma, M., Asare, M., Lakhan, R., Kanekar, A., Nahar, V. K., & Moonie, S. (2021). Can the Multi-Theory Model (MTM) of health behavior change explain the intent for people to practice meditation? *Journal of Evidence-Based Integrative Medicine, 26*, 1−12. https://doi.org/10.1177/2515690X211064582

Silverstein, M., Cong, Z., & Li, S. (2006). Intergenerational transfers and living arrangements of older people in rural China: Consequences for psychological well-being. *The Journals of Gerontology Series B: Psychological Sciences and Social Sciences, 61*(5), S256−S266. https://doi.org/10.1093/geronb/61.5.s256

Stahl, S. T., Beach, S. R., Musa, D., & Schulz, R. (2017). Living alone and depression: The modifying role of the perceived neighborhood environment. *Aging and Mental Health, 21*(10), 1065−1071. https://doi.org/10.1080/13607863.2016.1191060

Statista. (2022). *Percentage of households with internet use in the United States from 1997 to 2020*. https://www.statista.com/statistics/189349/us-households-home-internet-connection-subscription/.

Tandoc, E. C., Jr., Ferrucci, P., & Duffy, M. (2015). Facebook use, envy, and depression among college students: Is facebooking depressing? *Computers in Human Behavior, 43*, 139−146. https://doi.org/10.1016/j.chb.2014.10.053

Vanoh, D., Shahar, S., Yahya, H. M., & Hamid, T. A. (2016). Prevalence and determinants of depressive disorders among community-dwelling older adults: Findings from the towards useful aging study. *International Journal of Gerontology, 10*(2), 81−85. https://www.sciencedirect.com/science/article/pii/S1873959816300333

Verduyn, P., Lee, D. S., Park, J., Shablack, H., Orvell, A., Bayer, J., Ybarra, O., Jonides, J., & Kross, E. (2015). Passive Facebook usage undermines affective well-being: Experimental and longitudinal evidence. *Journal of Experimental Psychology. General, 144*(2), 480−488. https://doi.org/10.1037/xge0000057

Vogel, P., Grotherr, C., Von Mandelsloh, F., Gaidys, U., & Böhmann, T. (2021). Older adults' use of online neighborhood social networks: Perceptions, challenges and effects. *Proceedings of the 54th Hawaii International Conference on System Sciences*, 3913−3922. https://hdl.handle.net/10125/71090.

Vogels, A. E. (2022). *Pew Research Center. Digital divide persists even as Americans with lower incomes make gains in tech adoption*. https://www.pewresearch.org/fact-tank/2021/06/22/digital-divide-persists-even-as-americans-with-lower-incomes-make-gains-in-tech-adoption/.

Vu, H. Q., Law, R., & Li, G. (2019). Breach of traveller privacy in location-based social media. *Current Issues in Tourism, 22*(15), 1825−1840. https://doi.org/10.1080/13683500.2018.1553151

Wang, J., Mann, F., Lloyd-Evans, B., Ma, R., & Johnson, S. (2018). Associations between loneliness and perceived social support and outcomes of mental health problems: A systematic review. *BMC Psychiatry, 18*(1), 1−16. https://doi.org/10.1186/s12888-018-1736-5

Wilson, J., Heinsch, M., Betts, D., Booth, D., & Kay-Lambkin, F. (2021). Barriers and facilitators to the use of e-health by older adults: A scoping review. *BMC Public Health, 21*(1), 1−12. https://doi.org/10.1186/s12889-021-11623-w

World Health Organization. (2022a). *Aging and health*. https://www.who.int/news-room/fact-sheets/detail/ageing-and-health.

World Health Organization. (2022b). *WHO remains firmly committed to the principles set out in the preamble to the constitution?* https://www.who.int/about/governance/constitution.

Wu, H. Y., & Chiou, A. F. (2020). Social media usage, social support, intergenerational relationships, and depressive symptoms among older adults. *Geriatric Nursing, 41*(5), 615−621. https://doi.org/10.1016/j.gerinurse.2020.03.016

Xie, B., Watkins, I., Golbeck, J., & Huang, M. (2012). Understanding and changing older adults' perceptions and learning of social media. *Educational Gerontology, 38*(4), 282−296. https://doi.org/10.1080/03601277.2010.544580

Xu, W., & Larsson, A. T. (2021). Communication officers in local authorities meeting social media: On the production of social media photos of older adults. *Journal of Aging Studies, 58*, 100952. https://doi.org/10.1016/j.jaging.2021.100952

Yu, R. P., Ellison, N. B., & Lampe, C. (2018). Facebook use and its role in shaping access to social benefits among older adults. *Journal of Broadcasting and Electronic Media, 62*(1), 71−90. https://doi.org/10.1080/08838151.2017.1402905

Yu, R. P., Mccammon, R. J., Ellison, N. B., & Langa, K. M. (2016). The relationships that matter: Social network site use and social wellbeing among older adults in the United States of

America. *Ageing and Society, 36*(9), 1826−1852. https://doi.org/10.1017/S0144686X15000677

Zebhauser, A., Baumert, J., Emeny, R. T., Ronel, J., Peters, A., & Ladwig, K. H. (2015). What prevents old people living alone from feeling lonely? Findings from the KORA-age-study. *Aging and Mental Health, 19*(9), 773−780. https://doi.org/10.1080/13607863.2014.977769

Zhang, K., Kim, K., Silverstein, N. M., Song, Q., & Burr, J. A. (2021). Social media communication and loneliness among older adults: The mediating roles of social support and social contact. *The Gerontologist, 61*(6), 888−896. https://doi.org/10.1093/geront/gnaa197

Chapter 6

Social media, diversity, equity, and inclusion

Tiffiny R. Jones and Sely-Ann Headley Johnson
Wayne State University, Detroit, MI, United States

Learning objectives
1. To define diversity, equity, and inclusion (DEI).
2. To identify ways DEI can be implemented via social media to increase health outcomes.
3. To examine the health outcomes and health disparities revealed due to the emergence of the COVID-19 pandemic.
4. To analyze the ways in which social media affects mental health.

Identifying diversity, equity, and inclusion

Equity can be defined as the fair treatment, access to needed resources, opportunity, and advancement for all people, while simultaneously striving to identify and eliminate barriers that have prevented some groups from their full participation (Kapila et al., 2016). If we seek to improve equity, we need to increase fairness and justice embedded in the procedures and processes of institutions or systems, besides an equitable distribution of resources (Kapila et al., 2016). Equity can only be achieved, when systemic and institutional barriers to engaging in teams, groups, or organizations have been acknowledged and removed, or when opportunities and resources are evenly distributed. True equity exists when opportunities are equally available to all, regardless of individual differences (Lyman et al., 2022). In a world of growing technology advancement, equity must also be addressed in the digital platform.

Equity in the digital domain implies a situation where all individuals and communities have the technological capacity required for full participation in our society, economy, and democracy (National Digital Inclusion Alliance, 2021). It has also been established that digital equity is essential for employment, cultural participation, lifelong learning, access to essential services, and civic participation (National Digital Inclusion Alliance, 2021). In a world with an increasing reliance on virtual spaces and interactions, digital equity is closely related to, if not essential to, attaining health equity.

Healthy People 2030 explains that health equity involves attaining the highest level of health possible. To achieve health equity, everyone must be valued equally, while ongoing societal efforts are made to address inequalities, historical and contemporary injustices, and healthcare disparities (Office of Disease Prevention and Health Promotion, n.d). Healthcare disparities are also intricately linked with diversity and inclusion. Several organizations recognize the importance of explicitly addressing cultural diversity, which is defined by the American Psychology Association APA (2020) as having subcultures, societies, or communities, within an organization that differ substantially from one another. At its core, diversity means having or being composed of differing elements (Merriam-Webster, 2022). Although diversity is often used in reference to inherent diversity, we should expand this to include acquired diversity as well. Inherent diversity includes characteristics a person was born with, namely race, ethnicity, and gender, while acquired diversity includes those characteristics a person has acquired throughout their live. These acquired diversity characteristics include religion, military status, age, pregnancy status, disability, sexual orientation, socioeconomic status, education, language, and physical appearance. It is also important to incorporate diversity in life experiences, beliefs, values, perspectives, resources, relationships, and appearances (Kapila et al., 2016) to create a truly inclusive environment.

At its core, inclusion is about creating environments, where an individual or group can be and feel welcomed, supported, respected, and valued to fully participate in that group or organization (Kapila et al., 2016). An organization with an inclusive and welcoming climate embraces and values differences and shows respect in words as well as actions for all people. While an inclusive group is diverse, a diverse group is not always inclusive (Kapila et al., 2016). General inclusion is achieved, when every team member feels welcomed, respected, supported, and valued to fully participate (Lyman et al., 2022).

While general inclusion is essential to team dynamics, digital inclusion addresses activities vital to ensure that all individuals and communities (including the most disadvantaged) have access to or are able to properly use communication technologies and information. This often requires intentional strategies and investments to reduce and eliminate historical, institutional, and structural barriers to technology use and access (National Digital Inclusion Alliance, 2021).

It is difficult to talk about inclusion without discussing diversity because we need inclusivity in order to foster diversity. It is also challenging to discuss inclusion and diversity without addressing equity, as we cannot actually attain a diverse and inclusive environment without being equitable. The link between diversity, equity, and inclusion (DEI) has been well established (Lyman et al., 2022). If we want to achieve health equity and improve health outcomes, we need both public health and healthcare teams that are diverse, inclusive, and equitable to eliminating barriers that cause people to feel unappreciated, unheard, and unwanted. As diversity, equity, and inclusion overlap, we may require different strategies and solutions to address them all (Crews et al., 2021). While several institutions aim to address diversity within its workforce and collaborations, it is much less often addressed within a wider context going outside of the immediate workforce and into the surrounding communities that are often affected of institutional policies downstream (Crews et al., 2021). Structural diversity, while a key component in achieving equity, is only a small step in true equity and inclusion. Structural diversity involves including representation of socially marginalized groups within learners, faculty, and leaders in institutions (Crews et al., 2021). While this may seem like a big step, it is only one of many in the overarching goal of true health equity, diversity, and inclusion.

Psychological climate is also a component in which institutions have made moderate strides. Psychological climate involves persons to report experiences of discrimination, but this also is problematic considering that while discrimination may not be present in spaces, persons may report feeling unwanted. Creating an environment of inclusivity is vital to long-term DEI. Studies also show that when an environment of inclusivity and equity is fostered, their team members feel equally respected, valued, and heard, and as a result, leaders can ensure their organization better serves the needs of diverse patient populations (Lyman et al., 2022), which leads to better health outcomes and contributes to the attainment of health equity. More specifically, increasing DEI of employees in a setting leads to improved patient outcomes and satisfaction and contributes to the reduction of health disparities (Branche et al., 2022).

The overarching aim of Healthy People 2030 is to replace health disparities with health equity and improve health literacy to in turn improve health and well-being (Office of Disease Prevention and Health Promotion, n.d.). Tackling equity concerns requires an understanding of the root causes of outcome disparities in our society. While healthcare is often the target place for DEI, implementing these principles in public health can produce even greater returns considering far-reaching nature of public health. Improving health outcomes requires public health and healthcare teams that are diverse, equitable, and inclusive. Both healthcare and public health can set the standard for DEI.

How does health equity, diversity, and inclusion affect mental health?

Mental health can look different for everyone and is mediated by our socioeconomic factors including gender, race, disability, religion, culture, age, etc.

Other factors also play a significant role in mental health like lived experiences and our backgrounds. All these factors can lead to not only the presence of negative mental health issues but also govern the magnitude in which these mental health issues occur. Health equity, diversity, and inclusion can all shape the way mental health shows up in both individual and collective spaces.

Those who have access to mental health services or health insurance have more opportunity to improve their mental health and health and social outcomes associated with it. This access refers to having the opportunity to (1) identify healthcare needs, (2) seek healthcare services, (3) reach and obtain or to use healthcare services, and (4) actually have the need for services fulfilled (Levesque et al., 2013). In many instances, access poses a significant barrier to obtaining necessary mental health services. In efforts to improve access, the introduction of e-mental health services provides patients with fewer barriers to break down when seeking mental health resources and services. Even e-mental health has its issues with DEI. While it is meant to eliminate barriers, Stone and Waldron (2019) found that poor literacy was a significant issue in e-mental health resources. Incidentally, those with poor literacy had the greatest need for services (Stone & Waldron, 2019). This theme is translatable in almost every aspect of health. There is a need to create more accessible spaces to seek and obtain needed mental health services.

"Access" involves this push and pull feature as well as the supply and demand of services (Coombs et al., 2021). Access to mental health resources not only involves the ability to pay but also involves the ability to perceive a need, understand where to find the services, seek the services out, and actually engage in services (Coombs et al., 2021). This process can be daunting for many persons who do not understand or have never navigated the health system. In this society, it is more common to blame the victim for not knowing than it is to find and connect people to resources in ways that show them how to navigate a very confusing and arbitrary process. The saying "Give a man a fish and you feed him for a day. Teach him how to fish and you feed him for a lifetime" has so much importance even within the aspect of health. Because those in the greatest need did not create these systems, it is only fair to "show" how to navigate a system with so much unfamiliarity and many nuances.

Barriers to mental health service utilization

While access may be a structural issue that prevents individuals from utilizing mental health services and resources, mental health stigma plays a significant role as well. This stigma can prevent individuals from expressing challenges with mental health, expressing the needs for services, and from seeking care altogether (Shechtman et al., 2018) and has significant concerns surrounding mental health (Bharadwaj et al., 2017). According to Goffman (1963), stigma refers to an attribute that is deeply discrediting toward a group of people.

Mental health historically has not been a positive topic of conversation, and typically when discussed in public, it is often associated with negative consequences or negative health outcomes. Stigma reveals itself in various ways including through the public stigma, self-stigma, institutional stigma, negative or discriminatory attitudes some people may have regarding mental health or their own mental condition, and government and private organizational policies that intentionally or unintentionally limit opportunities for people with mental illness (APA, 2020). These forms of stigma can be overt or covert. They often reveal themselves as society "blaming the victims" of mental illness, individuals blaming themselves due to experiences with mental illness, or policies and laws in place that stigmatize individuals with mental health disorders/disclosures (Corrigan et al., 2014). This stigma is present among the healthcare professionals when seeking medical services (Knaak et al., 2017), creating even further barriers for those who are in need.

Religion also serves as a barrier to seeking mental health resources. Religion plays a significant role in seeking mental health services (Lukacho & Hankerson, 2015). In some cultures, individuals are urged to seek alternative ways to ease mental health concerns and illness. Alternatives to seeking mental health services include prayer, exorcisms, and religious counseling. In a nationally represented sample, Lukachko and Hankerson (2015) found that among African Americans, higher levels of religiosity was correlated with lower levels of mental help-seeking behavior, when compared to lower levels of religiosity. Though this may be true, other studies have found that religious involvement is actually a protective factor against psychological problems (Dein et al., 2012; Taylor et al., 2013). Several people who adhere to cultural and religious standards may view mental health issues through the scope of religion and not that of traditional mental healthcare professionals (Lukacho & Hankerson, 2015).

Culture may also serve as a deterrent to seeking mental health services. In American culture, the saying "boys will be boys" can be applied to situations from childhood to adulthood. There is mental health stigma present in both males and females; however, the gender roles created by society and culture shape the ways in which males and females engage in certain behaviors. Society influences the way the public views gender and gender roles, creating gendered stereotypes that make way for the performance of certain acceptable behaviors. Women are perceived as weaker, quiet, more reserved, etc., while men are stronger, louder, and more aggressive. These stereotypes support the reluctance for males to seek mental health services and resources. These stereotypes are then enforced and reinforced by social circles we occupy including friends, family, neighbors, coworkers, etc., creating a constant cycle of unresolved mental health issues. Breaking down and developing a better understanding of gender issues could help to address (mental) health equity but inclusion of a "whole" self in all spaces individuals occupy. While gendered culture in America can sometimes negatively influence mental health, workplace culture also diminishes mental health issues. The workplace should be a place of solace and well-being; it is often the opposite. Mental health issues in the workplace often go unaddressed. Negative outcomes of mental health issues in the workplace include reduced job performance, worse communication with coworkers, and even affects individuals physically (CDC, 2022a). In certain professions, individuals may choose not to disclose mental health information due to professional ramifications. Adams et al. (2010) conducted a study examining doctor's attitudes of stigma surrounding the label of depression and whether levels of perceived stigma affect patterns of help-seeking behaviors. The results show that 46.2% of respondents described feeling as though they had suffered from an episode of depression, and female doctors were significantly more likely to report that they suffered from depression with:

- 51.7% of women reporting having a history of depression,
- 89.9% of respondents reported that they would talk to friends or family if their health suffered because of strain or stress,
- 66% reported they would turn to a practitioner, and
- 7% would not seek help at all (Adams et al., 2010).

Barriers to help-seeking included concerns about letting colleagues down (70.6%), breach of confidentiality (50.1%), letting patients down (51.9%), lack of locum cover (44.2%), and career progression (14.0%) (Adams et al., 2010). These professional concerns are extremely important to consider, especially within the context of long-term career progression and health equity, diversity, and inclusion. Even highly educated persons are not immune to the effects of mental health stigma and inclusion.

Personal and environmental experiences can affect mental health in many ways. This can include experiences with poverty, education, and childhood trauma among many others. In many instances, individuals from structurally disadvantaged backgrounds (those who face poverty, racism, sexism, etc.) face many challenges with overall health including mental health. The city of Detroit filed for bankruptcy in 2013 (Chapter 9), as the largest municipal bankruptcy filing in US history (Turbeville, 2022). Though there were many factors that contributed to the bankruptcy, some included the long-term unemployment and reduction in population. Because of these issues, there are less resources for schools where most students from these structurally disadvantaged backgrounds reside. Instead, students are faced with dilapidated buildings and inadequate facilities where they must go to receive their education. Unfortunately, there are many things at play here. Education, poverty, racism, classism, and trauma all contribute to negative health and mental health outcomes. While the Detroit area has historically been a place of extreme diversity with very high populations of Blacks, Arab Americans, Polish, and Latinx, this diversity also brings a plethora of diverse mental health needs.

While these issues are widespread and many know that the issues exist, these same issues often do not look the same on social media. Social media has been used as a powerful tool to promote change and can certainly do the same in the moving forward changing the narrative surrounding mental health. Mental health stigma is a huge problem in American society. This leads to many people underreporting mental health illnesses compared with other illnesses (Bharadwaj et al., 2017). Creating more inclusive policies can significantly influence the presence of poor mental health. Implementing more federal, state, and local laws and workplace policies that address mental health challenges can cause a ripple effect redefining norm surrounding mental health. This can be felt by a shift in social media culture that is more accepting of imperfections and leaves behind the need to be perfect at all times. Pouring more resources into mental health awareness, an increasing number of educational institutions have already introduced "Mental Health Days" into their calendars, recognizing that mental health of students is vital to student success. A mental health day involves a person taking a day off from work or school or other responsibilities to focus on emotional and psychological well-being. This is also slowly being introduced into the workplace as more employers are also recognizing that mental health and job performance are inextricably linked.

Although challenging to validate the depth and breadth of the mental health problem in the United States, the current indicators of mental health include suicide rates, hospitalization rates, and utilization rates of health resources (Tannenbaum et al., 2009). Examining these mental health indicators helps researchers, educators, health providers, etc., get an idea of how mental health is affected by several variables enabling entities and institutions to properly and rigorously address this issue.

It has been well established that our lived experiences affect our mental health predisposition. This is the case for suicides among African American boys in the United States (Bridge et al., 2015; Lindsey et al., 2019), where Black boys aged 5—11 years have been dying by suicide at an increasing rate. Researchers speculate that the increased rates are attributed to multiple, intersecting risk factors, including historical and ongoing trauma (Marraccini et al., 2021). One way to mitigate the trauma Black boys experience is by addressing implicit bias and creating a culturally diverse environment. For example, a 2021 study found that school psychologists could help transform their schools into a safe and culturally affirming place for Black youth, if they recognized their own implicit biases, model antiracist practices, and listen to and recognize the strengths and diversity of Black youth (Marraccini et al., 2021). This finding is not only applicable to Black youth but also to all.

In the United States, international students from low socioeconomic backgrounds and first-generation ethnic minorities are at higher risk for developing mental health problems (Wynaden et al., 2005). In spite of this, international students report underuse of mental health services. For example, a 2005 study found that among Asian communities, help-seeking behaviors were hindered by shame and stigma (Wynaden et al., 2005). This is still true in the 21st century, where studies still find that acculturation, cultural barriers, and stigma attached to mental health problems are common factors that significantly contribute to Asian American college students' low mental health-seeking behaviors (Han & Pong, 2015; Shahid et al., 2021). Suicide rates are also high among veterans and people who identify as lesbian, gay, or bisexual (CDC, 2022a). Other risk factors for poor mental health include individual, social, and environmental factors. Individual factors include low self-esteem, substance abuse, or having a medical illness (Dattani et al., 2021). Social factors include exposure to violence, poverty, unemployment or high stress work environment, and environmental factors include poor access to basic services, discrimination, and inequalities (Dattani et al., 2021).

The Effect of Social Media on Health Equity, Diversity, and Inclusion: Social Media, The Twin Crises (COVID-19 and Racism), and Access to the Internet.

Countless diagrams outline the link between equity, diversity, and inclusion but have not addressed the role social media plays in navigating these outcomes. Social media is a powerful platform that has the power to create or destroy anything it allows to go "viral." The virality of social media is extremely widespread. The use of social media requires some sort of access to the internet and technology. Though many people may consider access to technology and internet commonplace, it is actually a source of disparity which has been made more noticeable during the novel COVID-19 pandemic where many people were forced to work from home. In the spring of 2020, many people, including students, were forced to adapt to a virtual learning space in an effort to slow the spread of the novel virus that causes COVID-19, while health systems around the world were pushed to their limits. While this sudden health crises encroached upon the lives of people all over the world, health, business, and educational entities were forced to adapt. While the world resorted to virtual interaction and communication, this posed a significant problem for students because many did not have access to high-speed internet in their homes. School districts and state and federal entities were forced to minimize the digital divide allowing students to attend school virtually. Before this was able to happen, this digital divide exacerbated learning conditions for students who did not have a computer in the home widening the gap for learning outcomes compared to students whose districts/schools had the infrastructure in place to provide students with resources needed to excel in school. Though most Americans have at least some access to a type of broadband internet, some areas of the country are still lagging far behind (mostly rural areas), and in many areas, students only have access to the internet while in school.

Social media has proven to be a powerful tool for change. Social media coupled with COVID-19 and the movement to uncover and combat racism unveiled the longstanding inequities and disparities present within our health and social systems. Many reasons exist as to why these health and social disparities persist. There are long-standing inequities within the social determinants of health that affect certain groups of people (CDC, 2020). During the onset of COVID-19, many victims were low-income, high-risk, and Black Americans. Factors like poverty and access to healthcare are interconnected and influence many health and quality of life outcomes (CDC, 2022b). Reasons that contributed to the high death rate of Black Americans included that they are more likely to live in areas with higher cases of new COVID-19 infections (Millet et al., 2020). Areas with higher social and economic inequities also have higher rates of COVID-19 infection (Lewis et al., 2020). Further, among people diagnosed with COVID-19, Black Americans (among other minorities) are more likely to live in areas with increased population density, more housing units, and/inadequate housing (Okoh et al., 2020). These high-deprivation areas have higher percentages of nonwhite residents, persons working in manual, essential, and public-facing sectors (U.S. Bureau of Labor Statistics, 2019). These areas also have more crowded housing and increased food and healthcare insecurity (CDC, 2020). Within many Black American groups, it is common for multiple generations to live in one household, therefore, increasing the likelihood of transmission to more vulnerable household members (CDC, 2020). Within employment, Black Americans are disportionately represented within essential work settings like healthcare facilities, farms, factories/warehouses, food and retail services, and public transportation, putting them in more vulnerable positions increasing their chances of being exposed to COVID-19 (Gould & Wilson, 2020).

Social media played a significant role in the COVID-19 pandemic because many people resorted to increased screen time and engagement with social media sites including Facebook, Instagram, TikTok, and Snapchat among many others. Health information about COVID-19 (whether it was true or not) was communicated through social media (Ren et al., 2020). Health entities did make significant strides in reaching various populations, but many failed to optimize the use of social media in ways that make things easier to understand for specific populations. Utilizing memes and other tools as a means to communicate important health information to specific populations can potentially reach a much broader audience especially when the content created is culturally sensitive and easy to understand (Headley et al., 2022).

Black Americans unfortunately were fighting two wars in the United States: the war against COVID-19 and the war against racism. The introduction of social media has played a critical role in making the daily lives of Black people more visible to the world. The world was able to see George Floyd killed by a Minneapolis, Minnesota police officer by kneeling on his neck for more than 8 min. The world was able to see Philando Castile fatally shot during a traffic stop by a police officer in Falcon Heights, Minnesota. The world was able to see many other instances of excessive force, racist and discriminatory practices toward Black Americans. The Black Lives Matter Movement is a great example of the power social media can have for social movements. The movement originated online using the hashtag #BlackLivesMatter in response to the acquittal of George Zimmerman for the murder of Trayvon Martin in 2013. The movement gained exponential traction in a matter of weeks unearthing the systemic racism that exists in the United States. The following year the movement gained even more attention after the murder of Michael Brown by a white police officer. After the murder of Michael Brown, the Black Lives Matter Movement brought attention to police brutality specifically toward African Americans. This movement however did not stop in the United States, but gained international attention, bringing even more awareness to racism and colorism that exists on a global scale. While racism has a negative effect on mental and physical health (Bacon et al., 2017; Hudson et al., 2013; Pascoe & Smart Richman, 2009), witnessing racism and discrimination can also have mental health consequences. This movement was a successful attempt to bring awareness to social issues through social media. It also unveiled the longstanding inequities present within our health and social systems.

Linking social media, mental health, and diversity equity and inclusion

Since 2005, social media usage among American adults has increased from 5% to about 72% in 2021 (Pew Research, 2021). A significant portion of adults also report using social media as a health information-seeking hub. Data indicate that people from some marginalized groups are also more likely to use a smartphone to access health-related information, and some groups even exclusively rely on their cell phones due to no broadband internet at home (Pew Research Center, 2015, 2019a). Given the rise in reliance on social media, mental health applications or apps are a promising approach to reduce the pervasive unmet mental health needs observed among marginalized groups, especially apps that consider issues of DEI (Ramos et al., 2021). Mental health applications also have the potential to address the shortage of mental health professionals, mitigate logistic service utilization barriers, decrease costs associated with care, and engage individuals in care who may not seek services otherwise (Kazdin, 2019; Ramos et al., 2021). In order to appropriately address this issue, it is important to ensure that application designers are utilizing practices that encompass DEI when designing mobile apps.

A 2021 study examining the number of diversities, equity, and inclusion considerations in apps developed after 2015 found that only 58% of apps evaluated considered at least one DEI criterion in the apps' development (Ramos et al., 2021). This lack of consideration for DEI variables may limit the ability of app-based interventions to serve marginalized communities, or worse, they create new disparities increasing barriers to needed health information and resources. Ignoring issues of DEI in health application development and evaluation could inadvertently create a new series of digital disparities (Alcaraz et al., 2017; Bridge et al., 2015; Friis-Healy et al., 2021; Veinot et al., 2018), where app interventions may only be feasible and effective among affluent users, and privileged populations are overrepresented (Ramos et al., 2021), reducing, even eliminating access to populations who demonstrate the greatest need.

Implications for public health practice

Education is often the simplest approach to improving health, health equity, and diversity/inclusion in health-related spaces. Providing an array of educational materials that reflect specific target audiences could improve adherence to health recommendations and overall health outcomes. It is far too commonplace to conceal history and personal experiences with mental illnesses. The virality of social media places various platforms in the perfect position to transform the status quo regarding mental health and experiences with mental illnesses. While many users display a sense of perfection while using social media platforms, the societal pressure to do so may sometimes be overwhelming. Social media platforms can be trailblazers in refuting perfection regarding lived experiences and mental health issues. Health and behavioral entities are not utilizing social media platforms to their maximum capacity. That said, neither are the creators behind these social media platforms. Unfortunately, we tend to view "health" in a very rigid way. Many people feel health has to look a certain way and feel a certain way, and if it does not, we are not doing it right. Public health entities should play a more active role in encouraging people to define health for themselves and allow them to formulate what "health" should look and feel like. This approach fosters a more diverse and inclusive meaning of health and urges people to view health from multiple lenses, as opposed to the very narrow ways in which many people traditionally view health.

The possibilities that exist for social media to be a part of health-based solutions are endless. However, this cannot be the sole driver in improving mental health challenges. Social media serves as more of a mediator among many factors at play. It can enhance health-based policies and decisions only after measures are put in place. Because social media has the potential to affect billions of people worldwide.

Because social media is available to people regardless of age, race, class, geography, etc., it can be used to increase access to health information to many high-risk populations. Below are other ways in which public health can address diversity and inclusion that are described in Table 6.1.

Social media is not going away, it will likely continue to proliferate as well as diversify. If we want to foster a more inclusive and equitable social media

TABLE 6.1 Ways to address diversity and inclusion.

Push for improved occupational policies regarding mental health	• Encourage more employers to implement mental health days • Amend worker benefits to include mental health services at low cost or no cost out of pocket expenses
Improved access to mental health services	• Out of pocket expenses for mental health services deter many people from seeking and utilizing these services • Once a need is recognized and mental health services are initiated, many people are forced to wait weeks before speaking with a provider. When services are inquired, the need is often immediate
Prioritize mental health and wellness	• Improving worker satisfaction on the job can significantly improve job performance
PH-social media collaborations	• Because public health has a broader audience, more and strengthened collaborations should be introduced to reach high-risk populations
Normalize struggles with mental health across racial/ethnic/cultural lines	• Social media gives the façade of perfection. Breaking down this norm via social media is essential in improving mental health, especially among adolescents and young adults
Community collaborations (churches, schools, and other agencies)—inclusivity—result in increased DEI	• Utilizing community organizations to engage high-risk populations should be a priority

culture, public health entities should take an active and direct leadership role on all social media platforms in ways that appeal to populations with the greatest need.

Chapter summary

DEI are facets that should be instilled into institutions. Unfortunately, this only recently has become the standard in medical and public health. The absence of DEI has led our world to disparate health and social outcomes that we are only just beginning to improve. Healthy People 2030 defines health equity as the attainment of the highest level of health for all people (Office of Disease Prevention and Health Promotion, n.d.). This can only be achieved when health and social institutions are ensuring that DEI is implemented in all spaces where everyone feels included and welcomed. These institutions can only achieve equity when systemic and institutional barriers are recognized, addressed, and removed. Further, achieving digital equity requires the same if not more intentional planning and collaborations. Digital equity is "a condition in which all individuals and communities have the technological capacity needed for full participation in our society, democracy, and economy" (National Digital Inclusion Alliance, 2021). Digital equity becomes increasingly important in employment, education, and community participation. Many factors affect progress toward equity including many social and historical factors. In ensuring equitable, diverse, and inclusive spaces, everyone in these spaces must feel equally valued and difficult conversations must be had addressing injustices that affect these spaces.

Mental health is a very important aspect in DEI and health DEI. Due to the trauma incurred as a result of various negative experiences, DEI severely affects mental health. Access is a key component in mental health. Access to mental health resources involves the ability to pay, to perceive a need, to understand where to find the services, seek the services out, and engage in services (Coombs et al., 2021). Due to barriers present not only in the mental health system but also social and community barriers as well, many persons are unable to resolve their mental health needs. Explicit efforts should be made to remove barriers and support DEI in mental health.

The emergence of COVID-19 caused most of the world to interact from a virtual space. Social media played an integral role in society during this time. Gaps in disparities present across many spaces were exposed, and systems were forced to address these disparities. While COVID-19 forced many people into virtual spaces, the issues of racism coupled with social media exposed the structural and institutional racism that has and continues to cause the inequities and disparities present within many systems and institutions. The inequities within the social determinants of health that affect certain groups of people have persisted for quite some time (CDC, 2020). Social media has the potential to minimize disparities present within certain groups of people. Social media also plays a role in awareness of social and health issues which can then be addressed using the same method they were brought to the light: social media. Social media may not always be effective, especially when there are so many people that lack access to internet. Though there are some initiatives that attempt to address this issue, COVID-19 forced many educational institutions, especially elementary and secondary schools that support underserved students and families.

Social media plays a significant role in mental health. Social media can be used as a tool to increase DEI and mental health. Social media occupies a critical position where it can be used to bridge the gap in health disparities for many underserved populations. Many adults use social media to seek health information. Mental health apps can be used to reduce the unmet mental health needs among marginalized groups, especially apps that consider issues of DEI (Ramos et al., 2021). Mental health apps can also address the shortage of mental health professionals, mitigate logistic service utilization barriers, decrease costs associated with care, and engage those in care who would normally not seek services otherwise (Kazdin, 2019; Ramos et al., 2021), creating a more inclusive environment.

Social media can be a great asset to public health practice when used in the right way. It is important that researchers, educators, and leaders understand how to use social media to communicate appropriate health messages to general populations. It is also important to impart principles of DEI into medical and public health institutions.

The following are other ways in which public health can address diversity and inclusion:

- Push for improved occupational policies regarding mental health
- Improved access to mental health services
- Prioritize mental health and wellness
- PH-social media collaborations
- Normalize struggles with mental health across racial/ethnic/cultural lines
- Community collaborations (churches, schools, and other agencies)—inclusivity—result in increased DEI

Definition of key terms

Diversity: The practice of involving people from various backgrounds including race/ethnicity, socioeconomic status, gender, sexual orientation, culture, etc.

Equity: The allocation of resources to marginalized populations that allow them to achieve equal outcomes.

Inclusion: Allowing all people to feel welcomed, recognized and appreciated in the spaces they occupy.

Review or discussion questions

1. What are other examples (not stated in this chapter) where social media has taken up worldwide issues?
2. How can health entities play a more active role in providing health messaging and communication to high-risk populations?
3. What are some health and social outcomes that occur due to mental health being ill-addressed?

4. How does health equity, diversity, and inclusion show up in social and educational spaces that affect mental and overall health?

Websites to explore

National Digital Inclusion Alliance: A website with resources that advance digital equity.

https://www.digitalinclusion.org/

Explore this website for support in reducing the digital divide and increasing digital inclusion. *What types of support does this organization offer? What are the organizations core values? Discuss where the NDIA's work could be most beneficial.*

TikTok: A popular social media website among adolescents and young adults.

https://www.tiktok.com/

Visit this website and explore the short-form video platform. *How was you experience with this relatively new platform? How can public health organizations utilize TikTok to reach younger populations?*

References

Adams, E. F., Lee, A. J., Pritchard, C. W., & White, R. J. (2010). What stops us from healing the healers: a survey of help-seeking behaviour, stigmatisation and depression within the medical profession. *International Journal of Social Psychiatry, 56*(4), 359–370.

Alcaraz, K. I., Sly, J., Fleisher, L., et al. (2017). The Connect Framework: a model for advancing behavioral medicine science and practice to foster health equity. *Journal of Behavioral Medicine, 40*, 23–38.

American Psychology Association. (2020). *APA dictionary of Psychology.* Retrieved 06/28/2022 from https://dictionary.apa.org/.

Bacon, K. L., Stuver, S. O., Cozier, Y. C., Palmer, J. R., Rosenberg, L., & Ruiz-Narvaez, E. A. (2017). Perceived racism and incident diabetes in the black women's health study. *Diabetologia, 60*(11), 2221–2225. https://link.springer.com/article/10.1007/s00125-017-4400-6#citeas.

Bharadwaj, P., Pai, M. M., & Suziedelyte, A. (2017). Mental health stigma. *Economics Letters, 159*, 57–60. https://doi.org/10.1016/j.econlet.2017.06.028

Branche, B., Black, K., Goh, M., Andino, J., & Ghani, K. R. (2022). Social media as a platform for diversity, equity and inclusion. *AUANews, 27*(4), 22–24.

Bridge, J. A., Asti, L., Horowitz, L. M., Greenhouse, J. B., Fontanella, C. A., Sheftall, A. H., Kelleher, K. J., & Campo, J. V. (2015). Suicide trends among elementary school-aged children in the United States from 1993 to 2012. *JAMA Pediatrics, 169*(7), 673–677. https://doi.org/10.1001/jamapediatrics.2015.0465

Centers for Disease Control and Prevention. (2020). *Covid-19 racial and ethnic disparities.* Centers for Disease Control and Prevention. Retrieved April 12, 2022, from: https://www.cdc.gov/coronavirus/2019-ncov/community/health-equity/racial-ethnic-disparities/index.html.

Centers for Disease Control and Prevention. (2022a). *Disparities in suicide.* https://www.cdc.gov/suicide/facts/disparities-in-suicide.html#age.
Centers for Disease Control and Prevention. (2022b). *Vaccination and case trends of COVID-19 in the United States. COVID data tracker. CDC COVID data tracker: Vaccinations and cases trends.*
Coombs, N. C., Meriwether, W. E., Caringi, J., & Newcomer, S. R. (2021). Barriers to healthcare access among U.S. Adults with mental health challenges: A population-based study. *SSM – Population Health, 15,* 100847. https://doi.org/10.1016/j.ssmph.2021.100847
Corrigan, P. W., Druss, B. G., & Perlick, D. A. (2014). The impact of mental illness stigma on seeking and participating in mental health care. *Psychological Science in the Public Interest, 15*(2), 37–70.
Crews, D. C., Collins, C. A., & Cooper, L. A. (2021). Distinguishing workforce diversity from health equity efforts in medicine. *JAMA health forum* (Vol 2,(12), e214820.
Dattani, S., Ritchie, H., & Roser, M. (2021). *Mental health.* Our World in Data. https://ourworldindata.org/mental-health#risk-factors-for-mental-health.
Dein, S., Cook, C., & Koenig, H. (2012). Current controversies and future directions. *The Journal of Nervous and Mental Disease, 200*(10), 852–855. https://doi.org/10.1097/NMD.0b013e31826b6dle
Friis-Healy, E. A., Nagy, G. A., & Kollins, S. H. (2021). It is time to REACT: opportunities for digital mental health apps to reduce mental health disparities in racially and ethnically minoritized groups. *Journal of Medical Internet Research Mental Health, 8*(1). https://mental.jmir.org/2021/1/e25456.
Gould, E., & Wilson, V. (2020). *Black workers face two of the most lethal preexisting conditions for coronavirus—racism and economic inequality.* Economic Policy Institute. https://www.epi.org/publication/black-workers-covid/.
Han, M., & Pong, H. (2015). Mental health help-seeking behaviors among Asian American community college students: The effect of stigma, cultural barriers, and acculturation. *Journal of College Student Development, 56*(1), 1–14. https://doi.org/10.1353/csd.2015.0001
Headley, S. A., Jones, T., Kanekar, A., & Vogelzang, J. (2022). Using Memes to Increase Health Literacy in Vulnerable Populations. *American Journal of Health Education.* https://doi.org/10.1080/19325037.2021.2001777
Hudson, D. L., Puterman, E., Bibbins-Domingo, K., Matthews, K. A., & Adler, N. E. (2013). Race, life course socioeconomic position, racial discrimination, depressive symptoms and self-rated health. *Social Science & Medicine, 97,* 7–14. https://doi.org/10.1016/j.socscimed.2013.07.031
Kapila, M., Hines, E., & Searby, M. (2016). *Why diversity, equity, and inclusion matter.* https://independentsector.org/resource/whydiversity-equity-and-inclusion-matter/.
Kazdin, A. E. (2019). Annual research review: expanding mental health services through novel models of intervention delivery. *Journal of Child Psychiatry, 60*(4), 455–472.
Knaak, S., Mantler, E., & Szeto, A. (March 2017). Mental illness-related stigma in healthcare: Barriers to access and care and evidence-based solutions. In , *Vol 30. Healthcare management forum* (pp. 111–116). Sage CA: Los Angeles, CA: SAGE Publications, 2.
Levesque, J. F., Harris, M. F., & Russell, G. (2013). Patient-centered access to health care: Conceptualising access at the interface of health systems and populations. *International Journal for Equity in Health, 12*(1), 1–9.
Lewis, N. M., Friedrichs, M., Wagstaff, S., Sage, K., LaCross, N., Bui, D., … Dunn, A. (2020). Disparities in COVID-19 incidence, hospitalizations, and testing, by area-level deprivation—Utah, March 3–July 9, 2020. *Morbidity and Mortality Weekly Report, 69*(38), 1369.

Lindsey, M. A., Sheftall, A. H., Xiao, Y., & Joe, S. (2019). Trends of suicidal behaviors among high school students in the United States: 1991–2017. *Pediatrics, 144*(5).

Lukachko, A., Myer, I., & Hankerson, S. (2015). Religiosity and mental health service utilization among African-Americans. *The Journal of Nervous and Mental Disease, 203*(8), 578–582. https://doi.org/10.1097/NMD.0000000000000334

Lyman, B., Parchment, J., & George, K. C. (2022). Diversity, equity, inclusion: Crucial for organizational learning and health equity. *Nurse Leader, 20*(2), 193–196.

Marraccini, M. E., Lindsay, C. A., Griffin, D., Greene, M. J., Simmons, K. T., & Ingram, K. M. (2021). A trauma-and justice, equity, diversity, and inclusion (JEDI)-Informed approach to suicide prevention in school: Black boys' lives matter. *School Psychology Review*, 1–24.

Merriam-Webster. (2022). *Diversity.* https://www.merriamwebster.com/diction/.

Millett, G. A., Jones, A. T., Benkeser, D., Baral, S., Mercer, L., Beyrer, C., & Sullivan, P. S. (2020). Assessing differential impacts of COVID-19 on black communities. *Annals of Epidemiology, 47*, 37–44.

National Digital Inclusion Alliance. (2021). *Definitions.* org/definitions/ https://www.digitalinclusion.

Office of Disease Prevention and Health Promotion. Objectives. Healthy People 2030. U.S. Department of Health and Human Services. https://health.gov/healthypeople/objectives-and-data/browse-objectives.

Okoh, A. K., Sossou, C., Dangayach, N. S., Meledathu, S., Phillips, O., Raczek, C., & Grewal, H. S. (2020). Coronavirus disease 19 in minority populations of Newark, New Jersey. *International Journal for Equity in Health, 19*(1), 1–8.

Pascoe, E. A., & Smart Richman, L. (2009). Perceived discrimination and health: A meta-analytic review. *Psychological Bulletin, 135*(4), 531.

Pew Research Center. (2015). *Racial and ethnic differences in how people use mobile technology.* Retrieved 06/25 from https://www.pewresearch.org/fact-tank/2015/04/30/racial-and-ethnic-differences-in-how-people-use-mobile-technology/.

Pew Research Center. (2019). *Digital divide persists even as lower-income Americans make gains in tech adoption.* https://www.pewresearch.org/fact-tan k/2019/05/07/digital-divide-persists-even-as-lower-income-americans-makegains-in-tech-adoption.

Pew Research Center. (2021). *Social media fact sheet.* https://www.pewresearch.org/internet/fact-sheet/social-media/?menuItem=f13a8cb6–8e2c-480d-9935-5f4d9138d5c4.

Ramos, G., Ponting, C., Labao, J. P., & Sobowale, K. (2021). Considerations of diversity, equity, and inclusion in mental health apps: A scoping review of evaluation frameworks. *Behaviour Research and Therapy, 147*, 103990.

Ren, S. Y., Gao, R. D., & Chen, Y. L. (2020). Fear can be more harmful than the severe acute respiratory syndrome coronavirus 2 in controlling the coronavirus disease 2019 epidemic. *World Journal of Clinical Cases, 8*(4), 652.

Shahid, M., Weiss, N. H., Stoner, G., & Dewsbury, B. (2021). Asian Americans' mental health help- seeking attitudes: The relative and unique roles of cultural values and ethnic identity. *Asian American Journal of Psychology, 12*(2), 138.

Shechtman, Z., Vogel, D. L., Strass, H. A., & Heath, P. J. (2018). Stigma in help-seeking: The case of adolescents. *British Journal of Guidance & Counselling, 46*(1), 104–119.

Stone, L., & Waldron, R. (2019). Great expectations and e-mental health: 'The role of literacy in mediating access to mental healthcare. *Australian Journal of General Practice, 48*(7), 474–479. https://search.informit.org/doi/10.3316/informit.516530301065149.

Tannenbaum, C., Lexchin, J., Tamblyn, R., & Romans, S. (2009). Indicators for measuring mental health: Towards better surveillance. *Healthcare Policy = Politiques de sante*, 5(2), e177−e186.

Taylor, R. J., Chatters, L. M., & Abelson, J. M. (2013). Religious involvement and DSM IV 12 month and lifetime major depressive disorder among African Americans. *The Journal of Nervous and Mental Disease*, 200(10), 856−862. https://doi.org/10.1097/NMD.0b013e31826b6d65

Turbeville, W. C. (2022). The Detroit bankruptcy: The causes of Detroit's bankruptcy and what the city's emergency manager can do to turn it around. *Demos*. https://www.demos.org/research/detroit-bankruptcy.

U.S. Bureau of Labor Statistics. (2019). *Labor Force Characteristics by Race and Ethnicity, 2018*. https://www.bls.gov/opub/reports/race-and-ethnicity/2018/home.h.

Veinot, T. C., Mitchell, H., & Ancker, J. S. (2018). Good intentions are not enough: how informatics interventions can worsen inequality. *Journal of the American Medical Informatics Association*, 25(8), 1080−1088.

Wynaden, D., Chapman, R., Orb, A., McGowan, S., Zeeman, Z., & Yeak, S. (2005). Factors that influence Asian communities' access to mental health care. *International Journal of Mental Health Nursing*, 14(2), 88−95.

Further reading

Kluch, Y., & Wilson, A. S. (2020). #NCAAInclusion: Using social media to engage NCAA student-athletes in strategic efforts to promote diversity and inclusion. *Case Studies Sport Management*, 9(S1), S35−S43.

Krys. (2019). *Belonging: A conversation about equity, diversity, and inclusion*. https://medium.com/@krysburnette/its-2019-and-we-are-still-talking-about-equity-diversity-and-inclusion-dd00c9a66113.

Substance Abuse and Mental Health Services Administration. (2021). *Key substance use and mental health indicators in the United States: Results from the 2020 national survey on drug use and health (HHS publication No. PEP21-07-01-003, NSDUH series H-56)*. Rockville, MD: Center for Behavioral Health Statistics and Quality. https://www.samhsa.gov/data/report/2020-nsduh-annual-national-report.

Section III

Social media and global exposure to research

Chapter 7

Ethical, privacy, and confidentiality issues in the use and application of social media

Amar Kanekar and Joseph Otundo
School of Counseling, Human Performance and Rehabilitation, College of Business, Health and Human Services, University of Arkansas, Little Rock, AR, United States

Learning objectives
After reading this chapter, you should be able to:
1. Define the terms media and social media.
2. Differentiate between privacy and confidentiality in social media-based platform usages, for example, Facebook, Twitter, etc.
3. Explain the terms beneficence, nonmaleficence, and justice as applicable to public health.
4. Discuss the role of ethics in social media usage.
5. Delineate major ethical issues related to social media usage.
6. Appraise the issues arising from the application of social media-based platforms in a research setting.
7. Discuss the literature on the application of privacy and confidentiality issues in social media-based public health studies.

Social media and its varied applications

We are currently living in times where social media has become as ubiquitous as print media in past times. Rarely would you find someone who is unfamiliar with any or all these such as Facebook, Twitter, YouTube, Instagram, and Snapchat. Before we explore the reasons for the rise of these platforms, we need to understand that the basis of having these is the invention of the "Internet." The rise of the Internet over the last several decades has given opportunities for human-to-human interaction via social media such that humans can easily communicate with each other via blogs, applications (called apps), video conferencing, and digital media (by which we mean platforms called social media (Vriens & Van Ingen, 2018)). Social media is not only a form of entertainment media for consumption, sharing, and producing digital information but also has been of immense application in recent times either for election campaigns, higher education marketing, sports, or as a self-branding tool. These applications of social media in a variety of fields show its growing importance.

Although "media" by definition means a plural term of the word "medium" of cultivation, conveyance, and/or expression (Media, n.d.) and "social" means marked by pleasant companionship with friends and associates (Social, n.d.) separately, "social media" can be defined as "forms of electronic communication (such as websites for social networking and microblogging) through which users create online communities to share information, ideas, personal messages and other content (such as videos)" (Social media, n.d.). This brings to "conversational media" which is "a group of web-based apps used for creating and transmitting content in the form of words, pictures, and multimedia" (Bensley & Brookins-Fisher, 2019).

The purpose of this book chapter is to discuss three specific parameters for social media platform usage: (a) ethics or ethical aspects in social media-based platform usage, (b) privacy issues while using social media-based platforms, and (c) confidentiality and issues revolving around it while using social media-based platforms. Although the authors have categorized these aspects of social media usage, they tend to overlap and cause complexity, while delineating their applications in either personal or professional lives.

To that end, "research ethics" is inbuilt into the application of social media-based platforms while designing and implementing research studies in the public health or health education field. The authors include two case study vignettes to demonstrate these complexities and offer possible approaches to navigate emerging and established issues arising from these applications.

Role of ethics in social media usage

In our current society, the role of ethics is valuable and often tested either when we make decisions for ourselves, or as a part of a group, organization, or society. Some of the common examples would be the role of vaccines and their allocation, end-of-life decisions, and using gene alteration of stem cells. These are controversial topics in society and ethical norms, beliefs, and dilemmas make addressing them a challenging lifelong process. Lines get blurred when the effects of human actions are weighed against what is morally and ethically correct or not when dealing with these issues and their outcomes in terms of the societal effects.

Are "ethics versus morality" interchangeable terms or is there a difference? Let us first see the definition of these two terms. Ethics is broadly defined as a theory or a system of moral values, principles of conduct governing an individual or a group (Ethic, n.d.), whereas "morality" or "morals" concerns what is right or wrong in human behavior, considered what is a right behavior by most people and agreeing with a standard of right behavior (Moral, n.d.). In the context of health education/promotion, ethics is the science of how choices are made, whereas morality sets standards for right or wrong in human behavior (Cottrell et al., 2023)

Why do we need to behave ethically as humans? One of the reasons we need to behave ethically is it provides us a sense of purpose and meaning (references) to one's life, particularly as one functions and thrives in a vibrant society. It's the right thing to do no matter what. There are numerous ethical theories such as *deontology* (also known as formalism or nonconsequentialism) where the primary reasoning is that the end does not justify means and *teleology* (also known as consequentialism) where the primary reasoning is that the end does justify the means. We will be seeing the application of these theories in the context of social media usage in the latter part of this section. But before we discuss the theories and their applications, particularly in terms of the use and application of social media in professional and research settings, we need to discuss the fundamental realms of ethics: (a) professionalism or professional ethics and (b) research ethics (Cottrell et al., 2023).

Professionalism and the role of professional ethics are extremely important, particularly when professionals use social media either for professional reasons or just for social purposes. Often professionals either get confused or are not fully versed in the balance between maintaining their identities in a professional environment versus a social environment and lines often get blurred. Maintaining and updating one's identity on a social media platform is as vital as in a professional platform (e.g., LinkedIn), as any kind of expression of morally inappropriate behavior whether, through sharing of information (textual, images, or videos), consumption of information (textual, images, or videos), or building of information (via groups or webpages) on social media-based platforms can adversely affect one's reputation and societal image. For example, a professional may post textual comments which could be religiously or politically offensive and hence may adversely affect the person's professional reputation and his work.

As per the Belmont Report which was the basis for the revision of 45CFR46-the common rule, the core ethical principles which govern ethical functions include beneficence and nonmaleficence, justice, and respect for persons (U.S Department of Health & Human Services, 2021). These principles when applied to social media usage can be discussed as follows:

Beneficence deals with maximizing the good and minimizing harm. There is an obligation to protect persons by creating and sharing content that provides maximum good and minimizes harm. In the context of social media usage, it would be expected that content creators of messages, stories, news, images, and pictures share the information being mindful that these do not harm readers in a reasonable manner, for example, sharing inappropriate and unrealistic body images and pictures of social gathering with excessive alcohol usage should be minimized or nonexistent (Tseng et al., 2019).

Nonmaleficence indicates doing no harm to the study participants. An area that gets challenging for social media platforms is advertisers which can make false advertising via social media and entice subscribers to buy things or sharing of fake news which can create inappropriate messaging and adds fuel to the fire (Relihan, 2018). Some of the strategies for counteracting this deal with banning and regulating false advertisements along with setting up diverse filters for regulating fake news (Kanekar & Thombre, 2019). It is strongly recommended that healthcare organizations, particularly public health organizations counter these practices with authentic evidence-based messages.

Justice, particularly, in the context of public health means fair deliberative procedures and equitable distributions of burdens and benefits. Social media-based studies inherently compromise justice to a major extent as those without a social media account do not get an opportunity to voice their opinions, thoughts, and attitudes. On the flip side, justice can be invoked by using social media as an advocacy tool to advocate for social justice (Fileborn, 2017). This is highly encouraged for public health justice.

Respect for persons can be initiated by having an informed consent document or a statement that introduces the social media-based study participants to the study and seeks their consent with a "yes" and "no" button to participate. Researchers and coinvestigators of a social media-based study should be aware that all posts seen on social media are not necessarily "public data" (most tweets on Twitter are "public" and some spaces and groups could be private [e.g., closed groups in Facebook and private one on one discussions in Twitter]) which need detailed informed consent either from the social media study participants or the administrator or gatekeeper of a social media-based group.

Role of privacy in social media usage

By definition, "privacy" means freedom from unauthorized intrusion and has an element of secrecy (Privacy, n.d.). When we think of social media applications such as Facebook, YouTube, Twitter, and Instagram, they all have "privacy" built into the user interface as part of the navigational settings. For instance, Facebook has quite detailed "privacy" settings which involve a variety of aspects such as "password" protection, 2-step authenticity verification, and content visibility to users as well as those who would access the content generated by users (such as availability of content to friends, friends of friends, and/or the general public). Similarly, a cursory glance at "YouTube" settings would indicate to media disseminators that one can keep one's playlists and subscriptions private or available to the public.

The social media platform "Twitter" has an extensive privacy setting where a Twitter user has adequate control over aspects of managing the information associated with tweets, managing the contacts and the visibility of the message, the consumption of the content along with the advertisements seen, and finally data sharing and connection with other businesses. The privacy policy of Twitter is very detailed and useful for anyone wanting to use Twitter for information consumption or information dissemination (Twitter, 2022).

Although all of the social media-based platforms have a privacy policy for intended users, it would be good and beneficial for researchers to train themselves and be aware of these policies prior to embarking on using social media platforms or applications as tools in the participant recruitment process and data collection. Ideally, it would be highly beneficial if the Institutional Review or Ethics review boards at institutions have some guidelines developed for maintaining privacy when it comes to social media-based research. This could be instituted in the manuals and websites designed by the Institutional Review Boards (IRBs) and/or via videos or modules which discuss the importance and implications of "privacy" when conducting social media-based research (for the participants as well as the researcher).

Sometimes it could be the role of a researcher or the study investigator to provide the participants with useful information on their rights as a participant in the social media-based study, for example, an investigator may ask the participant to familiarize themselves with the public versus private guidelines of a social media platform (such as Facebook or Twitter) before they commit to their participation in the study. Alternatively, the consent form for the proposed study could include specific language which attests that the participants are aware of whether their information would be either public or private before they engage actively with the social media platform.

Although participant privacy may be controlled at the researcher level along with oversight by the IRB, there can certainly be instances where there could be opportunities for risks. Thus, both the participants and nonparticipants could be affected such as third-party risks. For example, asking questions via social media-based blogs or platforms for a family-focused issue may protect research participants but not the participants' immediate family members. This can happen in a quantitative, qualitative, or mixed-methods research approach (Office for Human Research Protections, 2021). Therefore, it is suggested that the researchers or coinvestigators include a plan of how they would be addressing this, particularly when they use social media platforms for recruitment and data collection. This could be an additional protective layer for maintaining the "privacy" of participants.

The distinction between what would be "private" versus what would be "public" in research needs some discussion. "Private" information in the context of Internet-based information would be where an individual's behavior is reasonably expected to not be made public by the individual via observation or recording. This could be further clarified in terms of being a research participant where the identity of a participant can be readily ascertained via the associated information, that is, the information shared, and the identity can be linked.

If the individuals intentionally post textual materials or multimedia on the social media platforms, then it could be presumed to be public unless the platform has privacy or additional policies which preclude that (UA Little Rock Research Protection Program Policies and Procedures, 2018). Although, this is a bit easier when social networking sites mention what is public and what is private, often this information is hidden in deeply seated pages of the social media platform such as Facebook or Twitter and needs to be searched via the networking tools. Furthermore, a research participant may either forget that their profile is set up "publicly" or "privately" or "restricted for friends only" on platforms such as Facebook. This can create a lot of ambiguity for the participant as well as the researcher.

Let us see a couple of examples of how "public" versus "private" is distinguished on two of the popular social media platforms: Facebook and Twitter: Facebook collects a lot of information from its subscribers such as networks, information on product transactions, and even if a subscriber makes his data "private," it is still available to the company. Facebook policies clearly mention that "public" information can be seen by anyone, on or off our products, even if they don't have an account. These include Facebook username; any information shared with a public audience; information in individual profiles on Facebook (Facebook Help Center, n.d.); and content shared on Facebook Page, such as Facebook Marketplace. In addition, people that use Facebook and Instagram can provide access to or send public information to anyone on or off the company products, including other Meta Company Products, search results, or through tools and application programming interface (APIs). Public information can also be seen, accessed, reshared, or downloaded through third-party services such as search engines (like Google), APIs, and offline media such as TV, and by apps, websites, and other services that integrate with our Products (Meta Privacy Center, 2022). So, subscribers need to be aware of these policies when they share information and have not made a specific concerted effort to make their information "private." Another piece of confusion lies in the fact that though the Facebook subscribers can intentionally make their contributions "public," they do not assume that a researcher may use this information for research purposes, but it can be reasonably assumed by a researcher that this information is for "public" use unless the creators or sharers of this information object to it. Hence social media-based data collected via screen capture of Facebook profiles made "public" should and could be considered in the public domain for most cases. If a researcher or a coinvestigator of a research study involving screen capture data is unsure whether the "privacy" aspects of the study are violated, need to consult the IRB at their institution and/or seek oversight on the research process.

Twitter, another social media platform that is widely used for sharing content via short tweets, discussions, and other media sharing has a detailed "privacy" policy. Twitter privacy policy clearly states that "Most activity on Twitter is public, including your profile information and your display language when you created your account, and your Tweets and certain information about your Tweets like the date, time, and application and version of Tweet. The privacy policy states that subscribers may choose to publish their location in their Tweets or their Twitter profile. When subscribers share audio or visual content using the Twitter platform, the data generated is used for their services for example by providing audio transcription. The lists created by subscribers, people they follow and who follow them, and Tweets such as 'like' or 'retweet' are also public. If subscribers 'like', 'retweet', 'reply', or otherwise publicly engage with the Twitter advertisement services, the advertiser might thereby learn information about subscribers associated with the advertisement. Furthermore, broadcasts (including Twitter Spaces) created by subscribers to the Twitter platform are public along with the information about the date when it was created" (Twitter, 2022).

A subscriber's engagement with broadcasts, including viewing, listening, commenting, speaking, reacting to, or otherwise participating in them, either on Periscope (subject to your settings) or on Twitter, is public along with when they took those actions. On Periscope, for example, hearts, comments, the number of hearts received, and whether a live broadcast was watched or replayed. Any engagement with another account's broadcast will remain part of that broadcast for as long as it remains on the Twitter services. The information posted about you (as a subscriber) by other people who use Twitter services may also be public. For example, other people may tag a photo (if your settings allow it) or mention you in a Tweet. So again, when a researcher captures twitter data via screen capture or through "Twitter analytics" on several tweets or retweets, this information is presumed to be in the "public domain" unless it is specifically hidden by making specific private settings by the user (Twitter, 2022).

"Direct messages" is a Twitter tool that allows more control over privacy and allows users to have nonpublic conversations, protect their tweets, and/or host private broadcasts. Data collected via this tool are subject to "privacy laws" and will need an IRB approval or insight if researchers wish to use any of this for their research studies (Twitter, 2022).

As social media usage continues to rise among professionals, the lines between what is acceptable to share via social media versus professional media (such as LinkedIn-www.linkedin.com) continue to blur and cause confusion. Higher Education faculty is one of such "special groups" of individuals who struggle with maintaining this balance. Due to insufficient guidelines for social media policy usage across most of the institutions of higher learning across the world (Buraphadeja, & Prabhu, 2020), it is up to the individual faculty to find an appropriate balance between their professional and personal lives when it comes to sharing information, opinions, images, and other forms of multimedia via the social networking sites. E-professionalism in essence deals with maintaining a professional identity and expression of traditional professional paradigms through digital media (Cain & Romanelli, 2009).

Although Facebook, Instagram, and YouTube (owned by Google) are the social media platforms that have been very popular over the last few years, it is important to understand that "Snapchat" and "TikTok" are generally popular social media platforms, particularly for generating and sharing user-initiated and created video contents and avatars. These provided instant recognition and a brief sense of fame to the younger generation at the expense of losing their rights for content creation to the companies that own these (Johnston, 2020). Although YouTube videos are distinct from other social media outlets where an individual can make the video availability settings private versus public, data shared through Facebook can be shared to WhatsApp and Instagram as these are sister apps. So, although one can technically delete one's account on one of the platforms such as Facebook, Instagram, and/or WhatsApp, the data could be shared through another platform as per the terms and agreements of these platforms (belonging to one family). Furthermore, even if one deletes one's account, the information shared by others about you are not deleted (Karlis, 2019). This can have deleterious effects related to maintaining confidential information.

The use of social media in teaching is fraught with concerns from faculty as well as students. Faculty professionals are mainly concerned with privacy issues when sharing course information or having student interactions and are involved in mainly sharing YouTube videos passively. Videos subscribed via "YouTube" accounts can be made public by changing the preference in the settings and choosing "private subscriptions" via an individual account (https://www.youtube.com/account_privacy); similarly, while creating videos on YouTube, one needs to use the "YouTube creator studio" and pick the advanced setting of this tool, particularly if one is sharing educational videos to children (these are protected by the Children's Online Privacy Protection Act [COPPA]) (Fruhlinger, 2021). Hence, it is important to be aware of what information is collected or shared via "YouTube" usage as a social media tool, as governed by laws and regulations. While students have mixed feelings about the use of social media in the school context as seen in a recent study, which also reiterated the original use of social networking space more as a tool for social networking among family and friends (Dennen, & Burner, 2017).

Some of the recommendations involve being aware of one's professional identity and how to carefully cultivate one's presence on social networking sites such that the digital footprints left demonstrate mindfulness of positive social behaviors and engagement in ethical behaviors while creating and disseminating content and/or opinions via the social networking sites. This also extends to initiating and maintaining relationships with peers via social networking sites through communication (such as tweets) or through engagement via online support groups hosted by social media platforms such as Facebook and Twitter (Forbes, 2017). Professionals and students need to be mindful that most of the information shared via social media platforms can be used by anyone accessing their profiles through public search engines unless they have specifically requested that information to be hidden from a "public view."

> **Case study 1: Facebook-based research study**
>
> A researcher at a mid-western University wanted to conduct a research study about support mechanisms in a breast cancer support group (for participants who had a recent diagnosis or are in remission) that is hosted on Facebook. It is expected that participants in this support group would be sharing their thoughts, feelings, and opinions about breast cancer diagnosis and issues related to that. The researcher wants to approach this group for data collection as one of the researcher's primary research questions is what does the diagnosis of breast cancer mean to the participants?

Case study questions

1. Does a researcher need any kind of training to pursue such as research study?
2. What would be the approaches the researcher could take in conducting this social media-based research study?
3. What should the researcher be aware of from the research ethics point of view?
4. Should the researcher be concerned about issues of privacy or confidentiality in using or sharing this data?
5. What should be the role of the IRB for a study such as this?

   Possible solutions and approaches (Townsend & Wallace, 2016):

1. Other than the scientific training mandated by the IRBs at the researcher's respective institution, the researcher needs to be aware of the terms and policies of the social media platform which is being used as much as possible. It is also suggested that the researcher makes the participants of the study aware of the terms and policies of the social media platform, which is being used, particularly the policies related to public data use and privacy settings.
2. As a research approach, the researcher should first find out if the "online support group" on Facebook is a "closed group" or an "open group." A "closed group" usually is a password-protected group and has a group gatekeeper. The researcher needs to inform the "gatekeeper" of the study and research plans and then decide accordingly whether to be a participant in the group or just be an observer or both (as this could bias the researcher's findings, particularly for a qualitative study).
3. From the point of view of research ethics, it's important that the researcher asks either the group gatekeeper to provide informed consent to conduct

this study; alternately the group gatekeeper may seek permission via informed consent from all the group participants in this group (this is particularly important if the researcher wishes to publish or present any or all of the collected data at scientific meetings or via scientific publications).
4. Since the information accessed in this "online support group" is sensitive, utmost care for handling "privacy" and "confidentiality" of the data needs to be taken by the researcher. The researcher may either introduce oneself to the community members or check with them if anyone would like to "opt-out" of this study. The gatekeeper can also protect some of the group members who would not like to be a part of the study by restricting researcher access to the entire group. Any data collected need to be reported in aggregate. Furthermore, care should be taken to fully anonymize data if there are fewer participants such that the data cannot be linked to a specific individual.
5. The IRB along with the Research Compliance Officer is the primary board that makes sure that the researcher is compliant with the policies of informed consent, privacy, and confidentiality such that it protects the study participants from any potential or established harm. In this case, the IRB needs to make sure that this is clearly outlined by the researcher in the proposal to the IRB. An IRB oversight is needed throughout the process of the research study (if any aspects of the research are altered) and 3 years beyond the study completion as the study output data need to be protected and saved for three years of study completion.

Case study 2: Twitter-based research study (Townsend & Wallace, 2016)
A Professor at a Southern Research University was interested in studying pro- and antivaccination narratives in light of the vaccine initiative related to the COVID-19 pandemic. It was decided that the data would be collected via Twitter—as most data on Twitter are public and hence convenient to be collected. The researcher decides to collect data over the last two weeks using hashtags #covid vaccine #vaccine refusal #vaccine benefits. Some of the early concerns about this study are that this could be considered a controversial and hence sensitive topic and the researcher could have participants who are less than 18 years of age providing comments via tweets and hence anonymity concerns arise.

Case study questions

1. Does a researcher need any kind of training to pursue such as research study?

2. What should the researcher be aware of from the research ethics point of view?
3. What should be the role of the IRB for a study such as this?

Possible solutions:

1. Other than the scientific training mandated by the IRBs at the researcher's respective institution, the researcher needs to be aware of the terms and policies of the social media platform, which is being used as much as possible. It is also suggested that the researcher makes the participants of the study aware of the terms and policies of the social media platform, which is being used, particularly the policies related to public data use and privacy settings.
2. Since the data collected for this study were via tweets using hashtags, these data could be safely considered public. In case the researcher wanted to collect data via direct communications between participants, then that was "private data" and needed much more safety precautions, particularly since this could be considered sensitive data as well. There are concerns related to "privacy" and "confidentiality" in this case as well. Since we do not know the age of the tweet contributors (as they could be underaged and hence need protection from harm), it is important that the researcher provides a paraphrased version of the participant comments rather than direct quotes (as the actual participant quotes could be linked to their user profiles and a cause of harm). In case the researcher decides to use "direct quotes" then, informed consent from those participants needs to be taken. A research output should be in terms of emerging themes that are paraphrased, and Twitter handles removed.
3. The IRB along with the Research Compliance Officer is the primary board that makes sure that the researcher is compliant with the policies of informed consent, privacy, and confidentiality such that it protects the study participants from any potential or established harm. It is also expected that the IRB has its own policies for social media data usage and research conductance to facilitate the researchers' approach and conductance of the study. The IRB needs to make sure that this is clearly outlined by the researcher in the proposal to the IRB. An IRB oversight is needed throughout the process of the research study (if any aspects of the research are altered) and three years beyond the study completion as the study output data need to be protected and saved for three years beyond study completion.

Role of confidentiality in social media usage

Confidentiality is private information that is entrusted with confidence (Confidentiality, n.d.). From a researcher's perspective and often explained by the IRBs, "confidentiality" is when the researcher knows participants, irrespective of whether one can link a person to a set of answers. However, researchers do not present research results that identify participants. If the participant pool is small, it may be impossible to ensure confidentiality even if data are presented in the aggregate (University of Arkansas at Little Rock, n.d). Hence, confidentiality is maintained in any kind of research by reporting data in aggregate and it is challenging to maintain with a small pool of participants.

In the meantime, there are several questions to consider in "social media-based research": Will the data from the social media applications (apps) such as Facebook and Twitter be identifiable? What is the social media app's privacy policy? Does the app have access to research data? Do participants need to be trained on the use of social media apps? Do users know how to adjust security settings on their devices and apps? Should data in transit from the social media app to the researchers be encrypted? (University of Nevada Reno, 2021).

Data confidentiality is important in maintaining and transferring any data accessed using technology such as social media applications or social media platforms. Data confidentiality can be primarily maintained during the data collection phase as well as the data storage phase. In a data collection phase, researchers can ask social media users such as Facebook or Twitter to create their own "screen names" which could be possibly less identifiable. For instance, it can be linked to participant identities in a single database when downloaded. If the researcher has the participants involved in sharing opinions or other artifacts in a social media-based group, it is important to remind the research participants not to share personally identifiable information. It is recommended that the investigators send repeated announcements in a group-based social media study (Bull, 2011).

As an illustration, patients commenting on medical information on a Facebook page or sharing pictures of themselves where pictures of doctors or nurses are "tagged" can inadvertently transcend the boundaries of maintaining confidentiality. Similarly, a medical professional cannot share the private and protected medical information about a patient (which is considered confidential) via Facebook page comments although the patient himself/herself could do it based on the privacy setting set by the person on the social media platform (Medical Protection Society, n.d.). The above scenario highlights a very important aspect of the doctor-patient relationship and how social media-based platforms could jeopardize those if either party is not very cognizant of this.

Another profession where confidentiality is of paramount importance is the law. Lawyers and their clients need to make sure that no breach of confidentiality exists in the case. Because of this, case details should not inadvertently be disclosed via social media platforms or alternately used by a third party through accessing the social media pages of a client leading to unintended disclosure (Medical Protection Society, n.d.). Guidelines need to be established prior to professionals engaging with their clients on the handling of case-based information if shared via social media platforms.

Similarly, nursing professionals who craft professional messages or share videos or pictures via social media-based platforms need to be cognizant of the HIPAA guidelines. Hence, they ought to be careful when selecting social media-based platforms for their professional consultations and/or sharing professional messages. Precaution should be taken not to violate patient or client confidentiality. In addition, crafted messages should not contradict the mission or philosophy of the organizations they represent (Bradley University, n.d.).

Review of literature
Major ethical issues related to social media
Privacy

Data privacy is one of the major issues related to social media. Ethical issues associated with social media are basically divided into individual morality and information management (Turculet, 2014). Primarily, data privacy is dependent on the individual social media platform and individual settings or preferences (Hunter et al., 2018). Interactions on social media have often been an issue not only for the public but also for individuals. For instance, despite Facebook allowing users to post their personal or private details, it has been accused of infringing individual privacy. Specifically, allegations include sharing or selling members' information with other agencies or allowing other users to share personal information. Notwithstanding, individual postings have sometimes been pulled out for contravening Facebook community standards.

Likewise, among ethical issues related to privacy is the privacy document signed by social media platform users. More often, the policy is available on social media platforms, but many users rarely take the time to read and decide the level of privacy they need. For example, Facebook has a privacy policy that allows users to decide whether they want their profile and followers to be public, or only accessible to their friends, your friends' friends, or private. However, studies show that some users are either technologically limited or unable to comprehend the language used (Turculet, 2014). As a result, it may violate ethical issues since many users end up not setting up a privacy platform. To that effect, Facebook has been accused of sharing personal information with users. Despite the challenges, researchers have suggested several measures to enhance privacy: educating users with regard to personal privacy and social media privacy policy settings; development and use of programs that detect third-party users when browsing; and users' anonymity such as lack of personal image or data (Turculet, 2014).

Confidentiality

One of the key ethical requirements of IRBs is the protection and maintenance of participants' anonymity when using social media to conduct research. Even though researchers are required to exclude items asking for personal information, individual relational links can still be used to predict the personal attributes of the users (Zimmer, 2010). Even though it can be argued that whatever information individuals post is personal profiles, social media data such as personal pictures can still be used to erode confidentiality. In addition, personal information such as profile pictures of participants might as well result in researcher biases.

Can ethics and confidentiality Coexist?

The answer to this question is yes, but with conditions. IRBs are not only required to put measures in place but must also ensure that both ethics and confidentiality are implemented. American Speech-Language-Hearing Association (ASHA, 2022), there are several measures that IRBs can implement to support both ethics and confidentiality:

- Disseminate research findings without disclosing personal identifying information.
- Secure storage of research data.
- Anonymous responses.
- Removing and coding personal information.
- Obtaining electronic or written consent.

Trust

Trust is an impediment to social media. Trust tends to emerge from a lack of proper communication and a feeling of vulnerability (Turculet, 2014). Users tend to distrust social media platforms, thus making it a complex ethical issue. There are very little progress researchers can make when the participants lack trust. It is not only difficult to recruit participants, but it is also unethical to talk them into participating in a study when they lack trust. Ideally, individuals build trust from long-term interaction with other people (Turculet, 2014). However, with social media platforms, there is no face-to-face or personal interactions. As a result of social media platforms, many participants might not be willing to engage in online studies. In addition, false information and conspiracy theories have also contributed to mistrust witnessed when using social media. As observed earlier, trust emerges spontaneously through experience and mechanisms put in place to assure users that the platform is safe.

Application of privacy and confidentiality in public health studies using social media

Even though social media has become an integral part of public health studies, it has also been compounded with issues related to privacy and confidentiality. This part of the book chapter specifically explores privacy and confidentiality issues in social media use in public health.

Despite dynamic changes in social media use, public health researchers have an obligation to ensure that they observe ethical issues related to privacy. Protecting the privacy of research participants is very important. Researchers can maximize privacy by ensuring that they grasp the default settings as well as understand whatever the users have signed for (Hunter et al., 2018). Researchers can go a step further and take advantage of social media platforms and amend privacy risks. In such incidents, utmost confidentiality should be observed. Notwithstanding, some data are general or accessible in the public domain and hence require minimal levels of confidentiality. Public health researchers have successfully used social media for contact tracing and disease surveillance. Even though this information is important to the public, confidentiality must be adhered to, so as to give users confidence. For instance, when using Facebook, researchers can adjust the settings to hide personal information such as pictures. As observed earlier, most social media platforms have attempted to put measures in place to protect users. However, it is important for researchers to familiarize themselves with confidentiality information on each platform and reconcile with the ethical requirements guiding their research. In this case, a researcher would be in a position to determine which features to hide or remove.

Second, the users should be made aware of what they are signing. It has been reported that most social media policies and guidelines are so complex that users get confused. They require a detailed approach and frequent updates for them to be well utilized by the users. Thus, it is the responsibility of researchers to ensure that they provide adequate information to the users to be aware of issues associated with their privacy and confidentiality. A study exploring COVID-9 contact tracing on Twitter found that users are likely to share their personal information if they are aware of the intended use as well as have confidence that the information will be protected (Bhatt et al., 2022). To that extent, the researchers ought to explain all the small details about confidentiality and explain to the users why they are collecting the data.

Third, researchers should only collect information that is necessary and applicable in the final analysis (Nicholas et al., 2020). Questionnaires for research questions ought to be designed in a way to capture the key research questions or hypotheses and avoid gathering any data that are not required in the study. For example, researchers can leave out sensitive information such as usernames and public identifiers. But, if for any reason personal data are collected, then it should only be accessible to the main researcher. Another approach is saving personal information separate from the rest of the data. In other words, it should be made difficult to pair the data collected with the participant or user. In the unlikely event that there is a breach or accessibility to data, it will be impossible to compromise the personal information of the users.

In an event that a researcher uses direct quotes from social media platforms, it is recommended that the information be de-identified for the second time (Nicholas, 2020). For instance, a researcher can remove information that directly identifies the user such as the name, social media platform name, or pictures.

In the recent past, public health organizations and agencies have increased the use of electronic and social media platforms to conduct research, store information, and even submit information to coresearchers. However, these platforms, if not handled carefully, can compromise confidentiality by exposing the data to unauthorized users. To that extent, putting measures that secure and uphold confidentiality is an equally important part of the research process. The question that arises is how to balance individual and societal interests, especially when faced with an epidemic such as COVID-19. In an ideal situation, researchers are expected to assess the sensitivity of data, the possibility of maintaining confidentiality, and the risks associated with sharing or not sharing personal information (Harris, 2008). For example, sensitive data about a particular organization collected from employees might put them at risk. Such a situation demands that the information or data collected be kept in confidence. Remember, it is imperative for the users to be confident that the information they are sharing with the researcher is protected and will not expose them to undue risks.

One of the remedies is for public health researchers to establish routine disclosure protocols. This would include the appropriateness of disclosure, the integrity of the information being disclosed, the identity of the person receiving the information, and the security of the mode of data transmission (Myers et al., 2008). Education is another measure that should be incorporated into public health research. Even though there has been an outcry with regard to hacking, the real problem lies with the way the data are handled and secured by the researchers. There should be more training, increased surveillance, and accountability for data storage (Myers et al., 2008). In other words, all the persons involved in research should undergo frequent training and refresher courses. With the emergence of new social media platforms accompanied by dynamic changes, researchers have to be on top.

In order that public health experts to promote health behavior changes among patients, they have been encouraged to use social media. Nevertheless, this has led to the issue of confidentiality. Some scientists posit that social media be used for public conversations with regard to general public health issues rather than discussing individual patients (Crotty & Mostaghimi, 2014). Specifically, the patients or participants should be informed that the social media platform is not meant for clinical communication. Public conversations focus on general issues, for example, physical violence in the local community. Participants might discuss probable causes, without pinpointing suspects within the community.

In summary, social media platforms are very useful in public health research. The increased accessibility and use of social media platforms make it logical for researchers to utilize these platforms. At the same time, measures should be put in place to ensure that the user's privacy and confidentiality are implemented. Researchers should undertake training relevant to specific platforms to avoid issues emerging from breaches of ethical principles related to research.

Chapter summary

In summary, the use and applications of social media-based platforms have been common in current times for personal as well as professional usage. Since some of the social media-based platforms such as Facebook, Twitter, and LinkedIn can be used for personal as well as professional usage, they have blurred the lines between public versus private use of these platforms for sharing content, making comments, and taking part in conversations, and attending events and chats. This chapter has addressed the use and application of social media platforms from a lens of ethics, privacy, and confidentiality.

The authors have discussed tenets of beneficence, nonmaleficence, justice, and respect for persons while considering the use and application of social media-based platforms. Furthermore, detailed discussions about the public versus private mode of operationalization of platforms such as Facebook, Twitter, and YouTube are discussed. The information gleaned from this discussion can be applied not only for personal usage of these platforms but also for professional usage by faculty, staff, researchers, and allied professionals in the fields of health care, public health, and health education. The case vignettes are specifically designed to demonstrate the emerging issues when these platforms are used in conducting a research-based study.

Finally, the chapter wouldn't be complete without a brief literature review on issues of privacy, confidentiality, and trust in the usage and application of social media platforms. Additionally, the authors also throw some light on the relevant current literature which demonstrates the use and application of social media-based platforms in health education/public health.

The authors of this book expect that the review of the material in this chapter related to ethics, privacy, and confidentiality of social media usage generates a discussion among the readers and enhances their understanding of the complexity of issues when using social media for personal as well as professional usage. It is the expectation of the authors that this chapter generates a copious and enlightening discussion among the readers, including researchers, faculty, and students on generating guidelines for navigating the privacy and confidentiality aspects of social media-based platforms while designing and implementing research studies and health care interventions. It is also expected that IRBs at diverse institutions across the nation generate detailed guidelines for use of social media-based platforms while conducting research.

Questions for discussion

1. Describe the role of ethics in the use and application of social media.
2. Compare and contrast the privacy-related issues between Facebook and Twitter—two social media-based platforms.
3. Describe the role of confidentiality while using social media-based platforms.
4. Explain the role of trust in using social media-based platforms.
5. Apply a social media-based intervention addressing a public health issue.
6. Appraise the literature on the application of social media-based studies in public health/health education.

Important terms defined

Beneficence: Beneficence is defined as an act of charity, mercy, and kindness with a strong connotation of doing good to others including moral obligation (Kinsinger, 2009).

Confidentiality: The fact of private information being kept secret (Confidentiality, n.d.).

Ethics: The principles of conduct governing an individual or a group (Ethic, n.d.).

Facebook: This is an online social media and social networking service owned by American company Meta Platforms (Facebook, n.d.).

Institutional Review Board: The Institutional Review Board (IRB) is an administrative body established to protect the rights and welfare of human research subjects recruited to participate in research activities conducted under the auspices of the institution with which it is affiliated (Oregon State University, n.d.).

Justice: The maintenance or administration of what is just especially by the impartial adjustment of conflicting claims or the assignment of merited rewards or punishments (Justice, n.d.).

Media: A medium of cultivation, conveyance, or expression (Media, n.d.).

Nonmaleficence: It means an intention to avoid needless harm or injury that can arise through acts of commission or omission (Ethics of International Engagement & Service Learning, 2011).

Public: Of relating to or affecting all the people or the whole area of a nation or state (Public, n.d.).

Public Health: The art and science dealing with the protection and improvement of community health by organized community effort and including preventive medicine and sanitary and social science (Public Health, n.d.).

Privacy: Freedom from unauthorized intrusion (Privacy, n.d.).

Social media: Forms of electronic communication (such as websites for social networking and microblogging) through which users create online communities to share information, ideas, personal messages, and other content (such as videos) (Social media, n.d.).

Trust: Assured reliance on the character, ability, strength, or truth of someone or something (Trust, n.d.).

Twitter: Twitter is a free social networking site where users broadcast short posts known as tweets (Hetler, 2022).

YouTube: It is an American online video sharing and social media platform headquartered in San Bruno, California (YouTube, n.d.).

Websites to explore

Centers for Disease Control and Prevention, social media tools, guidelines, and best practices

http://www.cdc.gov/socialmedia/tools/guidelines/index.html.

The purpose of this website is to share guidelines and best practices to use social media by the Centers for Disease Control and Prevention. Please explore this website. Read the Facebook and Twitter Guide. What did you learn about the privacy and confidentiality policy? Review the Social Media Toolkit. Can you think of an application of this tool kit in designing at least two social media based public health campaigns? How would you evaluate this campaign?

Ethical dilemmas of social media and how to navigate them from Norwegian Business School

https://www.bi.edu/research/business-review/articles/2020/07/ethical-dilemmas-of-social-media–and-how-to-navigate-them/.

The above website from a Norwegian Business School discusses ethical dilemmas in navigating social media. Do you agree with these? Do you feel these are applicable in the US context as well? Do you have any additional ideas based on this chapter about navigating these?

Internet safety rules while using social media for teens

https://arkansasag.gov/education-programs/internet-safety/

This is a website demonstrating safety rules for teens engaging in social media. Please explore this website. Did anything on this website surprise you? What do you feel is missing from these rules in terms of safety? How many of these social media applications are you familiar with?

Issues in ethics: ethical use of social media

https://www.asha.org/practice/ethics/ethical-use-of-social-media/.

This Issues in Ethics statement is new and is consistent with the Code of Ethics (2016). The Board of Ethics reviews Issues in Ethics statements periodically to ensure that they meet the needs of the professions and are consistent with the American Speech Language and Hearing Association. Compare and contrast this with the code of ethics for at least two other organizations (e.g., National Commission for Health Education specialists, American Public Health Association, etc.) and state at least three aspects that were common and three aspects which were different.

Protecting student privacy on social media

https://www.commonsense.org/education/articles/protecting-student-privacy-on-social-media-dos-and-donts-for-teachers.

The purpose of this website is to share the do's and don'ts regarding student privacy on social media. Explore this website and read the do's and don'ts in detail. How many of these rules apply outside the school environment? Read the section on "Further reading" to enhance your understanding.

Public health guide to social media 101

https://www.rvphtc.org/wp-content/uploads/2019/05/MPHTC_SocialMediaGuide2015.pdf.

This is a Public Health Social Media training guide developed by Michigan Public Health Training Center. Please review this guide. Can you mention at least three ways in which public health professionals can benefit from this guide?

Social media research: public health versus privacy

https://www.ethicscenter.net/exploring-convergence-social-media-big-data-ethics/.

The above website and the associated video by Tim K. Mackey, MAS, PhD., Director, Global Health Policy Institute (www.ghpolicy.org) Associate Director, Joint Master's Program in Health Policy and Law & Associate Professor, UC San Diego, School of Medicine discuss ethical challenges for prescription drug abuse prevention in the social media environment. Please watch this video. Do you agree with the speaker's thoughts and points? If not, why? Are these points applicable for any other health behavior other than prescription drug abuse?

Social media and web 2.0 policy: US Department of Commerce

https://www.commerce.gov/about/policies/social-media.

This website presents a social media policy by the US Department of Commerce. Review this policy. Can you identify any strengths and weaknesses in this policy? Do you feel it's missing anything which needs to be added to it?

Theme issue: social media, ethics, and COVID-19 misinformation

https://www.jmir.org/themes/1142-theme-issue-social-media-ethics-and-covid19-misinformation.

This is a themed issue. Read any three articles from the year 2022. What were the strengths and weaknesses of these articles in terms of discussing privacy and ethical issues and their applications?

References

American Speech-Language-Hearing Association [ASHA]. (2022). https://www.asha.org/practice/ethics/confidentiality/.

Bensley, R. J., & Brookins-Fisher, J. (2019). *Community and public health education methods: A practical guide* (4th ed.). Jones and Bartlett Learning.

Bhatt, P., Vemprala, N., Valecha, R., Hariharan, G., & Rao, H. R. (2022). User privacy, surveillance and public health during COVID-19—An Examination of Twitter verse. *Information Systems Frontiers*, 1—16.

Bradley University. (n.d.) Pros and cons of social media for nursing professionals. https://onlinedegrees.bradley.edu/blog/social-media-in-nursing/.

Bull, S. (2011). *Technology-based health promotion*. Sage Publications Inc.

Buraphadeja, V., & Prabhu, S. (2020). Faculty's use of Facebook and implications for e-professionalism in Thailand. *Cogent Education, 7*(1). https://doi.org/10.1080/2331186X.2020.1774956

Cain, J., & Romanelli, F. (2009). E-professionalism: A new paradigm for a digital age. *Currents in Pharmacy Teaching and Learning, 1*(2), 66—70. https://doi.org/10.1016/j.cptl.2009.10.001

Confidentiality.(n.d.). Cambridge dictionary. Retrieved September 30th 2022 from https://dictionary.cambridge.org/us/dictionary/english/confidentiality.

Cottrell, R. R., Seabert, D. M., Spear, C. E., & McKenzie, J. F. (2023). *Principles of health education and promotion* (8th ed). Jones & Bartlett Learning.

Crotty, B. H., & Mostaghimi, A. (2014). Confidentiality in the digital age. *BMJ, 348*.

Dennen, V. P., & Burner, K. J. (2017). Identity, context collapse, and Facebook use in higher education: Putting presence and privacy at odds. *Distance Education, 38*(2), 173—192.

Ethic. (n.d.). Meriam-webster. Retreived February 27, 2023 from https://www.merriam-webster.com/dictionary/ethic#note-1.

Ethics of International Engagement and Service Learning. (2011). *Non-maleficence and beneficence*. http://ethicsofisl.ubc.ca/?page_id=172.

Facebook Help Center. (n.d.). What is public information on Facebook. https://m.facebook.com/help/203805466323736?ref=dp&_rdr.

Facebook (n.d.). Wikipedia. https://en.wikipedia.org/wiki/Facebook.

Fileborn, B. (2017). Justice 2.0: Street harassment victims' use of social media and online activism as sites of informal justice. *British Journal of Criminology, 57*, 1482—1501.

Forbes, D. (2017). Professional online presence and learning networks: Educating for ethical use of social media. *International Review of Research in Open and Distributed Learning, 18*(7), 175—190.

Fruhlinger, J. (2021). *COPPA explained: How this law protects children's privacy*. https://www.csoonline.com/article/3605113/coppa-explained-how-this-law-protects-childrens-privacy.html.

Harris, J. K. (2008). Consent and confidentiality: Exploring ethical issues in public health social network research. *Connections, 28*(2), 81—96.

Hetler, A. (2022). *Twitter. What is*. https://www.techtarget.com/whatis/definition/Twitter.

Hunter, R. F., Gough, A., O'Kane, N., McKeown, G., Fitzpatrick, A., Walker, T., & Kee, F. (2018). Ethical issues in social media research for public health. *American Journal of Public Health, 108*(3), 343—348.

Justice. (n.d.). Merriam-webster. https://www.merriam-webster.com/dictionary/justice.

Kanekar, A., & Thombre, A. (2019). *Fake medical news: Avoiding pitfalls and perils.* Family Medicine & Community Health. https://doi.org/10.1136/fmch-2019-000142

Karlis, N. (2019). *You just deleted Facebook. can you trust Facebook to delete your data?.* Salon website https://www.salon.com/2019/02/10/you-just-deleted-facebook-can-you-trust-facebook-to-delete-your-data/.

Kinsinger, F. S. (2009). Beneficence and the professional's moral imperative. *Journal of Chiropractic Humanities, 16,* 44−46.

Media(n.d.) Merriam-Webster. Retrieved February 27, 2023 from https://www.merriam-webster.com/dictionary/media.

Medical Protection Society. (n.d.) Casebook. Aspects of confidentiality: social media. https://www.medicalprotection.org/southafrica/casebook/casebook-may-2014/aspects-of-confidentiality-social-media.

Meta Privacy Center. (2022). Privacy policy: What is the privacy policy and what does it cover?. https://www.facebook.com/privacy/policy/?entry_point=data_policy_redirect&entry=0.

Moral. (n.d.) Merriam-Webster. https://www.merriam-webster.com/dictionary/moral.

Myers, J., Frieden, T. R., Bherwani, K. M., & Henning, K. J. (2008). Ethics in public health research: Privacy and public health at risk: Public health confidentiality in the digital age. *American Journal of Public Health, 98*(5), 793−801.

Nicholas, J., Onie, S., & Larsen, M. E. (2020). Ethics and privacy in social media research for mental health. *Current Psychiatry Reports, 22*(12), 1−7.

Office for Human Research Protections. (2021). Review of third -party research risk: Is there a role for IRBs https://www.hhs.gov/sites/default/files/2021-oew-summary-report.pdf.

Oregon State University. What is the Institutional Review Board (IRB)? https://research.oregonstate.edu/irb/frequently-asked-questions/what-institutional-review-board-irb.

Privacy. (n.d.). Merriam-webster. https://www.merriam-webster.com/dictionary/privacy.

Public Health. (n.d.). Merriam-webster. https://www.merriam-webster.com/dictionary/public%20health.

Public. (n.d.). Merriam-webster. https://www.merriam-webster.com/dictionary/public.

Relihan, T. (2018). *Social media advertising can boost fake news-or beat it.* https://mitsloan.mit.edu/ideas-made-to-matter/social-media-advertising-can-boost-fake-news-or-beat-it.

Social (n.d.) Merriam-webster. https://www.merriam-webster.com/dictionary/social.

Social media (n.d.). Merriam-webster. https://www.merriam-webster.com/dictionary/social%20media.

Townsend, L., & Wallace, C. (2016). *Social media research: A guide to ethics.* Economic and Social Research Council and the University of Aberdeen.

Tseng, T., Kanekar, A., Vogelzang, J. L., Hiller, M. D., & Headley, S. A. (2019). Commentary: Social media and the ethical principles of its use in public health and health education research. *American Journal of Health Studies, 34*(3), 155−161.

Turculet, M. (2014). Ethical issues concerning online social networks. *Procedia-Social and Behavioral Sciences, 149,* 967−972.

Twitter. (2022). *Twitter privacy policy.* https://twitter.com/en/privacy.

UA Little Rock Research Protection Program Policies and Procedures. (2018). *Office of research compliance institutional review board.* https://ualr.edu/irb/home/guidelines-and-regualtions/.

University of Arkansas at Little Rock. (n.d.). IRB Faq's https://ualr.edu/irb/home/irb-faqs/.

University of Nevada Reno. (2021). *Research integrity 410. maintaining data confidentiality.* https://www.unr.edu/research-integrity/human-research/human-research-protection-policy-manual/410-maintaining-data-confidentiality.

U.S. Department of Health and Human Services. (2021). *45 CFR 46. Office for human research protection.* https://www.hhs.gov/ohrp/regulations-and-policy/regulations/45-cfr-46/index.html.

Vriens, E., & Van Ingen, E. (2018). Does the rise of the internet bring erosion of strong ties? analyses of social media use and changes in core discussion networks. *News Media & Society, 20*(7), 2432−2449.

YouTube. (n.d.) Wikipedia. https://en.wikipedia.org/wiki/YouTube.

Zimmer, M. (2010). But the data is already public": On the ethics of research in Facebook. *Ethics and Information Technology, 12*(4), 313−325.

Trust (n.d.) Merriam-Webster. https://www.merriam-webster.com/dictionary/trust.

Further reading

Al-Bahrani, A., Patel, D., & Sheridan, B. J. (2017). Have economic educators embraced social media as a teaching tool? *Journal of Economic Education, 48*(1), 45−50.

Chapter 8

Applications of social media research in quantitative and mixed methods research

Rose Marie Ward[1], Mai-Ly N. Steers[2], Akanksha Das[3], Shannon Speed[4] and Rachel B. Geyer[3]

[1]*Department of Psychology, University of Cincinnati, Cincinnati, OH, United States;* [2]*School of Nursing, Duquesne University, Pittsburgh, PA, United States;* [3]*Psychology Department, Miami University, Oxford, OH, United States;* [4]*NIDA/NIAAA IRP—NIH, Translational Addiction Medicine Branch Lab, CPN Section, Biomedical Research Center, Baltimore, MD, United States*

Learning Objectives:

- Describe the important considerations for quantitative and mixed methods research in the social media arena.
- Discuss the methodological and statistical techniques that other researchers have utilized to examine health behavioral changes.
- Describe the challenges associated with conducting research within this domain.

Introduction

There are nearly four billion social media users worldwide (Statista, 2020). Across age and demographic groups (race, income, gender, and education), social media use continues to increase in the United States (Pew Research Center, 2021). Social media platforms allow for the creation and transfer of user-generated content (Wyrwoll, 2014). Kaplan and Haenlein (2010) suggest that there are two dimensions of social media—(1) media-related and (2) self-presentation. Media-related refers to the amount of intimacy and immediacy of the medium and the amount of social presence. Self-presentation (Goffman, 1959) refers to the individual's desire to influence other people's impressions of them. Due to these characteristics, social media represents a valuable source for quantitative and mixed methods research.

This chapter focuses on applications of quantitative and mixed methods research within the social media space. Quantitative refers to research that uses numbers to generalize information about groups of people or to explain specific phenomena (Muijs, 2010) using deductive research approaches, whereas qualitative research is defined as the quest to understand a particular social or human condition/issue and often involves inductive techniques (Creswell, 2007). Mixed methods include research that combines quantitative and qualitative aspects (Johnson et al., 2007). This chapter will provide examples from the alcohol literature to explore current applications of quantitative and mixed methods research, as alcohol-related content is common on social media too (Alhabash et al., 2018; Moreno, 2012). However, the methods discussed are applicable to a wide variety of other physical and mental health-related behaviors/outcomes.

First, several applications of using quantitative and mixed methods in social media research are discussed below. It is important to note that additional social media applications are being developed as social media research evolves. Next, challenges for quantitative and mixed methods research are described. Finally, we present future directions and opportunities for quantitative and mixed methods research.

Applications of social media research

Platform choice and across platforms use

A key aspect of quantitative and mixed methods social media research is the choice of platform (e.g., Snapchat, Facebook, WhatsApp, TikTok, YouTube). The choice of a platform influences the type of research that can be conducted because of the differing types of information that people are willing to share. For instance, different platforms are associated with various types of depictions of alcohol-related posts (Vranken et al., 2020). More specifically, Boyle et al. (2017) suggest that alcohol-related posts on Instagram contain more glamorous displays of alcohol use, whereas Snapchat posts are more likely to contain the negative consequences of alcohol consumption. Other researchers report those who use Snapchat and WhatsApp share more risky alcohol depictions, whereas Facebook and Instagram are used to share images of more moderate and socially accepted alcohol references (Hendriks et al., 2018). As a result, the quantitative and qualitative data gathered from each social media platform are not the same and can therefore lead to different interpretations or conclusions.

Posting across multiple platforms

In addition to tailoring material to the platform choice, some social media users embrace platform swinging (the practice of using several social media platforms; Tandoc et al. (2019). Within the practice of platform swinging, the user can present the same data in different ways (e.g., glamorous vs. more truthful) depending on the affordances and audience of the platform. In many of these platform-swinging practices, the user might present the same data (e.g., "I was so drunk last night") in different ways depending on the platform, even if audiences across the platforms overlap (i.e., the same friends follow them on Snapchat and Twitter). However, the message concerning the data shifts based on the characteristics of the platform (e.g., audience, platform's features, pictures vs. video vs. text, expectations of the platform).

Platform swinging provides an interesting opportunity for quantitative and mixed methods social media research. For example, researchers might consider examining how the message differs across platforms, how the followers react differently to the message depending on platform, or how the content on each social media platform ages over time (e.g., how a message on Facebook is perceived several years after the post compared to the same or similar post is perceived on TikTok). Quantitative researchers might examine platform swinging quantifying how viral (i.e., post that receives a large audience in a short amount of time). In contrast, a mixed methods research study might examine the perceived influence of the platform on the post by quantifying the number of engagements with the post and interviewing users about the post.

Self vs. other generated content

Beyond the platform of choice, who generates the content also impacts quantitative and mixed method research. For many people, the draw of social media is the ability to generate one's own content and gain an audience. This premise has given rise to people seeking notoriety via social media or becoming influencers (Gómez, 2019). The other key type of data available to quantitative and mixed methods research is the influence of other people's content (e.g., other friends, celebrities, influencers, and companies) on users' attitudes and behaviors. Therefore, where the content originates is a key aspect of quantitative and mixed methods research.

Overall, self-generated content has been found to be more strongly related to alcohol behaviors than other-generated content (e.g., sharing of alcohol memes). For example, in a quantitative study, self-generated posts (vs. being exposed to other people's posts) about alcohol correlated more with alcohol quantity and frequency, and risk of alcohol use disorders (Westgate et al., 2014). In addition, self-generated alcohol-related content relates to more positive attitudes about alcohol (Erevik et al., 2017; Geusens et al., 2020). Posting about alcohol also predicts alcohol-related negative consequences (Thompson & Romo, 2016). However, it is important to note that viewing other people's alcohol-related posts is also associated with alcohol consumption (Geusens et al., 2020; Mesman et al., 2020). In fact, a study found that self-generated alcohol-related posts moderated the association between alcohol-related posts viewed and drinking such that young people tended to drink more after seeing others' alcohol-related posts, particularly, if they did not post as much about alcohol themselves (Steers et al., 2021). In sum, generating (i.e., self-generated content) and viewing (i.e., other-generated content) posts with alcohol-related content relates to alcohol consumption behaviors.

Relationships with others

In addition to traditional survey methods, social media quantitative and mixed method research provides a snapshot as to how the user connects to others and their potential posting content and behaviors. Traditionally, this social network approach to research allows for the examination of the impact of one's friends or followers, and their friends' potential interaction with the users by way of network. The idea behind social network analysis is similar to "six degrees of separation from Kevin Bacon." Connections between people are considered "ties," and the people within a given network are dubbed "nodes." Within social network analysis, the researcher can quantify how central the person is in the network and the variety of ways they are connected to other people within the network (e.g., the connector between groups, one connection or pendant, the shortest distance from everyone else in the network). These connections (quantitative) or perceptions of the connections (quantitative or qualitative) can be used to explain data shared on social media or can be employed to predict associated behaviors. Applications of this method determine may examine how central each node is in the network or how many ties each node has. Examples are below.

For example, Kurten et al. (2022) examined "liking" behavior (i.e., engaging with a social media post and clicking on the "like" sign of approval) using a social network analysis approach. First, they determined that "liking" alcohol-related content on social media was more common than "liking" nonalcohol-related content. Second, "liking" behavior increased if it was reciprocated (e.g., I will "like" your post if you liked mine yesterday). Finally, "liking" behavior was related to greater alcohol consumption. For this social network study, the researchers were able to provide insight into the dynamics of posting about alcohol, engaging with other people's posts, followers engaging with one's posts, and subsequent drinking behaviors.

Quantitative and mixed methods researchers can use this method in a variety of ways. Social media platforms provide unique access to connection information (e.g., who the user is "friends" with or who the user follows). An example application of social network analysis to Facebook data is available through Sentinel Visualized (https://fmsasg.com/socialnetworkanalysis/facebook/). In addition, researchers can also generate a social network graph using the steps discussed in an article in Artificial Intelligence (https://ai.plainenglish.io/social-network-analysis-social-circles-of-facebook-611877849d). The article also provides basic social network analysis definition and ways to detect subgroups within a social network.

Engagement (likes, comments, and reposting)

Along these lines, engagement with social media content affords another opportunity for quantitative and mixed methods research. A recent meta-analysis suggests that level of social media engagement relates to subsequent behaviors (Curtis et al., 2018). Depending on the platform, social media users can engage with their own content and the content of others in a variety of manners. Social media users can engage with the content by "liking" it (e.g., LinkedIn, Twitter, Instagram, Facebook, YouTube), sharing it (e.g., Twitter, Facebook), commenting (e.g., TikTok, Twitter, Facebook, Instagram), sharing it on another platform, or a combination of all the aforementioned methods.

Engagement can be conveyed to one's friends/followers (e.g., a post a user comments on may be displayed in their friends/followers' feeds) and potentially also indicates one's approval or disapproval of the content. Social media platform engagement also communicates to others that people may condone the behaviors (e.g., seeing other people receive "likes" for their alcohol content may indicate to another user that heavy drinking is an acceptable or even a desirable behavior), thereby influencing users' perceived norms of the behaviors (e.g., people may be influenced to post alcohol content because they think it normal to drink and post based on what they see). When users engage with this material, norms become codified, which may lead to the maintenance or possible increase of alcohol-related posts and/or alcohol consumption within the network over time. For instance, in a mixed methods longitudinal study in which participants' alcohol-related posts to Facebook were assessed through coding participants' profiles by a team of research assistants, Steers et al. (2019) found that drinking more and perceiving friends as being more approving of drinking was prospectively predictive of posting more alcohol-related content.

In some cases, the ability to view or engage with the post is limited (e.g., Snapchat is a visual media platform in which the posts typically disappear within a certain timeframe), which relates to more risky alcohol content being shared. Engagement can also be restricted to a small group of people (e.g., users can create a Finsta ["fake instagram"] or limit the content of their Instagram stories to "close friends" so that only a select group of followers can see their content), thereby potentially increasing people's portrayals of their authentic self. Conversely, other platforms allow for social media users to engage with the posted content for a seemingly unlimited amount of time and search a users' profile for specific content (e.g., Twitter is a text-based microblog that users can check each other's profiles to see past content).

Another side of engagement with social media is the deletion of content. Reasons for deleting previously posted alcohol-related content on social media are still unknown. Anticipated regret (i.e., prior to decision-making, a person experiences negative emotions when thinking about how the outcome of a present situation could have been better if done differently; Bourgeois-Gironde, 2010; Somasundaram & Diecidue, 2017) relates to the removal of alcohol-related content (LoParco et al., 2022). However, the relationship between posting alcohol-related content and anticipated regret was moderated by age of the participant; wherein, among older participants, higher levels of anticipated regret were associated with deleting alcohol-related content. Quantitative and mixed methods research could further explore the removal of social media content and its relationship with other behaviors and attitudes.

Engagement represents a unique opportunity for quantitative and mixed methods researchers. Using social media content as data enables the research to quantify and explore the user's engagement with other social media content and the user's follower's engagement with their content. For instance, social media platforms provide a mechanism to quantify the number of followers or friends viewed or liked the user's information. This quantification allows for the assessment of impact of the user's post beyond the user's own perception of the impact. In short, the research no longer is limited to the user's assessment of impact.

Content (images, videos, and text)

Another application of social media research is with respect to the type of content displayed or generated. Across social media platforms, users can generate pictures, videos, and text-based content. Many mixed methods research studies use pictures, videos, and text generated for posts as a source for data. For example, in a mixed methods longitudinal study of 50 Black and/or Hispanic youth, researchers found that people who self-reported that they drank or had sex were more likely to post about alcohol and sex than those who did not engage in these offline behaviors; they also found themes in the content of the posts including using substances to relieve stress (Stevens et al., 2022). Given the sheer volume of content generated across platforms each day, researchers traditionally focus on one type of content or a limited number of platforms with the specified content. A brief selection of studies is discussed below.

Images

Given the ease of taking pictures on smartphones and subsequently posting pictures on social media sites, images are commonly used as a data source for quantitative and mixed methods research. On certain social media sites, the norm is for users' profiles to be public, which facilitates gathering images for research studies. For example, Geusens and Beullens (2021a) examined images on Instagram and reported that alcohol in the background of an image was not significantly related to binge drinking behaviors. More broadly, research suggests that the mere exposure to alcohol images on Instagram predicts future alcohol consumption (LaBrie et al., 2021). It is important to note that the social media research that uses images as a data source tends to be primarily qualitative (e.g., Hendriks et al., 2018, 2020). These studies have critical findings (e.g., most images about alcohol are typically positive and display social aspects of consuming).

Image analysis through machine learning now enables researchers to code a large number of images quickly (T.K et al., 2021); however, image-based alcohol-related content can be challenging to identify since what the user identifies as being alcohol-related (e.g., a drunk selfie in which the individual is clearly intoxicated but no alcohol is present in the photo) may not be detected as such through coding or an algorithm. Until researchers can refine techniques to more accurately identify image-based alcohol-related content on a large scale, growth in mixed methods and quantitative methods using images will be limited. Image analysis is a challenge and opportunity for quantitative and mixed methods social media researchers.

Videos

Whereas most social media sites have added the ability to share videos (e.g., Instagram, Facebook, Twitter), video sharing is central to the users' experience on certain platforms (e.g., Snapchat, YouTube, TikTok). In contrast to text-based social media posts, videos posted to social media provide context, tone, and the opportunity to hear directly from the user's voice. Videos can be shared publicly and privately to smaller audiences. Some social media sites (e.g., Snapchat) alert the user if the content is being saved, thereby letting the user know if the content has the potential to be shared more broadly than originally intended.

Three common video-based social media sites are Snapchat, YouTube, and more recently, TikTok. Boyle et al. (2017) determined that exposure to alcohol-related content videos on Snapchat was related to higher levels of alcohol consumption in the future. Primack et al. (2015) coded the most popular alcohol-related YouTube videos and determined that alcohol brand references and videos depicting active intoxication were common. In addition, they reported a higher level of "likes" in videos that portray alcohol use using humor. Similarly, Russell et al. (2021) examined the 100 most popular videos on TikTok which included the #alcohol hashtag using a mixed methods design. The researchers found most posts depicted alcohol consumption in a positive light with more than half of the videos illustrating rapid consumption of multiple drinks, humorous alcohol use, and bonding with others through drinking. Finally, negative associations with alcohol were rarely exemplified.

Videos on social media represent and challenge and opportunity for quantitative and mixed methods social media researchers. Videos can provide additional context to data collected through other sources. For example, researchers might be examining an event or major change in policy. Using videos from social media, the researcher can view aspects of the event through the data captured by the user attending the event. This potentially unfiltered glimpse of the event may provide additional context clues to the event or an additional angle to evaluate the event. In addition, the researcher can gain from any reaction that the user might provide during the video. The challenge to using videos to capture multiple perspectives on an event involves that the users may have only been aware of part of the event, the user may edit the content, and in large events, the vast amount of video content generate might prohibit research outside of machine learning applications.

Text-based

Several platforms provide text-based social media data (e.g., Facebook, Twitter). Given the ease of creating content directly on these platforms, some users post content in the moment and provide a "play by play" of their thoughts and feelings. Given the immediacy of content, some of the posts on these platforms provide very raw, uncensored data. Analysis of text-based content from social media has been more fully developed in comparison to visually based posts as there are mechanisms to retrieve and analyze the data fairly quickly, such as machine learning techniques (T.K, 2021). Studies using these approaches can easily survey millions of data points.

There are several examples of quantitative and mixed methods social media research in the text-based space. For example, participants granted researchers access to examine both their public and private Facebook posts. Information from posts was gathered posts through an application program interface (API). Using natural language processing, researchers were able to determine which words in Facebook posts might signal that the poster at risk for problematic alcohol use (e.g., words related to partying, swearing, anger; Jose et al. (2022). Stevens et al. (2020) also used natural language processing in their mixed methods approach to derive themes from substance use tweets which suggested that youth in the United States view substance use as positive and normative. Finally, Litt et al. (2018) employed mixed methods to analyze alcohol-related tweets from 186 young adults. They retrieved participants' tweets using the Twitter API and subsequently coded them for the presence of alcohol or not. Next, they used the coded data to derive a proportion of alcohol-related Twitter displays and found that this proportion was significantly and positively related to users' alcohol willingness, drinking, and consequences (Litt et al., 2018).

Text-based social media posts are an opportunity for text mining. Given the high volume of material being created every second of every day, it is possible to research nearly any topic using the text-based approach. Applications include examinations of sales, marketing, health surveillance, and more. However, the large volume of content provides a challenge to quantitative and mixed methods researchers as they must potentially separate the signal from the noise. In short, one challenge for this form of research is determining reliable information. For example, the primary author and colleagues (e.g., Ward et al. 2022) examine alcohol-induced blackouts on Twitter. As part of the process, the investigators needed to separate out Tweets concerning blackouts (loss of electrical power in select regions), blackout ops (references to a popular video game), blackouts at sporting events (everyone attending the event and wearing the color black), blackout Twitter (BIPOC people posting on Twitter), and media blackouts (suppression of the media by a government or media not addressing certain social issues). In addition, a challenge of text-based examinations in the context is often limited. In the aforementioned study, some Tweets said "blackout" and no additional information. It is difficult to use the text from these posts as it is often impossible to determine the purpose of the Tweet, the surrounding context, or the precipitating event. Even with these challenges, the sheer volume of text-based content generated allows for nuanced examinations (e.g., the timing of material shared or location of the user when they shared, how followers or the user react to the original post) and the ability to reach large groups of diverse users.

Reach (followers, friends of friends, viral, and influencers)

Another interesting application of social media research is the potential to examine the reach of the user. In contrast to face-to-face interactions, the potential reach can be determined based on metrics such as the number of followers or friends within their network. Other platform characteristics, such as the ability to share or retweet or make the content public, may extend the reach of a user's content beyond their known network; thus, researchers define the number of likes, comments, or views a post receives as being metrics of reach. In addition, it also allows users to have access to content from celebrities and companies. Moreover, the user has the possibility that the celebrity or company might engage with the content that they generated.

Given that certain users' self-generated content can sway the attitudes and behaviors of so many through their social media posts, researchers have begun to examine influencers (i.e., someone who generates social media content to guide the actions of others). Many social media users follow influencers (Gómez, 2019). Social media influencers promote brands and products (e.g., food choices; Byrne et al., 2017). In a mixed methods study, Hendriks et al. (2020) determined that Instagram influencers routinely post about alcohol; these posts tend to be positive; about one-fifth of influencers presented an alcohol brand in their posts; and influencers who disclosed their sponsorship in the post received fewer "likes" from their followers on those posts.

In addition, social media research has begun to examine the content that reaches several thousand followers to several million or viral posts (Hasan, 2022). In these posts, the user has the potential to become "internet famous." Quantitative or mixed methods research with regard to alcohol-related viral posts to date has been very limited.

Reach is a critical component of quantitative and mixed methods social media research. As mentioned above, there are a number of measures of reach. In this section, we will discuss viral news stories as an example. Viral news stories are the interesting application of social media research. Viral news stories have the potential to spread information about a news event faster and more broadly than traditional news outlets (Al-Rawi, 2017). For example, users might share or reshare content that was generated by traditional news media outlets. This sharing drives the content to their followers and potential users who might have been unaware of the news media story. Additionally, social media users have the opportunity to amplify certain stories that might be overlooked by traditional news outlets. For instance, social media users can draw attention to social movements and mobilize other users using their social media profile (e.g., Black Lives Matter; Mundt et al., 2018, p. 4).

Another form of this is hashtag (i.e., using the pound symbol in front of a word or series of words on social media) activism (Yang, 2016). Hashtag activism (e.g., using the # symbol in front of the name of a person of color who was killed through an interaction with the police) or the examination of hashtags, in general, provides the quantitative and mixed methods researcher a mechanism to group the social media content together. In short, the users have provided a preliminary coding for the social media content and a way to easily retrieve similar social media content. Some platforms (e.g., Twitter) provide the user with a listing of hashtags or topics that are trending (i.e., multiple people are posting on the same topic, person, or event).

Motivations for posting

Another venue for quantitative and mixed methods research is examining the motivations for sharing content online. Though social media use is pervasive, not all users create content; some users limit the type of content that they are willing to share, and others will never post content at all.

With respect to sharing alcohol-related content online, several researchers outline the motives for sharing alcohol-related content, including for entertainment, to cope, and/or celebrate. Hendriks et al. (2017) quantitative and qualitative findings suggest that the primary reason Dutch youth share alcohol-related content is for entertainment purposes. In a qualitative study, Riordan et al. (2019) examined tweets for themes concerning high-risk drinking or drinking to blackout and determined that users commonly tweeted about using blacking out as a well to celebrate or cope or they tweeted their intentions to blackout (prior to the event). Understanding the motivations behind particular posts can serve as a preliminary step toward developing internet-based interventions. Thus, further quantitative and mixed methods research is needed.

Change over time

An interesting aspect of quantitative and mixed methods social media research is the ease of examining research questions over time. Some social media platforms, such as Twitter, allow access to archival data of publicly available tweets spanning decades so that researchers may examine how users' posting trends about health behaviors have changed or remained over time. Other opportunities allow researchers to examine how much time users spend on social media and if their usage shifts over time or varies by platform. Finally, researchers can examine how engagement on the platforms evolves in response to modifications or additions to a platform's features (e.g., posting about alcohol following the introduction of the "stories" feature on Instagram in which the content disappears after 24 h).

Recently, several studies examined these characteristics with regard to alcohol use. More recently, Geusens and Beullens (2021b) extended this research by examining a triple spiral in which they hypothesized that the relationships between drinking, posting, and viewing alcohol-related content are reciprocal and prospectively predictive of one another at later time points (e.g., posting is not only predictive of drinking behaviors but drinking behaviors are also predictive of posting over time). They found that, across three time points, at the between-persons level, adolescents who drank more tended to post more frequently, and individuals who posted more frequently also tended to view more alcohol-related content. However, the results at the within level did not lend support for their model. In addition, Boyle et al. (2017) suggest that early exposure to alcohol-related posts during college relates to higher levels of drinking across the first year of college. Additional research is needed to examine user's posting, engagement, behaviors, and attitudes over time.

In addition to longitudinal studies of behaviors and attitudes, given that social media data are time-stamped, researchers can examine the trends regarding posting behaviors across a certain historical period. For instance, Riordan et al. (2022) examined the timing of alcohol-related tweets in the United States from January 2009 to January 2020 and uncovered that most alcohol and high-risk alcohol consumption (i.e., blackout) Tweets were written on the weekends, during evenings, and on major holidays (e.g., New Year's Eve). However, more general alcohol Tweets were posted on Cinco de Mayo, whereas more Tweets that were indicative of high-risk consumption were posted on Thanksgiving. Using social media to predict when future high-risk behaviors might occur is useful in informing and designing prevention efforts.

There are challenges and opportunities in quantitative and mixed methods social media research examining change over time. Among the challenges, there are some that are beyond the control of the researcher. Users can edit their content on some platforms, thereby changing the meaning or impact of the original post. Users can delete their content. Users can repost or share content that has been deleted. Users can delete their accounts or make their posts restricted to only a limited number of approved users. The opportunities may outweigh these challenges. For example, the researcher can potentially gain access to the user's social media content from the inception of the social media account. In addition, the researcher can continue to examine the contents of the social media account as long as the user maintains an active account and posts material (i.e., no end date to the study). The researcher can return to the social media account to verify or potentially provide an additional layer of analysis throughout the life of the study. The researcher can do all the aforementioned tasks with little to no additional effort from the user.

Experimental

In addition to longitudinal studies, quantitative and mixed method studies also employ experimental designs (i.e., quantitative methods of data collection aimed at controlling aspects of the research situation to enhance the validity of the findings) to examine how people respond to social media content. Experimental research provides opportunities to determine if certain aspects of the post or how much exposure to certain types of social media content related to the user's behaviors and attitudes. Moreover, some social media researchers use platforms to recruit participants and deliver interventions (e.g., Pechmann et al., 2015).

A few experimental studies examine the impact of exposure to alcohol-related content on social media on subsequent behavior. For example, Alhabash et al. (2015) conducted an experiment in which participants viewed a fictitious alcohol company's Facebook status update, in which the number of "likes" and "shares" were manipulated, was displayed alongside fake Facebook ads (a pro-alcohol ad, an antidrinking ad, or an ad for a bank). They determined that participants' intentions to engage with the alcohol company's status update (like, share, and comment) were predictive of their intentions to drink, particularly if the status update had a high amount of likes and shares. Alhabash et al. (2016) extended this research to a second experimental study in which exposure to alcohol ads (vs. ads for bottled water) increased the likelihood that participants would prefer a gift card to a bar versus a gift card to a coffee shop, particularly if they were a lighter drinker. Similarly, Mesman et al. (2020) manipulated the type of post the participant was exposed to and if the person posting the material was a close or distant friend. Results revealed that viewing other people's alcohol posts (vs. nonalcohol neutral posts) was related to higher intentions to use alcohol, greater willingness to use alcohol, and more positive attitudes toward alcohol. A close friend's (vs. distant friend's) alcohol posts were associated with more positive attitudes toward alcohol. Information from these experimental quantitative studies can inform future social media-based health interventions.

Experimental studies represent the "gold standard" for quantitative and mixed methods social media research. Experimental studies provide an opportunity for the researcher to narrow down potential influences on the outcome of interest. Quantitative and mixed methods researchers have more control over their variables. This control provides a mechanism for the researcher to examine potential causal influences. However, a concern with experimental studies is the potential selection bias (i.e., a nonrandom sample is used). If the users in the sample do not represent the population that the study is hoping to generalize to, the study may have a selection bias. Selection bias is an issue for social media research as social media use is widespread; yet, there are segments of the population that do not participate. Whereas selection bias is a concern, experimental studies by design are methods that be replicated with other samples, thereby increasing generalizability. Another limitation of experimental design is that some variables cannot be controlled (e.g., randomizing users to have certain experiences such as experiencing a traumatic event). Even considering these challenges and opportunities, experimental studies provide a powerful vehicle for examining social media data using quantitative and mixed methods.

Real-time

Another application of quantitative and mixed methods social media research is the ability to examine data in real time. Given that social media data are constantly being generated sometimes during major events (e.g., protests, elections, graduation) or calling attention to changes in situations (e.g., food truck locations, sales at stores), researchers can observe attitudes and behaviors as they unfold publicly. Moreover, many social media users post their material unedited and without thought that someone else might use it for research purposes. Therefore, the data seem to be more naturalistic and potentially provide more content than traditional surveys. For instance, when someone develops a post about an evolving situation, the researcher potentially has access to the user's posts prior to event and after the event. The researcher also has access to the amount of engagement with the post, the timing of posts, the description of the user in their profile, and other platform-specific characteristics (e.g., the location where the user made the post, images, and the ability to share a post). In short, researchers can often harness data embedded in the presentation of the post.

Due to the real-time nature of social media posts, some quantitative and mixed methods research can be used as a public health surveillance system. In a mixed methods study, Curtis et al. (2018) examined location-based data with regard to drinking levels. In short, they utilized 138 million tweets across 1384 US counties to predict drinking within those counties. The researchers argued that Twitter in conjunction with sociodemographic variables provided a better predictive model than sociodemographic variables alone; moreover,

Twitter data revealed that county-level drinking revolved around social drinking events (Curtis et al., 2018). Thus, Twitter data are useful in revealing more of the context surrounding drinking than sociodemographic variables along and could be harnessed to track public health trends.

Real-time data provide an opportunity for quantitative and mixed methods social media researchers. Given the "in the moment" creation of the content, the researcher can examine the spread of information and the interpretation of the information. For instance, the researcher can examine the time stamp of the original content and examine the number of shares, likes, and engagements with the post to quantify the spread of the information. The researcher can also observe other users' perceptions of the post as they become aware of the original user's post. A challenge for using data in real time is that it requires that the research have knowledge that the material is being created and approval from institutional review boards to use the content. In essence, the researcher is in the process of observing as the social media data unfolds. If real-time observation is possible (e.g., tracking social media posts around an election or referendum vote such as (Riordan et al., 2021)), the research has the potential to provide potential predictions based on the social media posts. The researcher also has the opportunity to preregister the study design and potential prediction.

Self-report vs. actual content

One concern about social media research applications includes whether the content of the social media posts is actually related to behavior. In many of the above studies, the content of the online posts was assumed to be indicative of or a proxy for the actual attitudes and behaviors of the users. However, it is unclear whether the actual *content* from social media sites versus the user's *perception* of their content on social media sites is more associated with attitudes and behaviors. The social media alcohol research provides some initial evidence concerning the importance of the user's perception of their posting behaviors.

In the alcohol social media quantitative and mixed methods research, several researchers examined social media posts and perceptions of one's posting behaviors in relation to their alcohol behaviors. Using a mixed methods design, researchers determined that self-reported posting behaviors to Instagram is not as accurate. Even though posting behavior was relatively infrequent (e.g., 80% of participants shared three or less alcohol posts in the last 12 months), participants tended to overestimate their posting behavior (Geusens & Beullens, 2021a). However, self-reported frequency of sharing alcohol-related content on Instagram was more related to the alcohol behavior than actual posting behavior on Instagram (Geusens & Beullens, 2021a). In another study, participants' perceptions of the frequency of their peers' pro-drinking posts were negatively associated with intentions to seek treatment; conversely, their perceptions of the frequency of their peers' posts depicting negative consequences of drinking or perceptions their peers posted frequently about their positive experiences in treatment/recovery were positively related to intentions to seek treatment (Russell et al., 2022). In short, users' perceptions of their and others' posting behaviors are more likely to be related to alcohol consumption behaviors and treatment seeking than actual posting behaviors.

Self-report vs. the measures easily accessible via each social media platform is an ongoing debate. A benefit of conducting quantitative and mixed methods social media research is the access to the measures (e.g., number of followers, time spent on the platform, number of likes). As reviewed above, the perception of the user of these measures seems to be more predictive in the alcohol literature. Additional research is needed to determine if these metrics or the perceptions of these metrics are more relevant for other behaviors or attitudes.

Social media use measures

A common discussion point for social media research is how to measure "use." How often does a user have to log on to the platform? How much time do they spend on the platform? How often do they post? How often do they choose not to post? How many platforms or functions of the platform does the user use? Early quantitative and mixed methods research asked users to estimate the amount of time spent on the platforms. As platforms expanded and the ease of use increased (from web-based to phone-based access points), the methods for estimating time spent on the platforms improved. Some smartphones provide a weekly report of how often and how much time a user spends on social media. In addition, social media applications have expanded from entertainment and social connection to education, marketing, and so much more. As a result, recent quantitative research provides measures of assessing use and posting behaviors sometimes irrespective of platform choice.

Use and posting behaviors are common quantitative social media research metrics. For example, Westgate et al. (2014) developed a 10-item scale to assess Facebook alcohol content posting behavior. Specifically, this measure assesses how often the user posts alcohol-related content on Facebook—including status updates, comments, and pictures of themselves and others. Other more recent assessments have been developed to determine social media users' likelihood for or reasons behind why people might refrain from posting alcohol-related content to social media. During the course of developing the former scale, Ward et al. (2022) uncovered that people who used social media more frequently and drank more were more likely to post alcohol-related content. By contrast, the latter measure found that people who actively abstain from posting alcohol-related content (which represents up to 50% of social media users) might do so out of fear that people might judge them negatively based on the content (Ward et al., 2022; in revision).

Challenges for quantitative and mixed methods research

Whereas the above sections provide some of the opportunities for quantitative and mixed methods social media research, there are a couple of challenges that warrant discussion. When conducting mixed methods or qualitative research, researchers need to develop detailed, all-encompassing codebooks. However, some of the content of social media posts are tailored to the user's assumed audience and may only clearly communicate to those in their audience (e.g., inside jokes or coded phrases). For instance, during sorority new member recruitment, posts concerning alcohol are commonly prohibited. Some users might circumvent this rule by posting pictures of themselves consuming alcohol from a popular water bottle. Although the researcher may not know this brand of water bottle signals to the intended audience that the user is consuming alcohol, the individual has clearly signaled they are drinking to their intended audience.

Moreover, given some of the changes in platform settings, it is becoming increasingly difficult to capture objective social media data for analysis. For example, Snapchat currently prohibits the downloading of content from their site using an API. Relatedly, it is often difficult for researchers to retrieve large volumes of videos and pictures from social media sites. Whereas several platforms which are primarily text-based have enabled studies of millions of users' posts, methods or access to the technology to harvest visual content are not as widely available.

Furthermore, platforms are also rapidly changing and evolving, and people may migrate from platform to platform. Early in social media research, much of the research used Facebook almost exclusively as a primary source of data. However, most young adults (18—24 years of age) have migrated to other platforms and report refraining from posting about alcohol or drug use to Facebook due to concerns that employers or their families will draw negative conclusions from the content (Stevens et al., 2022). The sites young adults choose to post to instead often depend on posting norms, features and functions, and audience composition of the platform. Moreover, the social media landscape is ever-expanding and evolving with regard to new platforms emerging and the pressure for more established sites to continue to advance in form and function to maintain their stronghold. By the time an article is published with the methods to access information from the social media site, it is possible that the rules of the site or the functions described in the site have changed. Taken together, some major challenges that researchers must overcome are how to generalize the findings from quantitative and mixed methods research across platforms and how to expand research from being platform-centric to user-centric since the same user may post dramatically different content (sometimes about the same event) to several platforms.

Future directions for quantitative and mixed methods social media research

Despite the challenges of quantitative and mixed methods social media research, the potential future directions leverage many opportunities. Discussed below are opportunities with regard to live-streaming, increased opportunities for tailoring, the impact of social norms, and augmented reality.

As live streaming continues to become mainstream, Twitch, Facebook Live, and other social media platforms provide an opportunity to not only observe real-time behavior but also interact with the user or observe other people's interactions with the user. As social media blends with web-based entertainment (e.g., users watching Netflix with friends while chatting on discord), the connections between users and the social part of social media become even more synchronous. Quantitative and mixed methods researchers have the potential to leverage these platforms to engage with research participants in ways that are not location bound (e.g., running a research participant through the experimental study via zoom) and to share the results of the research in new ways (e.g., live stream the experiment and get reactions and questions from the audience, which might include other researchers and observers).

Social media provides a venue for sharing of the approval or disapproval of attitudes and behaviors. By posting content, the user is sharing their views with their followers or other people that have access to their content. In short, they are describing their view of the norms of the behavior or attitude. Prior to social media, most social norm transmissions occurred via face-to-face interactions. Social media allows users to spread these beliefs to other people that they do not know and may never have met. The power of the spread of these messages represents one aspect of the future of quantitative and mixed methods social media research. This message transmission has been shown to share information concerning negative and positive health behaviors. The researcher might consider how this process might inform future social media interventions.

In addition, to live streaming and transmission of norms, augmented reality is a growing area. Augmented reality (i.e., the real-world environment is enhanced by computer-generated information such as the Pokemon Go game or Amazon's Echo glasses). As this technology continues to be embraced, the applications to social media research will expand. For example, researchers could observe via the video stream of the research participant's glasses; the researcher could provide visual stimuli to the participant that are not present in the research participant's reality; and the researcher could answer questions through the glasses in real time.

Conclusion

Quantitative and mixed methods research is evolving and expanding. In this chapter, we discussed several applications of using quantitative and mixed methods in social media research. In several sections, we described the challenges of quantitative and mixed methods research. Finally, we presented the future directions and opportunities for quantitative and mixed methods research. Each opportunity and challenge provide a fertile research opportunity in the social media research field. Perhaps the most exciting part of social media research is that we cannot even imagine how things will evolve in the next month, next 6 months, or next 5 years and how that will change how we conduct quantitative and mixed methods research.

Questions for discussion

1. What are some of the challenges and opportunities for quantitative and mixed methods social media research?
2. What are some examples of data sources for quantitative and mixed methods social media research? How might these data sources differ across different health behaviors?
3. What are some functions in different social media platforms which might change how a message is shared on social media? How do these functions impact how the post or message is received by other users?

Websites to explore

Facebook application using social network analysis

https://fmsasg.com/socialnetworkanalysis/facebook/

Explore this website to see an example of social network analysis using Facebook data. The website describes the users' experience by displaying graphically how their Facebook connections relate to each other. *What was a challenge described in the user's process? What are the advantages of using the API?*

Social network analysis—social circles of Facebook

https://ai.plainenglish.io/social-network-analysis-social-circles-of-facebook-611877849d59.

This website allows you to explore social network analysis of Facebook data. It introduces language commonly used to describe social network analysis. What are the advantages of using this type of data to describe social media usage? What are some of the challenges of this approach to quantitative and mixed methods social media research?

Definition of key terms

A list of key terms used in the chapter with definitions (to be used in the glossary).

Alcohol-related content to social media: Content that is posted to social media in which (1) alcohol is explicitly in an image (e.g., pictures of a cocktail); (2) alcohol is mentioned or implied in the text or images; (3) images in which alcohol is present in the background (e.g., alcoholic beverages on a table, photos at a club or bar); (4) posting of drink recipes; (5) articles involving alcohol; (6) drunk selfies (even if the individual is not holding an alcoholic beverage); (7) images in which the individual or others are clearly intoxicated; (8) memes about or involving alcohol; (9) sharing of others' alcohol-related content; and (10) text involving alcohol (e.g., mentions of partying, blacking out).

Experimental designs: Refer to quantitative methods of data collection aimed at controlling aspects of the research situation to enhance the validity of the findings.

Influencers: Include people who are internet celebrities or people who amass a following on one or more social media platforms.

Mixed methods research: Includes research that combines quantitative (numbers) and qualitative (words, text, images) aspects within the same project. It often uses inductive reasoning to draw conclusions.

Platform: A medium for sharing information, images, videos, etc., via the web (e.g., Snapchat, Facebook, WhatsApp, TikTok, YouTube). It allows for the dissemination of in the moment ideas and connection with other users.

Platform swinging: The practice of using several social media platforms to share the same or related information (Tandoc et al., 2019).

Quantitative research: Refers to research that uses numbers to generalize information about groups of people or to explain specific phenomena using deductive research approaches.

Viral: Refers to a post that receives a large audience in a short amount of time.

References

Alhabash, S., McAlister, A. R., Kim, W., Lou, C., Cunningham, C., Quilliam, E. T., & Richards, J. I. (2016). Saw it on Facebook, drank it at the bar! Effects of exposure to Facebook alcohol ads on alcohol-related behaviors. *Journal of Interactive Advertising, 16*(1), 44–58. https://doi.org/10.1080/15252019.2016.1160330

Alhabash, S., McAlister, A. R., Quilliam, E. T., Richards, J. I., & Lou, C. (2015). Alcohol's getting a bit more social: When alcohol marketing messages on Facebook increase young adults' intentions to imbibe. *Mass Communication and Society, 18*(3), 350–375. https://doi.org/10.1080/15205436.2014.945651

Alhabash, S., VanDam, C., Tan, P.-N., Smith, S. W., Viken, G., Kanver, D., Tian, L., & Figueira, L. (2018). 140 Characters of intoxication: Exploring the prevalence of alcohol-related Tweets and predicting their virality. *SAGE Open, 8*(4). https://doi.org/10.1177/2158244018803137

Al-Rawi, A. (2017). Viral news on social media. *Digital Journalism, 7*, 63–79.

Bourgeois-Gironde, S. (2010). Regret and the rationality of choices. *Philosophical Transactions of the Royal Society of London. Series B, Biological Sciences, 365*(1538), 249–257. https://doi.org/10.1098/rstb.2009.0163

Boyle, S. C., Earle, A. M., LaBrie, J. W., & Ballou, K. (2017). Facebook dethroned: Revealing the more likely social media destinations for college students' depictions of underage drinking. *Addictive Behaviors, 65*, 63–67. https://doi.org/10.1016/j.addbeh.2016.10.004

Byrne, E., Kearney, J., & MacEvilly, C. (2017). The role of influencer marketing and social influencers in public health. *Proceedings of the Nutrition Society, 76*(OCE3), E103. https://doi.org/10.1017/S0029665117001768

Creswell, J. W. (2007). *Qualitative inquiry and research design: Choosing among five approaches* (3rd ed.). Thousand Oaks, CA: Sage.

Curtis, B., Giorgi, S., Buffone, A. E. K., Ungar, L. H., Ashford, R. D., Hemmons, J., Summers, D., Hamilton, C., & Schwartz, H. A. (2018). Can Twitter be used to predict county excessive alcohol consumption rates? *PLOS ONE, 13*(4), e0194290. https://doi.org/10.1371/journal.pone.0194290

Curtis, B. L., Lookatch, S. J., Ramo, D. E., McKay, J. R., Feinn, R. S., & Kranzler, H. R. (2018). Meta-Analysis of the association of alcohol-related social media use with alcohol consumption and alcohol-related problems in adolescents and young adults. *Alcoholism: Clinical and Experimental Research, 42*(6), 978–986. https://doi.org/10.1111/acer.13642

Erevik, E. K., Torsheim, T., Vedaa, Ø., Andreassen, C. S., & Pallesen, S. (2017). Sharing of alcohol-related content on social networking sites: Frequency, content, and correlates. *Journal of Studies on Alcohol and Drugs, 78*(4), 608–616. https://doi.org/10.15288/jsad.2017.78.608

Geusens, F., & Beullens, K. (2021a). Perceptions surpass reality: Self-reported alcohol-related communication on Instagram is more strongly related with frequency of alcohol consumption and binge drinking than actual alcohol-related communication. *Drug and Alcohol Dependence, 227*, 109004. https://doi.org/10.1016/j.drugalcdep.2021.109004

Geusens, F., & Beullens, K. (2021b). Triple spirals? A three-wave panel study on the longitudinal associations between social media use and young individuals' alcohol consumption. *Media Psychology, 24*(6), 766–791. https://doi.org/10.1080/15213269.2020.1804404

Geusens, F., Bigman-Galimore, C. A., & Beullens, K. (2020). A cross-cultural comparison of the processes underlying the associations between sharing of and exposure to alcohol references and drinking intentions. *New Media & Society, 22*(1), 49–69. https://doi.org/10.1177/1461444819860057

Goffman, E. (1959). *The presentation of self in everyday life*. New York, NY: Doubleday Anchor Books.

Gómez, A. R. (2019). Digital fame and fortune in the age of social media: A classification of social media influencers. *aDResearch: Revista Internacional de Investigación en Comunicación, 19*, 8–29.

Hasan, R., Cheyre, C., Ahn, Y.-Y., Hoyle, R., & Kapadia, A. (2022). The impact of viral posts on visibility and behavior of professionals: A longitudinal study of scientists on twitter. *Proceedings of the International AAAI Conference on Web and Social Media, 16*(1), 323–334. Retrieved from https://ojs.aaai.org/index.php/ICWSM/article/view/19295.

Hendriks, H., Gebhardt, W. A., & van den Putte, B. (2017). Alcohol-related posts from young people on social networking sites: Content and motivations. *Cyberpsychology, Behavior, and Social Networking, 20*, 428–435. https://doi.org/10.1089/cyber.2016.0640

Hendriks, H., Van den Putte, B., Gebhardt, W. A., & Moreno, M. A. (2018). Social drinking on social media: Content analysis of the social aspects of alcohol-related posts on Facebook and Instagram. *Journal of Medical Internet Research, 20*(6), e226. https://doi.org/10.2196/jmir.9355

Hendriks, H., Wilmsen, D., van Dalen, W., & Gebhardt, W. A. (2020). Picture me drinking: Alcohol-related posts by Instagram influencers popular among adolescents and young adults. *Frontiers in Psychology, 10,* 2991. https://doi.org/10.3389/fpsyg.2019.02991

Johnson, R. B., Onwuegbuzie, A. J., & Turner, L. A. (2007). Toward a definition of mixed methods research. *Journal of Mixed Methods Research, 1*(2), 112−133. https://doi.org/10.1177/1558689806298224

Jose, R., Matero, M., Sherman, G., Curtis, B., Giorgi, S., Schwartz, H. A., & Ungar, L. H. (2022). Using Facebook language to predict and describe excessive alcohol use. *Alcoholism: Clinical and Experimental Research, 46*(5), 836−847. https://doi.org/10.1111/acer.14807

Kaplan, A. M., & Haenlein, M. (2010). Users of the world, unite! The challenges and opportunities of social media. *Business Horizons, 53*(1), 59−68. https://doi.org/10.1016/j.bushor.2009.09.003

Kurten, S., Vanherle, R., Beullens, K., Gebhardt, W. A., van den Putte, B., & Hendriks, H. (2022). Like to drink: Dynamics of liking alcohol posts and effects on alcohol use. *Computers in Human Behavior, 129,* 107145. https://doi.org/10.1016/j.chb.2021.107145

LaBrie, J. W., Trager, B. M., Boyle, S. C., Davis, J. P., Earle, A. M., & Morgan, R. M. (2021). An examination of the prospective associations between objectively assessed exposure to alcohol-related Instagram content, alcohol-specific cognitions, and first-year college drinking. *Addictive Behaviors, 119,* 106948. https://doi.org/10.1016/j.addbeh.2021.106948

Litt, D. M., Lewis, M. A., Spiro, E. S., Aulck, L., Waldron, K. A., Head-Corliss, M. K., & Swanson, A. (2018). #drunktwitter: Examining the relations between alcohol-related Twitter content and alcohol willingness and use among underage young adults. *Drug and Alcohol Dependence, 193,* 75−82. https://doi.org/10.1016/j.drugalcdep.2018.08.021

LoParco, C. R., Lowery, A., Zhou, Z., Leon, M., Galvin, A. M., Lewis, M. A., & Litt, D. (2022). Age as a moderator of the association between anticipated regret and the posting and deleting of alcohol-related content on social networking sites among adolescents and young adults. *Health Behavior Research, 5*(2). https://doi.org/10.4148/2572-1836.1111

Mesman, M., Hendriks, H., & van den Putte, B. (2020). How viewing alcohol posts of friends on social networking sites influences predictors of alcohol use. *Journal of Health Communication, 25*(6), 522−529. https://doi.org/10.1080/10810730.2020.1821130

Moreno, M. A. (2012). Associations between displayed alcohol references on Facebook and problem drinking among college students. *Archives of Pediatrics & Adolescent Medicine, 166*(2), 157−163. https://doi.org/10.1001/archpediatrics.2011.180

Muijs, D. (2010). Doing quantitative research in education with SPSS. In E. R. Babbie (Ed.), *The practice of social research* (12th ed.). Wadsworth Cengage.

Mundt, M., Ross, K., & Burnett, C. M. (2018). *Scaling social movements through social media: The case of Black Lives Matter.* Social Media + Society. https://doi.org/10.1177/2056305118807911

Pechmann, C., Pan, L., Delucchi, K., Lakon, C. M., & Prochaska, J. J. (2015). Development of a Twitter-based intervention for smoking cessation that encourages high-quality social media interactions via automessages. *Journal of Medical Internet Research, 17*(2), e50. https://doi.org/10.2196/jmir.3772

Pew Research Center. (2021). *Social media fact sheet.* Pew Research Center. https://www.pewresearch.org/internet/fact-sheet/social-media/.

Primack, B. A., Colditz, J. B., Pang, K. C., & Jackson, K. M. (2015). Portrayal of alcohol intoxication on YouTube. *Alcoholism: Clinical and Experimental Research, 39*(3), 496–503. https://doi.org/10.1111/acer.12640

Riordan, B. C., Merrill, J. E., & Ward, R. M. (2019). Can't wait to blackout tonight": An analysis of the motives to drink to blackout expressed on Twitter. *Alcoholism: Clinical and Experimental Research, 43*(8), 1769–1776. https://doi.org/10.1111/acer.14132

Riordan, B. C., Merrill, J. E., Ward, R. M., & Raubenheimer, J. (2022). When are alcohol-related blackout Tweets written in the United States? *Addictive Behaviors, 124*, 107110. https://doi.org/10.1016/j.addbeh.2021.107110

Riordan, B. C., Raubenheimer, J., Ward, R. M., Merrill, J. E., Winter, T., & Scarf, D. (2021). Monitoring the sentiment of cannabis-related tweets om the lead up to New Zealand's cannabis referendum. *Drug and Alcohol Review, 40*, 835–841.

Russell, A. M., Davis, R. E., Ortega, J. M., Colditz, J. B., Primack, B., & Barry, A. E. (2021). #Alcohol: Portrayals of alcohol in top videos on TikTok. *Journal of Studies on Alcohol and Drugs, 82*(5), 615–622. https://doi.org/10.15288/jsad.2021.82.615

Russell, A. M., Ou, T.-S., Bergman, B. G., Massey, P. M., Barry, A. E., & Lin, H. C. (2022). Associations between heavy drinker's alcohol-related social media exposures and personal beliefs and attitudes regarding alcohol treatment. *Addictive Behaviors Reports, 15*, 100434. https://doi.org/10.1016/j.abrep.2022.100434

Somasundaram, J., & Diecidue, E. (2017). Regret theory and risk attitudes. *Journal of Risk Uncertainty, 55*, 147–175. https://doi.org/10.1007/s11166-017-9268-9

Statista. (2020). *Number of social network users worldwide from 2017 to 2025 (in billions)*. https://www.statista.com/statistics/278414/number-of-worldwide-social-network-users/.

Steers, M.-L. N., Neighbors, C., Wickham, R. E., Petit, W. E., Kerr, B., & Moreno, M. A. (2019). My friends, I'm #sotallytober: A longitudinal examination of college students' drinking, friends' approval of drinking, and Facebook alcohol-related posts. *Digital Health, 5*, 1–11. https://doi.org/10.1177/2055207619845449

Steers, M.-L. N., Ward, R. M., Neighbors, C., Tanygin, A. B., Guo, Y., & Teas, E. (2021). Double vision on social media: How self-generated alcohol-related content posts moderate the link between viewing others' posts and drinking. *Journal of Health Communication*, 1–7. https://doi.org/10.1080/10810730.2021.1878311

Stevens, R., Bonett, S., Kenyatta, K., Chittamuru, D., Bleakley, A., Jingyi Xu, J., Wang, Y., & Bush, N. (2022). On sex, drugs, and alcohol: A mixed-method analysis of youth posts on social media in the United States. *Journal of Children and Media*, 1–18. https://doi.org/10.1080/17482798.2022.2059537

Stevens, R. C., Brawner, B. M., Kranzler, E., Giorgi, S., Lazarus, E., Abera, M., Huang, S., & Ungar, L. (2020). Exploring substance use tweets of youth in the United States: Mixed methods study. *JMIR Public Health and Surveillance, 6*(1), e16191. https://doi.org/10.2196/16191

Tandoc, E. C., Lou, C., & Min, V. L. H. (2019). Platform-swinging in a poly-social-media context: How and why users navigate multiple social media platforms. *Journal of Computer-Mediated Communication, 24*(1), 21–35. https://doi.org/10.1093/jcmc/zmy022

Thompson, C. M., & Romo, L. K. (2016). College students' drinking and posting about alcohol: Forwarding a model of motivations, behaviors, and consequences. *Journal of Health Communication, 21*(6), 688–695. https://doi.org/10.1080/10810730.2016.1153763

T.K, B., Annavarapu, C. S. R., & Bablani, A. (2021). Machine learning algorithms for social media analysis: A survey. *Computer Science Review, 40*, 100395. https://doi.org/10.1016/j.cosrev.2021.100395

Vranken, S., Geusens, F., Meeus, A., & Beullens, K. (2020). The platform is the message? Exploring the relation between different social networking sites and different forms of alcohol use. *Health & New Media Research, 4*(2), 135–168.

Ward, R. M., Dumas, T. M., Lewis, M. A., & Litt, D. M. (2022). Likelihood of posting alcohol-related content on social networking sites – measurement development and initial validation. *Substance Use & Misuse, 57*(7), 1111–1119. https://doi.org/10.1080/10826084.2022.2064505

Ward, R. M., Steers, M.-L. N., Guo, Y., Teas, E., & Crist, N. (2022). Posting alcohol-related content on social media: Comparing posters and non-posters (under review) *Alcohol.*

Westgate, E. C., Neighbors, C., Heppner, H., Jahn, S., & Lindgren, K. P. (2014). I will take a shot for every 'like' I get on this status": Posting alcohol-related Facebook content is linked to drinking outcomes. *Journal of Studies on Alcohol and Drugs, 75*(3), 390–398. https://doi.org/10.15288/jsad.2014.75.390

Wyrwoll, C. (2014). User-generated content. In C. Wyrwoll (Ed.), *Social media: Fundamentals, models, and ranking of user-generated content* (pp. 11–45). Springer Fachmedien Wiesbaden. https://doi.org/10.1007/978-3-658-06984-1_2

Yang, G. (2016). Narrative agency in hashtag activism: The case of #BlackLivesMatter. *Media and Communication, 4*(4), 13–17. https://doi.org/10.17645/mac.v4i4.692

Chapter 9

Applications of social media in qualitative research in diverse public health areas

Geetanjali C. Achrekar[1] and Kavita Batra[2,3]
[1]*GVM's College of Commerce and Economics, Ponda, Goa, India;* [2]*Department of Medical Education, Kirk Kerkorian School of Medicine at UNLV, University of Nevada, Las Vegas, NV, United States;* [3]*Office of Research, Kirk Kerkorian School of Medicine at UNLV, University of Nevada, Las Vegas, NV, United States*

Learning objectives
The main objectives of the present study are as follows:
1. To examine the nexus between social media and qualitative research.
2. To identify various beneficial applications of social media in diverse public health areas from a qualitative standpoint.
3. To identify novel methods through which social media platforms (e.g., Facebook) can be used to conduct qualitative research.
4. To describe the type of qualitative data available through Facebook and ways of performing qualitative research.
5. To discuss advantages and challenges associated with the use of different forms of data acquisition on/through Facebook in qualitative research.
6. To discuss some potentially harmful effects of social media in public health applications.
7. To identify policy measures for the optimum use of social media for the public health promotion and development globally.

Introduction
Social media has opened a floodgate of online interactions between physicians, medical professionals, nurses, health workers, patients, hospitals, medical institutions, research centers, pharma companies, and state health organizations (Safko & Brake, 2009). Social media offers new mechanisms of communications

in the sphere of public health systems, which may facilitate the emergence of new public health policies and systems in the long-term period (Safko & Brake, 2009). Social media applications helped both private as well as public health organizations to have a multidirectional communication flow with patients, physicians, customers, clients, health workers, and general public. Social media applications include but are not limited to social networking platforms, blogs, microblogs, chatting boards, bookmarks, and media forums to share information (Stern, 2010). Social networking services (SNSs) had been defined by Kamel Boulos in 2007 as collaborating and mediating surroundings, where desktops and android devices help to establish wide networks and connectivity, through which new forms of information get disseminated (Chen & Wang, 2021; Spallek et al., 2010). Some examples of social media and its types are wikis (e.g., Wikipedia), social networking sites (e.g., Facebook, LinkedIn), media-sharing sites (e.g., YouTube, SlideShare), blogs and microblogs (e.g., Blogger, Twitter), immersive worlds (e.g., Second Life), and 3-D virtual globes (e.g., Google Earth). The uses of social media in health care services have been very well captured by Moorhead and Colleagues (2013) as follows:

- Social media can provide information on a wide range of health issues in a timely manner.
- Internet can help patients to find answers to any health-related problems using search engines like Google.
- Public health professionals can use social media tools like Twitter and Facebook to interact with patients and conduct online consultations.
- Public health organizations can cost-effectively use social media as a tool of health intervention for health promotion and campaigns related to diseases.
- Social media can also help netizens to reduce the social stigma associated with certain illnesses.
- Pharma companies, hospitals, and pharmacies can market their products and services through social media.
- Last but not the least, social media can help in qualitative modes of inquiry, which will be largely discussed in the upcoming sections of this chapter.

Social media and qualitative research

The term "social media" is constantly evolving and is being used to describe tools based on Internet applications. Social media basically is designed for individuals and communities to enable them to congregate and interact; to exchange data, ideas, personal messages, photos, videos and other media information; and, in some situations, to collaborate with other people. The world's most popular social media platforms include Facebook, WhatsApp, Twitter, Instagram, LinkedIn, YouTube, etc. (Chen & Wang, 2021; Grajales et al., 2014). A standard classification (Kotler et al., 2010) divides social media into two major categories based on its primary purpose: (a) expressive social media, in which users share text, videos, pictures, and music; and (b) collaborative social media, in which users work together to solve problems. In general, users share knowledge and content and collaborate toward a common goal. Social media provides a one-of-a-kind opportunity. From a unidirectional or bidirectional information flow, there is a chance to alter the government—citizen interaction and exchange into a process of many-to-many communication (Kutsikos, 2007). Citizen-centric schemes and social media are effective tools for increasing public participation (Agostino, 2013).

Over the last 2 decades, there has been a plethora of social media studies in the field of public health, using qualitative research techniques and methods. Most of these are high-quality journal articles with important findings and observations and have mobilized primary data through personal interviews, focus groups, interviews using open-ended questions, and sample surveys. Another important method used by researchers using qualitative approach is content analysis, which uses posts on various social media platforms like Facebook and Twitter to draw meaningful conclusions about events and findings (Chen & Wang, 2021; Franz et al., 2019; Snelson, 2016).

O'Dwyer and Bernauer (2014) define qualitative research as a method that aims to unfold new knowledge within the existing natural surroundings and complexities. Qualitative research seeks to get deep insights into people and the environment they are living in, by posing questions or evolving questions as a study progresses, with the intent of adding knowledge to the existing body of theories. Wocott (1988) has aptly defined qualitative research by citing Margaret Mead who says the question in qualitative research is not "Is this case representative?" But rather "What is this case representative of?" Participants are the key focus of qualitative research and need to be selected in tandem with the research questions (P. 209). Qualitative research studies have played a significant role in bringing out the value addition social media has made in various branches of knowledge.

Social media platforms offer an information-rich opportunity to have a broader reach to diverse demographic groups, which are otherwise hard to locate (Franz et al., 2019). In the evolving landscape of social media, Facebook is the most dominant component (Franz et al., 2019). A sizable portion of people social lives is being recorded on Facebook, which make it a useful and rich source of qualitative data for researchers (Franz et al., 2019; Snelson, 2016). Historically, several studies used Facebook data as well as users (as study participants) for research activities and also for conducting or implementing behavioral interventions (Beullens & Schepers, 2013; Kent et al., 2016; Pederson et al., 2015). However, till today, little is known on how qualitative data from Facebook users can be utilized for health services research, which will require some additional strategies beyond traditional qualitative data collection strategies, such as individual or group interviewing, focus groups, etc., to unveil the deepest meaning, values, beliefs, and assumptions of people participating in the research

TABLE 9.1 Types of user-generated (Textual) data and its potential sources.

Type of user-generated data	Forms	Additional features
Posts	It can be appeared as a status update (most frequently used), newsfeed	Posts generated by the user; posts tagged in (posts where other users tag the user)
Comments	A response to the Facebook post or a response on the comment itself	Comments along with likes and reactions
Messages	Privately sent to another Facebook user and this message does not appear on a user's Facebook timeline or newsfeed	Like posts and comments, this can be accompanied with image, video, or emoticon (emoji)

Source: Franz, D., Marsh, H. E., Chen, J. I., & Teo, A. R. (2019). Using facebook for qualitative research: A brief primer. *Journal of Medical Internet Research, 21*(8), e13544. https://doi.org/10.2196/13544.

(Franz et al., 2019). Facebook is a rich source of qualitative data, including user-generated data, images, reactions, and text (Franz et al., 2019). There are 3 types of user-generated textual data and potential sources of these data on Facebook, which are described below (Table 9.1).

There are 3 ways in which the qualitative research methods can be applied through social media: active analysis, passive analysis, and research self-identification (Eysenbach & Till, 2001).

Active analysis involves the active engagement of research members with Facebook users, whereas passive analysis involves investigation of patterns and interactions using the existing data (Cheung et al., 2017; Kent et al., 2016). In research self-identification, Facebook can be used for recruiting research subjects, focus groups, and web-based interviews (Pederson et al., 2015). There are multiple ways through which Facebook data can be obtained for research, which are described by previous studies and indicated in Table 9.2 (Abramson et al., 2015; Eysenbach & Till, 2001; Kant et al., 2016; Kosinski et al., 2015; Kramer et al., 2014; Tower et al., 2015).

Methods/tools

Facebook data can be used for thematic and content analysis (Snelson, 2016). Selection of the analytical tool depends upon the quantity of the data as well qualitative approach (Franz et al., 2019). For instance, NVivo can be used to conduct active and passive analysis of large data (Franz et al., 2019). Another software called the Linguistic Inquiry and Word Count (aka Luke) can be used

TABLE 9.2 Methods of data acquisition on/through Facebook.

Options	Ways	Advantages	Challenges
#1	Partner with Facebook	• Access to large amount of Facebook data • Leveraging Facebook data processing systems to do thematic and topic analysis	• Resource intensive in terms of money and time • Setting up collaboration with Facebook
#2	Using ppublicly available data (using appropriate data extraction methods through the Facebook pages and groups)	• Access to large amount of data related to health topics • Informed consent will not be required	• Social desirability bias • Privacy concerns
#3	Creating or monitoring a Facebook page or group	• Allows customization of Facebook group pertaining to a particular health topic • Promotes engagement of the participants as it happens among targeted group • Offers a secure environment for the participants	• Difficult to navigate old information as recent interactions appear at the top pushing old interactions at the bottom • Hawthorne effect (activities among group can be different as subjects are aware that they are being observed)
#4	Private messages through a web-based portal	• Secure environment	• Inability to observe interactions

Source: Franz, D., Marsh, H. E., Chen, J. I., & Teo, A. R. (2019). Using facebook for qualitative research: A brief primer. *Journal of Medical Internet Research*, 21(8), e13544. https://doi.org/10.2196/13544.

for the text analysis that calculates the percentage of words in a given text that fall into any linguistic, psychological, and topical categories (Kramer et al., 2014). Other qualitative software include ATLAS.ti, which can be utilized to code HTML files of individual's Facebook downloads (Franz et al., 2019).

Social media has a very important role to play in a contemporary society in which one in every four individuals have access to internet and smart devices

(Pew Research Center, n.d). Social media has envisaged all the fields of life and health care sector. Medical industry has gradually accepted the use of social media for the numerous applications it has, which can potentially benefit all the stakeholders in the health and medical fields if used judiciously. This has been evidenced especially in the developed countries with positive results (though limited) to patients, physicians, health professional and health organizations as illustrated in the forthcoming sections of this article (Grindrod et al., 2014).

Applications of social media in diverse public health areas

Social media is important in all businesses, including medicine and public health. Health care organizations and medical facilities have begun to look into the function of social media in the connection between patients and physicians, as well as how it relates to various health information systems (Grindrod et al., 2014). Public health authorities are utilizing social media to interact with the public and build on relationships, a thing that was difficult in the past due to time and long-distance barriers. Today, state health agencies and community partnerships have been initiated through social media channels, and emerging issues in public health have been addressed. Leaders in public health have realized that social media platforms and social media-driven initiatives can be used to create, implement, and evaluate various public health care outreach programs (Peck, 2014). One such example is the COVID-19 pandemic, where public health outreach programs were instituted to improve the public health and patient outcomes. Building efficient treatment plans with patients and interacting with future patients using social media platforms offer numerous advantages (Chauhan et al., 2012). Social media has helped to equalize and widen access to information to millions of people irrespective of age, gender, caste, race, and personality type. Social media can provide a valuable source of peer-to-peer interaction, an online cost-effective source to provide psychological and emotional support to individuals especially those with various health conditions (Scanfeld et al., 2010). Hwang and Colleagues. (2010) in their study observed that inspiration, motivation, and sharing common experiences prove to be important social support features of social media sites. However, there are numerous drawbacks to adopting social media in health care that professionals in the medical field should be aware of (Lambert et al., 2012). Health care professionals (HCPs) have access to a variety of social media technologies, including social networking sites, blogs, video, virtual reality, microblogs, wikis, media-sharing sites, and gaming environments (Dizon et al., 2012; George et al., 2013). These tools can be used to boost performance or to improve professional networking, education, organizational effectiveness, patient care, patient education, and public awareness on health-related programs (Von Muhlen & Ohno-Machado, 2012). However, there are risks and dangers to patients and health care providers from the spread of low-quality information, harm to one's professional reputation, and security breaches with respect to patient confidentiality, breach of personal—professional boundaries as well as licensing and legal difficulties (Bernhardt et al., 2014; Farnan et al., 2013; Fogelson et al., 2013; MacMillan, 2013; Pirraglia & Kravitz, 2013). There are numerous health care institutions, as well as other professional associations, which have given directives in this regard (Moorhead et al., 2013).

Application of social media by health care professionals

HCPs use social media to share medical information among peers, to debate health care-related policy issues, to promote desirable health behaviors among patients, to interact with public, and to educate patients, caregivers, students, and colleagues (McGowan et al., 2012). HCPs can use social media to potentially improve health outcomes, develop a professional network, increase personal awareness of news and discoveries, motivate patients, and provide health information to the community (Korda & Itani, 2013). Health researchers utilize social media for a variety of purposes, including health-related research, professional development, and doctor—patient contact and offline services (Korda & Itani, 2013).

Social media can help health-related research in two ways: it can supply rich data to learn about patients' health experiences, and it can help researchers recruit participants as discussed in the preceding sections. Moreover, researchers can assess patients' understanding of the disease and coping strategies (Guidry et al., 2016), identify their concerns about the disease (Jiang, 2020), understand their doubts to health behavior change (Oser et al., 2019), identify symptoms related to the disease (Erikainen et al., 2019), and assess patients' state after recovery (Koutrolou-Sotiropoulou et al., 2016) by analyzing their social media conversations. Because patients' self-reported sickness experiences aren't always reported to and recorded by doctors, studying patient conversations on social media could help health researchers and clinicians (McDonald et al., 2019).

Physicians often join online communities for their own professional health-related research, where they can read news articles, listen to experts, research medical developments, consult colleagues regarding patient issues, and participate in conferences and seminars to expand their network (Househ, 2013). They can share practical guidelines, discuss practice management challenges, make referrals, disseminate their research, market their practices, or engage in health advocacy. A sample survey of over 4000 physicians conducted by the social media site QuantiaMD reported that both personal and professional use of social media among physicians is increasing. Some HCPs have set up their own members-only sites, such as "Sermo," "Doximity," and "Medical Directors Forum," where member physicians can discuss, post messages, exchange expert advice, get alerts, and have virtual libraries to mention a few applications. Microblogs have emerged as a very important social channel of brief communication. For instance, Twitter users can tweet a message of up to 140 characters with hashtags and hyperlinks. Some famous hashtags include #MDChat and #Health20 (Ventola, 2014).

Social media can be used by health professionals and researchers for personal growth, such as learning, collaboration, and job progress. They can use social media to collaborate on research projects and practices (Cherrez-Ojeda et al., 2020), access and share trending research findings and medical knowledge (Mascia et al., 2021; Yüce et al. et al., 2021), broaden their access to funders and publishers (Bamat et al., 2018), conduct a job search (Helm et al., 2016), follow medical conferences remotely (Loeb et al., 2020, market their team and services, and discuss interesting or difficult cases with colleagues. According to the previous studies, different social media platforms play different roles in the professional development of health professionals and researchers. Pinterest is primarily used for health care—quality education (Pizzuti et al., 2020). Twitter is primarily used for conference news and information, and LinkedIn is primarily used for career advancement and professional growth (Pizzuti et al., 2020). Nearly 85% of medical professionals agreed that social media can be effectively used for educational purposes, and 71% of health professionals, researchers, and businesspeople in the urology discipline agreed that social media can help in one's career development (Loeb et al., 2020). Social media has been actively used by health professionals to engage in mobile learning and to interact with peers. Benetoli and Colleagues (2015) found that Facebook was accessed by pharmacists in rural areas quite often through their smart devices, for connecting with colleagues to break feelings of loneliness and boredom (Benetoli et al., 2017).

Online health services and doctor—patient communication have become easier due to social media platforms like Facebook. Responding to patient inquiries submitted on social media (Birnbaum et al., 2017), delivering online consultation (Tang et al., 2019), and proactively providing advice and health information to social media followers (Benetoli et al., 2017) are all examples of how health professionals use social media for building their doctor—patient contact relationship. Furthermore, social media can be utilized to supplement traditional health services. Health practitioners can utilize social media to inform patients after examining their results, on how to improve medication compliance (Hermansyah et al., 2019), obtain feedback from patients about their health services, and collect data from patients after discharge to inform future practices. Encouragement of social media interactions between health professionals and patients may promote both patients' well-being and patient—physician relationships (Liu et al., 2020). Furthermore, social media may be utilized to deliver medical services, such as appointments, medical inquiries, personal information management, and medical charge payment, which can improve patient experiences and promote accessibility to medical treatment (Shen et al., 2019). The major social media applications for health professionals can be summarized in Fig. 9.1 below.

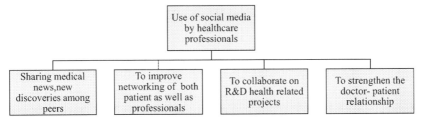

FIGURE 9.1 Applications of social media by health care professionals.

As health professionals and researchers utilize social media more frequently, there are concerns and obstacles on the way. Several studies have raised concerns about patient privacy and confidentiality and health professionals' lack of social media awareness. Studies have looked at issues like preserving content creators' intellectual property and dealing with unfavorable patient feedback. The majority of these articles advocated for the creation of a policy or regulation (Chen & Wang, 2021; Ventola, 2014).

Social media and public

Given the advancement of technology and its use, patients are increasingly using internet to find health information. Social media has transferred the earlier model of one-way communication, where health professionals and health service centers would inform the public about health-related issues, to a multilevel communication where health professionals and patients can simultaneously exchange their views, opinions, knowledge on blogs, Facebook, YouTube, and Twitter (Kenny & Johnson, 2016). The potential of social media to engage directly with patients could significantly improve the professional relationship between patients and the health care field. According to a study published in the *Journal of Medical Internet Research*, about 80% of internet users in the United States of America explore and research about their personal health online, by navigating and browsing through various health and medical websites, blogs, and YouTube channels and joining online community platforms which share their health issues (Suarez-Lledo & Alvarez-Galvez, 2021). As a result, medical experts may help these millions of individuals better understand health issues and how to cope with them if they maintain a regular presence on social media. Patients can access information directly from their doctors, nurses, and public health specialists via the internet in this manner (Peluchette et al., 2016). Individuals can share their updates and postings with their friends, family, and followers. Patient involvement can be aided by using social media in health care because it keeps them informed and updated on a regular basis (Peluchette et al., 2016). The public can use social media to follow and share news about trending health issues like COVID-19, find information on daily health behaviors like fitness, health, food choices

(Lambert et al., 2019), seek health advice for their own health concern like pregnancy or breast feeding-related information (Zhu et al., 2019), request a second opinion about their illness (Schwartz & Woloshin, 2019), access health care law, and follow social media accounts. For people with various information demands, social media has become a key source of information (Zhong et al., 2021). Golder and Colleagues (2015) in their systematic review of 170 journal articles on effective use of social media in public health, found that 88.8% of the articles found that social media has a positive influence on patients' behavior and health results by way of major health improvements using patient—doctor interactions via information technology.

Social media gives more freedom to people to express themselves on public health issues that are considered as a social "taboo," while staying anonymous. These issues may range from problems related to sexual health, HIV-AIDS, risky sexual behavior (with the same sex), alcohol, tobacco or substance abuse to human trafficking, harassment of youth, and depression. Social media maintains a high level of confidentiality, thereby making online netizens more comfortable in expressing their views on issues which are labeled as stigmatized traditionally (Norman & Yip, 2012). People diagnosed with chronic diseases or those who have successfully recovered from ailments can join similar online communities and share information on their illness experiences such as a brain tumor journey, treatment options, medical costs, diet, and fitness routines, which were followed during the treatment and their outcomes (Song & Paradice, 2020). Such online social networking patient groups help other similar patients to understand and cope with their own health situations positively. Social media sites enable users to seek health information, advice, and also validate the same from doctors and health advisors. Social media networks help patients connect emotionally with each other through online communities and encourage and empower each other (Pretorius et al., 2019). The major social media applications for public and patients can be summarized in Fig. 9.2 below.

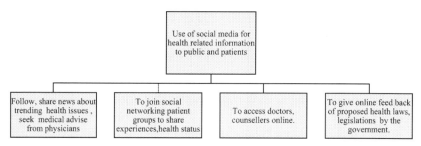

FIGURE 9.2 Applications of social media in health by public and patients.

Social media and public health organizations

Most of the studies in this area have been found to be on the interaction between public health bodies and citizens or patients. The use of social media platforms like Twitter and Facebook have been successful for evaluating public health services used by citizens, by conducting online feedback surveys and asking online suggestions. According to Bertot and Colleagues' (2016) classification framework of social media and public health organizations, public (government) health bodies use social media for seeking democratic participation by citizens. Public health bodies tend to use social media as a means to disseminate a government decision related to public health more often. For instance, a US-based study found the public health department used social media to educate the public on the state's proposed e-cigarette regulation to label e-cigarettes as tobacco product and to monitor their consumption and get feedback on the same. The above findings reiterate similar other e-government research findings on the capability of social media to engage in useful dialogue and constructive suggestions from citizens, while monitoring the misuse of social media (Cox et al., 2016).

Public health authorities must make use of social media platforms to educate and increase public health literacy among net users, as the latter tend to be easily misled by false and misleading posts from fake sources (Bin Naeem & Kamel Boulos, 2021). Studies suggest regulation and supervision by state machinery and uploading actual facts and figures about any disease outbreak or health-related information on official government websites will go a long way in building the trust between people and public health organizations and its systems. Since many Facebook posts during the COVID-19 pandemic contained misinformation on the nature, symptoms, vaccination programs, and its side effects, spreading fear and panic in the minds of people, social media users need to be more responsible (Cheung et al., 2017).

Thackeray et al. (2012) in their study on the extent and type of social media applications by state public health departments observed that 60% of the state health departments used at least one social media application for public communication, 86.7% of the health departments had a Twitter account, 56% had a Facebook account, and 43% had a YouTube subscription of the state health department. The latter used social media mainly to circulate health and disease-related data and factsheets to the people, and few active online users posted queries on health services. The study suggested that state health departments need to play a proactive role through social media channels and increase the interactions with online users. Public health organizations can use social media interventions to condition individuals' behavior to improve lifestyles and metabolic indicators of noncommunicable diseases (Balatsoukas et al., 2015; Gabarron et al., 2018). Online public health programs through social media can be started to increase exercise levels and cut down sugar and fat intake of patients (Jane et al., 2017; Richardson et al., 2010). Photos, videos, locations, and people networks can be used for public health monitoring and locating vulnerable groups living in backward and remote areas (Abramson et al., 2015; Chen & Schulz, 2016; Yeung, 2018).

According to the Global Web Index research data and Datareportal (2022) global overview, nearly 58.4% of the world's population use social media and this massive growth spurt has been evident typically during the COVID-19 pandemic. Most of the national governments and health authorities used social media tools, such as Facebook and Twitter to spread information on the virus outbreak and the preventive and safety norms to be followed by people to control the situation (Datareportal, 2022). Usage of social media by the World Health Organization (WHO) through its Twitter handle and official Facebook page during H1N1 2009 and the COVID-19 pandemic to educate people online was received very favorably across the world (Al-Dimour et al., 2020).

Among lower-income countries where the budgetary spending on health is a small share of the gross domestic product (GDP), social media can offer a cost-effective solution to spread public awareness on health issues. World organizations like WHO, World Bank, and non-governmental organizations (NGOs), and state-level health authorities in Chile, Mexico, and some Latin American countries have used social media for nutrition promotion, and health campaigns with limited funds (Mendoza-Herrera et al., 2020). In another study, it was found out that Facebook and Twitter are two popular platforms among youth and middle-aged people globally, and several government bodies have been utilizing these social networks for public messaging and awareness (Mendoza-Herrera et al., 2020). Social media has gained popularity even among senior citizens. as it helps them to reduce their social isolation, feelings of loneliness, improving their social connectivity with family and friends, and also to get quick medical assistance in times of emergency (Fu & Xie, 2021; Merchant et al., 2011). Online social platforms like "Patientlike me" and "Daily Strength" have many patients and users sharing online health-related information. Social media has the potential to help health policy planners to design age-specific policies in the future (Chen & Schulz, 2016).

Health care organizations like hospitals and pharmaceutical firms use social media for communicating with patients, promoting their products, services, providing customer support systems, and educating the public (Peck, 2014). With the increasing use of social media, pharmacies are communicating actively with wide WhatsApp groups of clients and customers simultaneously. They are conducting online consumer surveys, about the services pharmacies are offering in order to help clients and patients to identify themselves as part of the value addition chain of the pharmacy community. Many pharmacies are taking to social media platforms to increase their outreach to people regarding their health products, medical services, discounts, news events, and health information. Several large pharmacies and insurers have undertaken pilot projects that provide prescription refill and reminders regarding appointments via social media text messaging. In a study, it was found that 12.5% of health

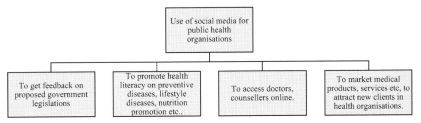

FIGURE 9.3 Applications of social media for public health organizations.

care bodies were able to draw and admit new patients mainly through promotion of their brand on social media (Househ, 2013). Additionally, even NGOs and not for profit health organizations have used social media platforms to engage stakeholders from communities by conduct of webinars, podcasts, and meetings using the online mode (Househ, 2013). In fact, the latter also have active blogs run by citizen groups which can be very handy for public health authorities during medical emergencies and disaster management situations.

In the United States, the Centre for Disease Control (CDC) as well as the U.S. Preventive Services Task Force use their own Facebook page and Twitter handle to share important information, contact numbers which can be shared fast on social media sites. The nature and form of health-related interaction between individuals and public health organizations have undergone a massive transformation since the past decade. Social media channels such as Facebook and LinkedIn are gaining popularity as a means of health information exchange, in the advanced countries like the United States of America, the United Kingdom, Norway, and Switzerland (Fox & Jones, 2009). The major social media applications for public health organizations can be summarized in Fig. 9.3 below.

Pitfalls of social media

Data reliability questionable: Health-related data and information posted on Facebook or Twitter are not always authentic or from reliable sources, since most of them are people posts. Social media is finally an informal source, where the information posted maybe of varying quality and consistency. The health-related information may not be referenced or validated from an official source. The WHO in this context has been evolving a monitoring mechanism, wherein privacy settings of account users should enable every account holder to limit his or her exposure to suspicious outsiders on a social media site. In fact, it is trying to start an official domain suffix that would strictly use validated health information and content of websites on this domain will be monitored by WHO officials (Childs & Martin, 2012).

Patient and health care professional personal boundary crossed: In a study, it was observed that patients send friend requests online to their own physicians which could be embarrassing for the latter as for medical professionals' patients are just subjects who have to be treated to cure a sickness (Peluchette et al., 2016). HCPs should discourage personal online chats with patients and request them to restrain themselves to the health problem in question. At the same time, patient's privacy has to be respected (Peluchette et al., 2016). Posting of inappropriate and unethical content about patients on social sites by medical students has been observed. A study of medical blogs posted by medical professionals found that in 47% of the 271 patient samples studied had the identities revealed on the social media sites. Patient privacy information needs to be protected by taking a patient's consent form before taking online web-based surveys or studies. There has to be a regulatory mechanism and a social media privacy guide to be followed by health professionals (Chen & Wang, 2021).

Health concerns: Social media can have negative effects on personal and professional lives of people in the health sector. Balatasoukas and Colleagues (2015) in their study pointed out that viral spread of false medical news, fake news, misinformation posted by quakes, social isolation, cyberbullying among teenagers, headaches, poor eyesight, stress and anxiety among youngsters, addiction to smart phones, and loss of self-confidence are some of harmful fall outs of excessive use of social media channels.

Privacy concerns: Social media can breach privacy of patients and their health problems (Pretorious et al., 2019). Some patients are uncomfortable discussing their health issues online with their health care provider. Therefore, it is important to take patients' consent before using their personal information on a social site.

Licensing revoked: Inappropriate use of social media by HCPs warrants disciplinary action and even suspension of the doctor. Health care institutions must include the effects and penalties for misuse of social media by their staff, in their Human Resources (HR) policy (Ventola, 2014). Medical professionals must be made aware of the legal requirements of state medical boards with reference to online chats, posts, and other forms of communications to ensure they do not commit any breach of rules that may lead to their license suspension. Nursing boards abroad caution their nurses for violations with regard to online disclosure of patients' personal health status and impose penalties ranging from letters of concern to license suspensions. Many HCP students in the United States have been found posting unprofessional content about their patients without the latter's consent, on social media, and this needs to be stopped by the institutions (Chauhan, 2012).

Guidelines for ethical use of social media in public health

One major risk of messaging on social media sites like Twitter is that the original message may become distorted through retweets and we can't be sure of how others may change the original meaning (Gough et al., 2017). Public health information on public domains needs to be posted responsibly, accurately, and timely. Sound information must be safeguarded and protected through mechanisms that will reply retweets if posts carry distorted messages. Health care organizations face social media risks on various grounds like violation of patient privacy, ignoring patient consent, violating physician and patient boundaries, and posting unethical, unwarranted material on social media alerts.

Some of the points all health organizations must incorporate in their social media policies and employee policies are as follows (Ventola, 2014):

- Leaking out confidential data of patients and health organizations should be banned.
- Any form of visible discrimination based on gender, caste, color, or race must be forbidden.
- Employees' time and purpose spent on social media during working hours need to be monitored.
- The penal action and consequences of violating organizational norms related to social media and its misuse at workplace must be defined in employee policies.
- Medical staff and health employees have to declare any conflict of interest at their organization.
- Even medical students and researchers must respect their patient's privacy and may post or publish patient-related information only after the voluntary consent of the patient.

Similarly, HCPs, including doctors, physicians, nurses, pharmacists, etc., are required to follow some guidelines with respect to the use of social media in their professional life. Some of the following guidelines are as follows (Ventola, 2014):

- Only information from reliable sources must be shared.
- Copyright laws of the state must be respected and adhered to.
- Securing a legal license for one's health profession practice.
- Physicians should try to avoid online friend requests from their patients.
- Instead, direct patients should be encouraged to join safe patient community social networking sites. Patients' identity should be undisclosed, and voluntary consent of the patient to be taken before sharing any related information on social media sites.
- The safest and most secure privacy settings should be made available to all the social networking site users.
- Health professionals need to verify that their credentials are correctly mentioned on social media channels.
- Correct self-identification on one's professional sites.

High-quality research surveys should be evolved to assess the impact of all forms of social media on health outcomes to find out the efficacy of usage of social media in enhancing the public health services. The state public health authorities should make it mandatory for health professionals to adopt social media as a communication channel in their interactions with patients, typically in the remote and backward areas of the country, Simultaneously, there should be strong regulatory systems to layout ethical use of social media in different health areas. To achieve public health outcomes, social media should be used more objectively, ethically to spread correct information on public health areas by all the stakeholders in the relevant fields. More focus on action research, "learning by doing" and sharing lessons among peers across the globe will be beneficial to all concerned (Ventola, 2014; Von Muhlen & Ohno-Machado, 2012).

Conclusions and future directions

The social media sites and platforms (if used judiciously) can offer several ways to promote individual as well as public health outcomes. Social media can have positive impact on shaping public awareness and education. One such example was the COVID-19 pandemic, which highlighted the role of social media in the emergency preparedness. However, strict guidelines in disseminating accurate information will be necessary to prevent the infodemic. Future studies can be planned to improve the dissemination of these guidelines to all potential users to prevent pitfalls associated with the social media discussed in this chapter.

Questions for discussion

- What strategies can increase social media usage among various public health professionals?
- Is social media as an information channel risker than offline communication for public health policy makers?
- What kind of public health training outreach programs should be promoted to increase the safe usage of social media among the public?
- What is the scope of social media application for medical experiments, trials, and research and development in the future?

Websites to explore

1. Twitter best practices and tips for physician users.

URL: https://www.hopkinsmedicine.org/fac_development/_documents/course-offerings/Twitter%20Best%20Practices.PDF.

Description: Visit this link to learn how to build and optimize your profile, to monitor health care community, and common hashtags used in Medicine.

Suggested activity for the readers: Read all the instructions to make your profile and start making the connections using these instructions.

2. Guidelines for the use of social media

URL: https://hr.umich.edu/sites/default/files/voices-social-media-guidelines.pdf.

Description: This describes the implications of "friending," "linking," and "following."

Suggested activity for the readers: Read all the instructions and locate individual's posting guidelines.

3. Google Trends

URL: https://trends.google.com/trends/?geo=US.

Description: Google Trends provides access to a largely unfiltered sample of actual search requests made to Google. It's anonymized (no one is personally identified), categorized (determining the topic for a search query), and aggregated (grouped together).

Suggested activity for the readers: Open Google Trends, search a term, in the top right of the chart, click download. Investigate data.

Key terms

Blog: Blog is an interactive website or its part maintained by either individual or organizations to ensure regular entries of comments and events.

Content analysis: This refers to a research tool used to determine the presence of certain words, themes, or concepts within some given qualitative data (i.e., text).

Emoticon or emoji: This is a graphical facial expression that is embedded in the text communication and provides emotional information in the lieu of traditional face-to-face interactions.

Gross domestic product: GDP is defined as the standard monetary measure of the market value of all the final goods and services being produced and sold in a specific period of time by countries.

Hawthorne effect: It refers to a type of reactivity in which individuals modify an aspect of their behavior in response to their awareness of being observed.

HTML: The HTML is a Hypertext Markup Language, which is a code to structure a webpage and its content.

Public health policy: A set of guidelines or measures which appropriate state authorities frame and implement for the promotion of good health standards and betterment of its people living in the society.

Qualitative research: It is a form of research that tries to gain deep insights into human behavior, attitudes, interests, interactions, beliefs, and experiences about reality through descriptive, nonquantitative data, typically using focus groups, case study method, interviews, observations, and content analysis to mention a few techniques.

Social media: A set of tools designed for individuals and communities to enable them to congregate and interact; to exchange data, ideas, personal messages, photos, videos, and other media information; and, in some situations, to work with other people.

Tags: Names or key words added to the posts or pictures.

Social desirability bias: Social desirability is a form of response bias, which occurs when respondents give answers to questions that they believe will make them look good to others, concealing their true opinions or experiences.

References

Abramson, K., Keefe, B., & Chou, W. Y. (2015). Communicating about cancer through facebook: A qualitative analysis of a breast cancer awareness page. *Journal of Health Communication, 20*(2), 237−243. https://doi.org/10.1080/10810730.2014.927034

Agostino, D. (2013). Using social media to engage citizens: A study of Italian municipalities. *Public Relations Review, 39*(3), 232−234. https://doi.org/10.1016/j.pubrev.2013.02.009

Al-Dmour, H., Masa'deh, R., Salman, A., Abuhashesh, M., & Al-Dmour, R. (2020). Influence of social media platforms on public health protection against the COVID-19 pandemic via the mediating effects of public health awareness and behavioral changes: Integrated model. *Journal of Medical Internet Research, 22*(8), e19996. https://doi.org/10.2196/19996

Balatsoukas, P., Kennedy, C. M., Buchan, I., Powell, J., & Ainsworth, J. (2015). The,role of social network technologies in online health promotion: A narrative review of theoretical and empirical factors influencing intervention effectiveness. *Journal of Medical Internet Research, 17*(6), e141. https://doi.org/10.2196/jmir.3662

Bamat, N. A., Manley, B. J., Harer, M. W., & Roland, D. (2018). Social media for pediatric research: What, who, why, and. *Pediatric Research, 84*(5), 597−599. https://doi.org/10.1038/s41390-018-0140-7

Benetoli, A., Chen, T. F., & Aslani, P. (2015). The use of social media in pharmacy practice and education. *Research in Social and Administrative Pharmacy, 11*(1), 1−46. https://doi.org/10.1016/j.sapharm.2014.04.002

Benetoli, A., Chen, T. F., Schaefer, M., Chaar, B., & Aslani, P. (2017). Do pharmacists use social media for patient care? *International Journal of Clinical Pharmacy, 39*(2), 364−372. https://doi.org/10.1007/s11096-017-0444-4

Bernhardt, J. M., Alber, J., & Gold, R. S. (2014). A social media primer for professionals: Digital dos and don'ts. *Health Promotion Practice, 15*(2), 168−172. https://doi.org/10.1177/1524839913517235

Bertot, J., Estevez, E., & Janowski, T. (2016). Universal and contextualized public services: Digital public service innovation framework. *Government Information Quarterly, 33*(2), 211−222. https://doi.org/10.1016/j.giq.2016.05.004

Beullens, K., & Schepers, A. (2013). Display of alcohol use on facebook: A content analysis. *Cyberpsychology, Behavior and Social Networking, 16*(7), 497−503. https://doi.org/10.1089/cyber.2013.0044

Bin Naeem, S., & Kamel Boulos, M. N. (2021). COVID-19 misinformation online and health literacy: A brief overview. *International Journal of Environmental Research and Public Health, 18*(15), 8091. https://doi.org/10.3390/ijerph18158091

Birnbaum, M. L., Rizvi, A. F., Confino, J., Correll, C. U., & Kane, J. M. (2017). Role of social media and the Internet in pathways to care for adolescents and young adults with psychotic disorders and non-psychotic mood disorders. *Early Intervention in Psychiatry, 11*(4), 290−295. https://doi.org/10.1111/eip.12237'

Chauhan, B., George, R., & Coffin, J. (2012). Social media and you: What every physician needs to know. *Journal of Medical Practice Management, 28*(3), 206–209.

Chen, Y. R., & Schulz, P. J. (2016). The effect of information communication technology interventions on reducing social isolation in the elderly: A systematic review. *Journal of Medical Internet Research, 18*(1), e18. https://doi.org/10.2196/jmir.4596. PubMed: 26822073.

Chen, J., & Wang, Y. (2021). Social media use for health purposes: Systematic review. *Journal of Medical Internet Research, 23*(5), e17917. https://doi.org/10.2196/17917

Cherrez-Ojeda, I., Vanegas, E., Felix, M., Mata, V. L., Jiménez, F. M., Sanchez, M., Simancas-Racines, D., Cherrez, S., Gavilanes, A. W. D., Eschrich, J., & Chedraui, P. (2020). Frequency of use, perceptions and barriers of information and communication technologies among Latin American physicians: An Ecuadorian cross-sectional study. *Journal of Multidisciplinary Healthcare, 13*, 259–269. https://doi.org/10.2147/JMDH.S246253

Cheung, Y. T. D., Chan, C. H. H., Wang, M. P., Li, H. C. W., & Lam, T. H. (2017). Online social support for the prevention of smoking relapse: A content analysis of the WhatsApp and facebook social groups. *Telemedicine Journal and E-Health: The Official Journal of the American Telemedicine Association, 23*(6), 507–516. https://doi.org/10.1089/tmj.2016.0176

Childs, L. M., & Martin, C. Y. (2012). Social media profiles: Striking the right balance. *American Journal of Health-System Pharmacy, 69*(23), 2044–2050. https://doi.org/10.2146/ajhp120115

Cox, E., Barry, R. A., & Glantz, S. (2016). E-Cigarette policymaking by local and state governments: 2009-2014. *The Milbank Quarterly, 94*(3), 520–596. https://doi.org/10.1111/1468-0009.12212

Datareportal. (2022). *Digital 2022*. Global Overview Report. Available at: https://datareportal.com/reports/digital-2022-global-overview-report.

Dizon, D. S., Graham, D., Thompson, M. A., Johnson, L. J., Johnston, C., Fisch, M. J., & Miller, R. (2012). Practical guidance: The use of social media in oncology practice. *Journal of Oncology Practice, 8*(5), e114–e124. https://doi.org/10.1200/JOP.2012.000610

Erikainen, S., Pickersgill, M., Cunningham-Burley, S., & Chan, S. (2019). Patienthood and participation in the digital era. *Digital Health, 5*, 1–10. https://doi.org/10.1177/2055207619845546

Eysenbach, G., & Till, J. E. (2001). Ethical issues in qualitative research on internet communities. *BMJ (Clinical Research ed.), 323*(7321), 1103–1105. https://doi.org/10.1136/bmj.323.7321.1103

Farnan, J. M., Snyder Sulmasy, L., Worster, B. K., Chaudhry, H. J., Rhyne, J. A., Arora, V. M., & American College of Physicians Ethics, Professionalism and Human Rights Committee, American College of Physicians Council of Associates, and Federation of State Medical Boards Special Committee on Ethics and Professionalism. (2013). Online medical professionalism: Patient and public relationships: Policy statement from the American College of physicians and the federation of state medical boards. *Annals of Internal Medicine, 158*(8), 620–627. https://doi.org/10.7326/0003-4819-158-8-201304160-00100

Fogelson, N. S., Rubin, Z. A., & Ault, K. A. (2013). Beyond likes and tweets: An in-depth look at the physician social media landscape. *Clinical Obstetrics and Gynecology, 56*(3), 495–508. https://doi.org/10.1097/GRF.0b013e31829e7638

Fox, S., & Jones, S. (2009). *The social life of health information*. http://www.pewinternet.org. Reports/2007/Information-Searches-.

Franz, D., Marsh, H. E., Chen, J. I., & Teo, A. R. (2019). Using facebook for qualitative research: A brief primer. *Journal of Medical Internet Research, 21*(8), e13544. https://doi.org/10.2196/13544

Fu, L., & Xie, Y. (2021). The effects of social media use on the health of older adults: An empirical analysis based on 2017 Chinese general social survey. *Healthcare (Basel, Switzerland), 9*(9), 1143. https://doi.org/10.3390/healthcare9091143

Gabarron, E., Årsand, E., & Wynn, R. (2018). Social media use in interventions for diabetes: Rapid evidence-based review. *Journal of Medical Internet Research, 20*(8), e10303. https://doi.org/10.2196/10303

George, D. R., Rovniak, L. S., & Kraschnewski, J. L. (2013). Dangers and opportunities for social media in medicine. *Clinical Obstetrics and Gynecology, 56*(3), 453–462. https://doi.org/10.1097/GRF.0b013e318297dc38

Golder, S., Norman, G., & Loke, Y. K. (2015). Systematic review on the prevalence, frequency and comparative value of adverse events data in social media. *British Journal of Clinical Pharmacology, 80*(4), 878–888. https://doi.org/10.1111/bcp.12746. PubMed: 26271492.

Gough, A., Hunter, R. F., Ajao, O., Jurek, A., McKeown, G., Hong, J., Barrett, E., Ferguson, M., McElwee, G., McCarthy, M., & Kee, F. (2017). Tweet for behavior change: Using social media for the dissemination of public health messages. *JMIR Public Health and Surveillance, 3*(1), e14. https://doi.org/10.2196/publichealth.6313

Grajales, F. J., Sheps, S., Ho, K., Novak-Lauscher, H., & Eysenbach, G. (2014). Social media: A review and tutorial of applications in medicine and health care. *Journal of Medical Internet Research, 16*(2), e13. https://doi.org/10.2196/jmir.2912

Grindrod, K., Forgione, A., Tsuyuki, R. T., Gavura, S., & Giustini, D. (2014). Pharmacy 2.0: A scoping review of social media use in pharmacy. Research in social and administrative. *Pharmacy, 10*(1), 256–270. https://doi.org/10.1016/j.sapharm.2013.05.004

Guidry, J., Zhang, Y., Jin, Y., & Parrish, C. (2016). Portrayals of depression on Pinterest and why public relations practitioners should care. *Public Relations Review, 42*(1), 232–236. https://doi.org/10.1016/j.pubrev.2015.09.002

Helm, J., & Jones, R. M. (2016). Practice paper of the academy of nutrition and dietetics: Social media and the dietetics practitioner: Opportunities, challenges, and best practices. *Journal of the Academy of Nutrition and Dietetics, 116*(11), 1825–1835. https://doi.org/10.1016/j.jand.2016.09.003. PubMed: 27788767.

Hermansyah, A., Sukorini, A. I., Asmani, F., Suwito, K. A., & Rahayu, T. P. (2019). The contemporary role and potential of pharmacist contribution for community health using social media. *Journal of Basic and Clinical Physiology and Pharmacology, 30*(6), 1. https://doi.org/10.1515/jbcpp-2019-0329. PubMed: 31800395.

Househ, M. (2013). The use of social media in healthcare: Organizational, clinical, and patient perspectives. *Studies in Health Technology and Informatics, 183*, 244–248.

Hwang, K. O., Ottenbacher, A. J., Green, A. P., Cannon-Diehl, M. R., Richardson, O., Bernstam, E. V., & Thomas, E. J. (2010). Social support in an Internet weight loss community. *International Journal of Medical Informatics, 79*(1), 5–13. https://doi.org/10.1016/j.ijmedinf.2009.10.003

Jane, M., Hagger, M., Foster, J., Ho, S., Kane, R., & Pal, S. (2017). Effects of a weight management program delivered by social media on weight and metabolic syndrome risk factors in overweight and obese adults: A randomised controlled trial. *PLOS ONE, 12*(6), e0178326. https://doi.org/10.1371/journal.pone.0178326

Jiang, T., Osadchiy, V., Mills, J. N., & Eleswarapu, S. V. (2020). Is it all in my head? Self-Reported psychogenic erectile dysfunction and depression are common among young men seeking advice on social media. *Urology, 142*, 133–140. https://doi.org/10.1016/j.urology.2020.04.100

Kenny, P., & Johnson, I. G. (2016). Social media use, attitudes, behaviours and perceptions of online professionalism amongst dental students. *British Dental Journal, 221*(10), 651–655. https://doi.org/10.1038/sj.bdj.2016.864

Kent, E. E., Prestin, A., Gaysynsky, A., Galica, K., Rinker, R., Graff, K., & Chou, W. Y. (2016). "Obesity is the new major cause of cancer": Connections between obesity and cancer on facebook and twitter. *Journal of Cancer Education: The Official Journal of the American Association for Cancer Education, 31*(3), 453−459. https://doi.org/10.1007/s13187-015-0824-1

Korda, H., & Itani, Z. (2013). Harnessing social media for health promotion and behavior change. *Health Promotion Practice, 14*(1), 15−23. https://doi.org/10.1177/1524839911405850

Kosinski, M., Matz, S. C., Gosling, S. D., Popov, V., & Stillwell, D. (2015). Facebook as a research tool for the social sciences: Opportunities, challenges, ethical considerations, and practical guidelines. *The American Psychologist, 70*(6), 543−556. https://doi.org/10.1037/a0039210

Kotler, P., Kartajaya, H., & Setiawan, I. (2010). *From products to customers to the human spirit: Marketing 3.0*. John Wiley & Sons.

Koutrolou-Sotiropoulou, P., Lima, F. V., & Stergiopoulos, K. (2016). Quality of life in survivors of peripartum cardiomyopathy. *American Journal of Cardiology, 118*(2), 258−263. https://doi.org/10.1016/j.amjcard.2016.04.040

Kramer, A. D., Guillory, J. E., & Hancock, J. T. (2014). Experimental evidence of massive-scale emotional contagion through social networks. *Proceedings of the National Academy of Sciences of the United States of America, 111*(24), 8788−8790. https://doi.org/10.1073/pnas.1320040111

Kutsikos, K. (2007). *Distribution-collaboration networks (DCN): A systems-based model for developing collaborative e-government*.

Lambert, K. M., Barry, P., & Stokes, G. (2012). Risk management and legal issues with the use of social media in the healthcare setting. *Journal of Healthcare Risk Management, 31*(4), 41−47. https://doi.org/10.1002/jhrm.20103

Lambert, M., Chivers, P., & Farringdon, F. (2019). In their own words: A qualitative study exploring influences on the food choices of university students. *Health Promotion Journal of Australia, 30*(1), 66−75. https://doi.org/10.1002/hpja.180

Liu, Q. B., Liu, X., & Guo, X. (2020). The effects of participating in a physician-driven online health community in managing chronic disease: Evidence from two natural experiments. *MIS Quarterly, 44*(1), 391−419. https://doi.org/10.25300/MISQ/2020/15102

Loeb, S., Carrick, T., Frey, C., & Titus, T. (2020). Increasing social media use in urology: 2017 American urological association survey. *European Urology Focus, 6*(3), 605−608. https://doi.org/10.1016/j.euf.2019.07.004

MacMillan, C. (2013). Social media revolution and blurring of professional boundaries. *Imprint, 60*(3), 44−46.

Mascia, D., Rinninella, E., Pennacchio, N. W., Cerrito, L., & Gasbarrini, A. (2021). It's how we communicate! Exploring face-to-face versus electronic communication networks in multidisciplinary teams. *Health Care Management Review, 46*(2), 153−161. https://doi.org/10.1097/HMR.0000000000000246'

McDonald, L., Malcolm, B., Ramagopalan, S., & Syrad, H. (2019). Real-world data and the patient perspective: The Promise of social media? *BMC Medicine, 17*(1), 11. https://doi.org/10.1186/s12916-018-1247-8

McGowan, B. S., Wasko, M., Vartabedian, B. S., Miller, R. S., Freiherr, D. D., & Abdolrasulnia, M. (2012). Understanding the factors that influence the adoption and meaningful use of social media by physicians to share medical information. *Journal of Medical Internet Research, 14*(5), e117. https://doi.org/10.2196/jmir.2138. http://www.jmir.org/2012/5/e117/

Mendoza-Herrera, K., Valero-Morales, I., Ocampo-Granados, M. E., Reyes-Morales, H., Arce-Amaré, F., & Barquera, S. (2020). An overview of social media use in the field of public health nutrition: Benefits, scope, limitations, and a Latin American experience. *Preventing Chronic Disease, 17*, E76. https://doi.org/10.5888/pcd17.200047

Merchant, R. M., Elmer, S., & Lurie, N. (2011). Integrating social media into emergency-preparedness efforts. *New England Journal of Medicine, 365*(4), 289−291. https://doi.org/10.1056/NEJMp1103591

Moorhead, S. A., Hazlett, D. E., Harrison, L., Carroll, J. K., Irwin, A., & Hoving, C. (2013). A new dimension of health care: Systemic review of the uses, benefits, and limitations of social media for health care professionals. *Journal of Medical Internet Research, 15*(4), e85. https://doi.org/10.2196/jmir.1933

Norman, C. D., & Yip, A. L. (2012). Ehealth promotion and social innovation with youth: Using social and visual media to engage diverse communities. *Studies in Health Technology and Informatics, 172*, 54−70.

O'Dwyer, L., & Bernauer, J. (2014). *Quantitative research for the qualitative researcher*. SAGE.

Oser, T. K., Minnehan, K. A., Wong, G., Parascando, J., McGinley, E., Radico, J., & Oser, S. M. (2019). Using social media to broaden understanding of the barriers and facilitators to exercise in adults with type 1 diabetes. *Journal of Diabetes Science and Technology, 13*(3), 457−465. https://doi.org/10.1177/1932296819835787

Peck, J. L. (2014). Social media in nursing education: Responsible integration for meaningful use. *Journal of Nursing Education, 53*(3), 164−169. https://doi.org/10.3928/01484834-20140219-03

Pedersen, E. R., Helmuth, E. D., Marshall, G. N., Schell, T. L., PunKay, M., & Kurz, J. (2015). Using facebook to recruit young adult veterans: Online mental health research. *JMIR Research Protocols, 4*(2), e63. https://doi.org/10.2196/resprot.3996

Peluchette, J. V., Karl, K. A., & Coustasse, A. (2016). Physicians, patients, and facebook: Could you? Would you? Should you? *Health Marketing Quarterly, 33*(2), 112−126. https://doi.org/10.1080/07359683.2016.1166811

Pew Research Center. (n.d.). *Internet/Broadband Factsheet*. Available at: https://www.pewresearch.org/internet/fact-sheet/internet-broadband/.

Pirraglia, P. A., & Kravitz, R. L. (2013). Social media: New opportunities, new ethical concerns. *Journal of General Internal Medicine, 28*(2), 165−166. https://doi.org/10.1007/s11606-012-2288-x

Pizzuti, A. G., Patel, K. H., McCreary, E. K., Heil, E., Bland, C. M., Chinaeke, E., Love, B. L., & Bookstaver, P. B. (2020). Healthcare practitioners' views of social media as an educational resource. *PLOS ONE, 15*(2), e0228372. https://doi.org/10.1371/journal.pone.0228372

Pretorius, C., Chambers, D., & Coyle, D. (2019). Young people's online help-seeking and mental health difficulties: Systematic narrative review. *Journal of Medical Internet Research, 21*(11), e13873. https://doi.org/10.2196/13873. https://www.jmir.org/2019/11/e13873/

Richardson, C. R., Buis, L. R., Janney, A. W., Goodrich, D. E., Sen, A., Hess, M. L., Mehari, K. S., Fortlage, L. A., Resnick, P. J., Zikmund-Fisher, B. J., Strecher, V. J., & Piette, J. D. (2010). An online community improves adherence in an internet-mediated walking program. Part 1: Results of a randomized controlled trial. *Journal of Medical Internet Research, 12*(4), e71. https://doi.org/10.2196/jmir.1338

Safko, J., & Brake, D. K. (2009). *The social media bible*. Hoboken: Wiley.

Scanfeld, D., Scanfeld, V., & Larson, E. L. (April 2010). Dissemination of health information through social networks: Twitter and antibiotics. *American Journal of Infection Control, 38*(3), 182−188. https://doi.org/10.1016/j.ajic.2009.11.004

Schwartz, L. M., & Woloshin, S. (2019). Medical marketing in the United States, 1997-2016. *JAMA, 321*(1), 80−96. https://doi.org/10.1001/jama.2018.19320

Shen, L., Wang, S., Chen, W., Fu, Q., Evans, R., Lan, F., Li, W., Xu, J., & Zhang, Z. (2019). Understanding the function constitution and influence factors on communication for the wechat official account of top tertiary hospitals in China: Cross-sectional study. *Journal of Medical Internet Research, 21*(12), e13025. https://doi.org/10.2196/13025

Snelson, C. L. (2016). Qualitative and mixed methods social media research: A review of the literature. *International Journal of Qualitative Methods, 15*(1), 1−15. https://doi.org/10.1177/1609406915624574

Song, J., Xu, P., & Paradice, D. B. (2020). Health goal attainment of patients with chronic diseases in web-based patient communities: Content and survival analysis. *Journal of Medical Internet Research, 22*(9), e19895. https://doi.org/10.2196/19895

Spallek, H., O'Donnell, J., Clayton, M., Anderson, P., & Krueger, A. (2010). Paradigm shift or annoying distraction: Emerging implications of web 2.0 for clinical practice. *Applied Clinical Informatics, 1*(2), 96−115. https://doi.org/10.4338/ACI-2010-01-CR-0003

Stern, J. (2010). *Social media metrics*. Hoboken: Wiley.

Suarez-Lledo, V., & Alvarez-Galvez, J. (2021). Prevalence of health misinformation on social media: Systematic review. *Journal of Medical Internet Research, 23*(1), e17187. https://doi.org/10.2196/17187

Tang, W., Ren, J., & Zhang, Y. (2019). Enabling trusted and privacy-preserving healthcare services in social media health networks. *IEEE Transactions on Multimedia, 21*(3), 579−590. https://doi.org/10.1109/TMM.2018.2889934

Thackeray, R., Neiger, B. L., Smith, A. K., & Van Wagenen, S. B. (2012). Adoption and use of social media among public health departments. *BMC Public Health, 12*, 242. https://doi.org/10.1186/1471-2458-12-242

Tower, M., Blacklock, E., Watson, B., Heffernan, C., & Tronoff, G. (2015). Using social media as a strategy to address 'sophomore slump' in second year nursing students: A qualitative study. *Nurse Education Today, 35*(11), 1130−1134. https://doi.org/10.1016/j.nedt.2015.06.011

Ventola, C. L. (2014). Social media and health care professionals: Benefits, risks, and best practices. *P and T, 39*(7), 491−520.

Von Muhlen, M., & Ohno-Machado, L. (2012). Reviewing social media use by clinicians. *Journal of the American Medical Informatics Association, 19*(5), 777−781. https://doi.org/10.1136/amiajnl-2012-000990

Wocott, H. F. (1988). Ethnographic research in education. In R. M. Jaeger (Ed.), *Complementary methods for research in education* (pp. 187−212). American Educational Research Association.

Yeung, D. (2018). Social media as a catalyst for policy action and social change for health and well-being: Viewpoint. *Journal of Medical Internet Research, 20*(3), e94. https://doi.org/10.2196/jmir.8508

Yüce, M.Ö., Adalı, E., & Kanmaz, B. (2021). An analysis of YouTube videos as educational resources for dental practitioners to prevent the spread of COVID-19. *Irish Journal of Medical Science, 190*(1), 19−26. https://doi.org/10.1007/s11845-020-02312-5

Zhong, Y., Liu, W., Lee, T. Y., Zhao, H., & Ji, J. (2021). Risk perception, knowledge, information sources and emotional states among COVID-19 patients in Wuhan, China. *Nursing Outlook, 69*(1), 13−21. https://doi.org/10.1016/j.outlook.2020.08.005

Zhu, C., Zeng, R., Zhang, W., Evans, R., & He, R. (2019). Pregnancy-related information seeking and sharing in the social media era among expectant mothers: Qualitative study. *Journal of Medical Internet Research, 21*(12), e13694. https://doi.org/10.2196/13694

Further reading

Eghdam, A., Hamidi, U., Bartfai, A., & Koch, S. (2018). Facebook as communication support for persons with potential mild acquired cognitive impairment: A content and social network analysis study. *PloS One, 13*(1), e0191878. https://doi.org/10.1371/journal.pone.0191878

Chapter 10

Role of social media in research publicity and visibility

Sely-Ann Headley Johnson and Tiffiny R. Jones
Wayne State University, Detroit, MI, United States

Objectives

- Identify various social media-friendly research dissemination metric resources.
- Describe common traditional research dissemination techniques.
- Interpret the versatility of social media in health.
- Formulate ways in which social media can benefit the field of public health and research dissemination.

Introduction

Research dissemination patterns

Research dissemination is a planned process, which considers target audiences, settings in which research findings are to be received, and communicating/interacting with wider audiences in ways that facilitate research uptake and understanding to influence decision-making processes and practice (Wilson et al., 2010).

Traditionally research has been disseminated in journals read by academics and professionals in their respective fields. It is also common for certain groups of researchers to publish in collections of journals that sometimes require additional fees for other scholars and the general public to access them (Millar & Lim, 2022). For example, several nonopen access journals, such as the American *Journal of Public Health*, *Journal of Infectious Diseases*, *Journal of Exposure Science and Environmental Epidemiology*, and others require a fee for readers to access the published research. This is a huge barrier to the general public, who are interested in the topics published in these journals. For example, a parent whose child, has contracted a rare infectious disease wants to read as much as possible about new trials and medicines to help his/her child, would not be able to do much research since the highly specialized articles require a cost to access them. Another example is a former faculty member or a retired academician, who now wants to read up on the latest developments in public health, but due to lack of resources, he/she can no longer maintain subscription fees to major organizations, such as APHA, which grants access to the *American Journal of Public Health* and *other journals*. This results in people in this group being unable to keep up with the latest research on COVID-19, monkeypox, and other developing health issues that could pose a threat to the general public. While journals need to maintain infrastructure and retain high quality staff and other administrative features, the cost of doing so must be balanced against the danger of creating knowledge deficits or knowledge black holes. These occurs when people who want to make informed decisions but are not given the opportunity to access the information they need to make informed decisions. One solution to this knowledge black hole is creating membership streams that the average person can afford. For example, membership that results in two free articles per week, or the opportunity to be a reviewer in exchange for access to 20 articles per month. This shift will also offer an opportunity for research impact to be measured in new, unconventional ways.

The impact of research, particularly in academia, has conventionally been measured quantitatively by the number of publications, the impact factors of journals that the research was been published in, h-index, and the number of citations an article receives. This type of assessment is slowly becoming outdated as more researchers turn to social media platforms to disseminate their findings and, in turn, rely on newer publication metrics. Below are some new ways of disseminating research.

Novel publication metric trends

More recently, as a result of advancements in digital technology, novel dissemination metrics have been pioneered by institutional repositories and journal websites, such as Altmetric and PlumX Metrics. These indices reflect digital footprints and provide a more comprehensive overview of the interest that the published research has had in terms of citations, including clinical, patent and policy documents, usage, captures, mentions, and social media tags, likes or mentions (Millar & Lim, 2022). This is because these metrics aim to provide a more comprehensive indication of the impact of research outputs, within the online environment and complement the traditional bibliometrics. In addition to monitoring citation-based metrics, Altmetric monitors citations on electronic platforms such as Wikipedia, blogs, social media, and bookmark reference managers like Mendeley (Altmetric, 2022). PlumX Metrics takes this a bit further by providing insight on how people are interacting with research products, through five metric categories, namely; citations, usage, captures, mentions, and social media. These five categories track a myriad of electronic footprint such as the number of downloads, views, favorites, likes, and various types of citations such as policy, patent and clinical citations (Plum Analytics, 2022). These novel metrics have the ability to give a more comprehensive view of the impact research has in our communities.

As such, there should be no surprise that social media platforms have been known to play a role in the academic promotion and tenure process (Deeken et al., 2020) as well as help improve more traditional metrics of academic and publication success. Similarly, studies revealed that the number of tweets an article gains correlates with citations in a variety of fields, specifically, psychiatry, gastroenterology, and professional health educators (Deeken et al., 2020). Even journals have benefited from this new shift. This is evidenced by the fact that a robust presence on Twitter is associated with an increased impact factor for pediatric urology, plastic surgery, and radiology journals. This is partly due to the fact that dissemination via social media is associated with increased readership, as documented by larger numbers of article views and downloads. It is evident that social media coupled with the new research metric tools clearly has the added advantage of being able to measure dissemination and indicate the potential influence and impact of research that would ordinarily not be available with traditional citation-based metrics.

As technology grows and publication metrics are in flux, research dissemination has also changed. Over the last two decades, several journals have moved online, and currently, digital dissemination with social media as a platform is on the rise (Ashcraft et al., 2020). Although some researchers still use blogs, and wikis to disseminate their work, others prefer to share their research with other scholars via professional social platforms such as LinkedIn, Adacemia.edu, and ResearchGate. The rise of social media usage has also created a new channel for disseminating research to the general public, because researchers can also reach the layperson as a well as other scholars via TED Talks, and exclusively reach the lay audience via YouTube. More recently, we see the shift toward television through shows, shorts, and (Ashcraft et al., 2020), movies or docuseries on Netflix and Prime Video. These new dissemination habits have contributed to new cross-disciplinary collaborations that aid the creation of new research, publication, and funding opportunities. This new avenue also creates several new challenges and opportunities for improved health outcomes among various groups, as they rely on electronic sources for health information.

Using social media for health information

Approximately 73% of US adults use social media, specifically, the Pew Research (2021) reports that 81% of adults report using YouTube and 61% of adults report using Facebook. Among all social media users, three-quarters use social media at least once daily, and almost 50% of adults report that information found through social media influences their health decisions (Pew Research, 2021). This creates an opportunity for researchers to reach critical audiences with their research and interventions. For example, the Rescue Agency, an organization committed to helping government agencies and nonprofits create award-winning campaigns that drive health behavior change, partnered with several community-based organizations to create health campaigns and disseminate them on TikTok and Snapchat, where at-risk youth received lifesaving information (Rescue Agency, 2022). One such example is the "A dose of truth" campaign that was designed to help change behavior in the heat of the moment by giving youth critical truth that made them question the safety of the source of their drugs in addition to information about what an overrode truly looks like. By showing youth that an overdose can be very subtle and quiet, as opposed to dramatic and loud like in TV shows, lives were saved because our youth were better informed of the truth behind what an overdose looks like. This is the same concept behind stickers like "can't wake, don't wait" as a call to action to use naloxone and dial 911 when an overdose is suspected (Rescue Agency, 2022). The genius behind these life saving campaigns was the dissemination platform. We are living in an era where social media is a valuable venue for reaching and engaging with various populations on their smartphones.

Social media as dissemination avenues

The abundant use of smartphones coupled with the rapid emergence of free, widely used social media platforms has accelerated the dissemination of information on a scale and speed that would have been unimaginable just a few years ago. Consequently, researchers are seeking innovative ways to disseminate research findings (Deeken et al., 2020) as quickly as possible, while also securing funds for subsequent projects. This is evident by the fact that many scientists are using crowdfunding strategies available through social media to improve research funding opportunities (Jang et al., 2019).

One critical component of securing funding is ensuring the audience has the ability to understand the purpose and impact of research. A great way to ensure complex research is understood by the intended audience, is by creating visual abstracts.

Visual abstracts

Visual abstracts were created by journals and researchers to save time and increase the ability to share complex abstracts (Millar & Lim, 2022), with health educators and communities. Visual abstracts have an infographic style format, coupled with a truncated, limited word summary of the research abstract detailing the key question, methodology, findings, and take-home message. Visual abstracts are so simple they can be shared on social media platforms, which in turn increase the interest and impact within the research community and target population. Due to its practicality and effectiveness, visual abstracts are being increasingly introduced to medical journals and organizations to help disseminate valuable research findings (Millar & Lim, 2022).

This effective summary method saves time and avoids translation or simplification errors by journalists or lay audiences. Due to their succinctness, these image-focused summaries provide clinicians and researchers with a snapshot of current research findings and provides a guide for selecting articles with further in-depth examinations. These sharable files are convenient and suitable for educators and researchers (Millar & Lim, 2022). Although not a replacement for the full article, visual abstracts act as a "taster" to entice a wide audience to examine, retrieve, and read the full article in greater depth (Millar & Lim, 2022).

Some healthcare professionals have indicated that they prefer visual infographic formats rather than conventionally written abstracts when communicating via social media and viewing online journals (Millar & Lim, 2022). These visual abstracts can also be valuable increasing the spread and consumption of factual health information on social media. Researchers can choose to create visual abstracts that the media can circulate, thus reducing the need for media personnel to interpret research findings for the lay audience.

Challenges and advantages of disseminating research on social media

As social media becomes an increasingly important venue for disseminating and constructing health and scientific knowledge, health communication on social media comes with several caveats, including the distortion of truth, and the spread of inaccurate information (Tambuscio et al., 2015). This has been a real issue for decades, but recent years have highlighted the true danger of this problem.

In the past, health and science communication had been traditionally unidirectional, where experts released information after long periods of research. Today, health and science news outlets pick up research results and translate them into knowledge, which was and still is essential for the public understanding of science and policy decision processes (Davies & Horst, 2016). Since few people have first-hand connections to or understanding of complex scientific findings, mass media have functioned as important conduits in spreading scientific information (Schafer, 2012) to the general public. In 1990, years before the explosion of social media, Hilgartner (1990) noted that the news outlet popularization process often oversimplifies and distorts scientific knowledge, but at the same time also provided an undeniably useful resource for public discourse.

Despite increasing warnings about inaccurate information online, before the COVID-19 pandemic, little was known about how social media would contribute to the widespread diffusion of unverified health information (Jang et al., 2019). During the pandemic, it was evident that social media has huge power and possessed the ability to influence attitudes, beliefs, norms, and behaviors. While this power can be used to influence better health outcomes, we also know that social media can undermine public health efforts (Schillinger et al., 2020).

Social media has such a massive impact on public opinion, that the Community Preventive Services Task Force (CPSTF) recommends health communication campaigns use multiple channels of dissemination, one of which must be social media, in addition to providing free or reduced-price health-related products to help produce intended behavior change (Community Preventative Services Task Force CPSTF, 2010).

Effect of social media on health outcomes in spite of research findings

As more people turn to online venues such as e-news, websites, blogs, and social media networks for health information, they are also using this information to gauge public opinion regarding divisive health topics such as the vaccine-autism controversy (Betsch et al., 2012; Goldstein et al., 2015). We can see this trend clearly if we observe the relationship between vaccination rates and attitudes toward vaccination.

Studies found that antivaccination beliefs either directly or indirectly predict under immunization rates by monitoring people's perceived vaccination risks (Betsch et al., 2010; Brewer et al., 2007; Gust et al., 2004). Additionally, a 2017 study found that antivaccine tweets coincided with vaccine-related news events and clusters geographically in areas with high concentrations of women who recently gave birth, households with high income levels, or men with no college degree (Tomeny et al., 2017).

Antivaccination beliefs represent a spectrum of diverse thoughts and attitudes toward elements and characteristics of vaccines, which have the potential to manifest themselves though awide range of negative attitudes ranging from being completely against vaccines to expressing hesitancy about them (Gust et al., 2005). These beliefs are often driven by distrust of government entities, pharmaceutical industry (Larson et al., 2014), medical professionals, lack of perceived need, and doubts about vaccine safety and potential side effects (Chen & DeStefano, 1998; Chen & Hibbs, 1998; Smailbegovic et al., 2003). While questioning the authenticity and safety of health decisions is a normal part of health literacy and the health decision-making processes, the challenge is social media users often share sensational and novel information that is often inaccurate and worse harmful.

This is particularly important because healthcare providers remain trusted sources for information on vaccines for many parents (Kennedy et al., 2011), while numerous parents turn to social media for information instead. Whether the information source acts as a detriment or asset to high-quality health decisions is completely contingent upon the validity of the information posted and the person receiving and processing the information. This is where media literacy and health literacy are needed to filter through the massive amount of information on social media.

Linking health literacy and media literacy

The Health Resources and Services Administration (HRSA) defines health literacy as the degree to which individuals have the capacity to obtain process and understand basic health information needed to make appropriate health decisions (HRSA, 2019) or to promote and maintain good health (Sørensen et al., 2012) for themselves and others (Office of Disease Prevention and Health Promotion, n.d.). Because there is a direct correlation between health literacy and positive health outcomes, it is essential that public health efforts to continue improving health literacy of target populations. As technology has improved, people have turned away from reading as the primary source of information.

Studies show that regardless of their health literacy skills, all adults were more likely to receive health information from radio, TV, friends, and family instead of printed written materials (U.S. Department of Health and Human Services DHHS, 2008). People also prefer receiving health information in a manner that offers interactive elements and resources for relatives while also providing an opportunity for online contact with others (Wollmann et al., 2021). Studies also reveal that picture-based instructions promote a better understanding of how to take medication and decrease medication errors among patients (DHHS, 2010). This is in line with research on the efficacy of memes and other simple-to-understand images in communicating succinct health messages to vulnerable populations (Headley et al., 2022).

Deciphering facts from inaccuracies online require more than obtaining and processing health information. In order to increase health literacy via social media, people need media literacy; the ability to effectively gain access to, analyze, evaluate, and generate messages across a variety of contexts (Livingstone, 2004). Media literacy is a critical key to increasing health literacy because it gives people the confidence and ability to identify credible sources of information posted on social media. Studies found that those who had media literacy were positively associated with fact-checking behavior for health information (Lee & Ramazan, 2021).

Someone who have media literacy will not take information presented at face value, they will double check it before sharing it. If we want technologically savvy and health literate populations, it is important that we start to teach media literacy in schools. Bergsman & Carney (2008) believe that when done correctly, users will be able to critically examine media messages that can influence a receiver's perceptions and practices and even help drive change on a social and policy scale. This is still true today.

Engaging policy makers through social media

It has been well established that policy change is one of the most effective ways of changing behavior and improving the health of the general population, especially when fueled by evidence-based research (Purtle et al., 2015). Major examples include cessation of indoor smoking, T-21, and seatbelt legislation. In order for research to contribute to and fuel policy change, it is critical that research findings be carefully and effectively disseminated to politicians. As such, scientists have sought to understand how research may influence policy (Dodson et al., 2015; Purtle et al., 2018), especially because of the increased availability of funding for policy-specific research (Purtle et al., 2015).

In order to build healthy and durable working relationships with policymakers, while maximizing the relevance of research products for policy work, researchers must first understand the policy process (Ashcraft et al., 2020) and make efforts to adhere to timelines and consider succinctness and practicality in their recommendations. The steps needed to enact policy are impacted by personal narratives, public opinion, special interest groups, expressed needs of constituents, and the news media. All of these interest parties have access to social media and therefore are all influenced by the narratives painted by their social media sphere of influence (Ashcraft et al., 2020).

Social media can also play a role in facilitating the dissemination of relevant reports electronically, thus creating the opportunity for more succinct information to be delivered in less time it takes to schedule a meeting. This opportunity offers authors the opportunity to creatively deliver information via video, visual presentation, or social media post tagging the policy maker him/herself, and motor all of the above from their smartphones.

Disseminating research on social media and the rise of enterprise social media

Since 2005, smartphone usage and social media usage among American adults have increased from 5% to about 72% in 2021 (Pew Research, 2021). As such, people have grown accustomed to the convenience of information, which has contributed to social media being increasingly implemented in work organizations as tools for communication among employees. Over time, new dynamic social media tools have started to propagate organizations (Leonardi et al., 2013).

Several modern knowledge management systems like Evernote, Slack, Canva, and Lessonly allow people to post files (documents, images, videos, etc.) that others can search and read at their leisure. As new technology is developed and introduced, people want the ability to do these things at the same time and in the same place as they do on social media. As such, Leonardi et al. (2013) define enterprise social media as web-based platforms that allow employees to do four specific tasks:

(1) Communicate with individual colleagues or broadcast messages to an entire organization;
(2) Indicate or reveal particular colleagues as communication partners;
(3) Post, edit, and sort text and files linked to themselves or others; and
(4) View messages, connections, text, and files communicated, posted, edited, and sorted by anyone else in the organization at any time of their choosing.

Enterprise social media is designed to also increase productivity and facilitate collaboration. Studies found that if employees appropriately use two routes to disseminate the knowledge compared with only using one of the two routes, increases dissemination speed and widens its scope (Zhu et al., 2022).

Implications for public health practice

A particular issue that public health may face is making health/healthy look appealing to all audiences—this is a challenge for the technologically rich social media era we currently live in and should be seriously addressed as social media has an unimaginable amount of sway over our population. Due to the proliferation and rapid sharing of information on social media, public health entities need to create credible social media channels and share reliable information if accurate health information will be shared in an effort to improve health outcomes. Since the majority of adults rely on social media platforms for daily updates, it is imperative that dedicated public health social media experts are posting relevant information at least daily.

Due to the widespread use of social media, health professionals should be assigned to monitor social media platforms in specific regions for region-specific antihealth beliefs that are spreading and then counteract them with factual visual bits of information. Public health entities should also reach younger populations who are more malleable to current trends on social media. In some instances, current trends can be extremely dangerous, while others are healthy. It is the role of public health to use social media as a tool to respond, with valid and reliable information that ensures public safety and well-being.

Since most research studies are written for experts within a particular field, the readability rating is high, and most of the general population is unable to understand what the article is saying. If the studies are controversial, news outlets often summarize and interpret research studies for readers. What often happens is the summary is not as accurate or as focused on the conclusion of the original authors. We saw that during the height of the COVID-19 pandemic when CDC guidelines were confusing to the general public, several news outlets had differing interpretations of the CDC guidelines, which results in more chaos and worst, people not being able to make informed health decisions due to incomplete or inaccurate information, which then contributed to an increased number of infections due to incorrect interpretation of guidelines.

One solution to this proliferation of modified summary of high literacy level documents is for PH entities to release a "layperson-friendly" version of recommendations of publications. This "layperson-friendly" version could be similar to an abstract, but written at a fourth or fifth grade literacy level with easy-to-understand graphics and a short accompanying video. These materials can be easily shared on social media, and thus contribute to the proliferation of accurate health education materials.

Finally, in public health and health education, there are several courses required so that once completed, the student will become proficient at the particular subject. Requiring an "information access equity" course or workshop would greatly benefit future researchers in translating vital research results and findings to those who the research is often meant to benefit. This would not only increase the level of practical skills of future researchers but also increase collaborations among educational institutions and community organizations.

Conclusions

As researches, it is time to determine which social media platform the target audience is using the most, and use that for dissemination venues and a component of interventions. Areas such as using video games to help find and lead youth to mental health or crisis management services are yet to be studied. In an effort to make research access and dissemination more equitable, researchers and educators should be trained to distribute study and research results to general populations. The virality of social media makes it perfect for timely research dissemination. Creating community-based partnerships with an emphasis on social media, can help target populations become consumers of researcher.

Chapter summary

Research dissemination is an important step in the scientific process. Many researchers aim to publish research findings into academic journals. In these journals, other researchers have access to current trends and ideas. This however poses a problem for the general public who do not have access to these academic journals or who would face significant challenges in reading material in academic language. Research dissemination is often a process that involves the publishing of research for other researchers to read and utilize when conducting their own research. The impact of research within academia is typically measured by the number of publications, impact factors of the journals, and the number of citations an article receives. This process can be amended to include the dissemination of research to general audiences, especially within the target communities. Novel trends in research dissemination and publication metrics point to social media as a gatekeeper. Many people now use social media as a source of health information. It is time to start disseminating research on social media as this has huge implications for public health practice.

Definition of key terms

Knowledge black holes: A situation where people who want to make informed decisions are not given the opportunity to access the information they need to make informed decisions.

Media literacy: The ability to effectively gain access to, analyze, evaluate, and generate messages across a variety of contexts.

Review or discussion questions

1. How can research metric websites work to make their platforms more accessible to the general population?
2. What are ways researchers can directly address knowledge black holes in the general public?
3. In what other ways can researchers make their research more practical?
4. Is there a need for print media in the age of social media? Explain.

Websites to explore

Academia.edu: An open-sourced research repository
https://www.academia.edu/
Activity for readers: Visit this open-sourced library with millions of academic research papers to answer any question you may have on a specific topic.

ResearchGate: A website to find and connect to researchers
https://www.researchgate.net
Activity for readers: Explore this social networking website to share your research and collaborate with other researchers. *How can ResearchGate help to bridge the gap in research dissemination? What are some ways ResearchGate supports novice researchers?*

YouTube: A popular online video-sharing website
https://www.youtube.com/
Activity for readers: Explore this social media platform to view video content. *How is YouTube different than other social media platforms? What are some ways that health organizations are utilizing this platform?*

References

Altmetric. (2022). What does Altmetric do?. https://www.altmetric.com/. (Accessed 2 August 2022).

Ashcraft, L. E., Quinn, D. A., & Brownson, R. C. (2020). Strategies for effective dissemination of research to United States policymakers: A systematic review. *Implementation Science, 15*(1), 1−17.

Bergsma, L. J., & Carney, M. E. (2008). Effectiveness of health-promoting media literacy education: A systematic review. *Health Education Research, 23*(3), 522−542.

Betsch, C., Brewer, N. T., Brocard, P., Davies, P., Gassmaier, W., Haase, N., Leask, J., Renkewitz, F., Renner, B., Reyna, V. F., Rossman, C., Sachse, K., Schachinger, A., Siegrist, M., & Stryk, M. (2012). Opportunities and challenges of Web 2.0 for vaccination decisions. *Vaccine, 30*(25), 3727−3733.

Betsch, C., Renkewitz, F., Betsch, T., & Ulshofer, C. (2010). The influence of vaccine-critical websites on perceiving vaccination risks. *Journal of Health Psychology, 15*(3), 446−455.

Brewer, N., Chapman, G., Gibbons, F., Gerrard, M., McCaul, K., & Weinstein, N. (2007). Meta-analysis of the relationship between risk perception and health behavior: The example of vaccination. *Health Psychology, 26*(2), 136−145.

Chen, R. T., & DeStefano, F. (1998). Vaccine adverse events: Causal or coincidental? *Lancet, 351*, 611−612.

Chen, R. T., & Hibbs, B. (1998). Vaccine safety: Current and future challenges. *Pediatric Annals, 27*(7), 445−455.

Community Preventative Services Task Force (CPSTF). (2010). *Health communication and social media marketing: Campaigns that include mass media and health-related product distribution.* https://www.thecommunityguide.org/media/pdf/Health-Communication-Mass-Media.pdf.

Davies, S. R., & Horst, M. (2016). *Science communication: Culture, identity and citizenship.* London, UK: Palgrave MacMillan.

Deeken, A. H., Mukhopadhyay, S., & Jiang, X. (2020). Social media in academics and research: 21st-century tools to turbocharge education, collaboration, and dissemination of research findings. *Histopathology, 77*(5), 688−699.

Dodson, E. A., Geary, N. A., & Brownson, R. C. (2015). State legislators' sources and use of information: Bridging the gap between research and policy. *Health Education Research, 30*(6), 840−848.

Goldstein, S., MacDonald, N. E., & Guirguis, S. (2015). Health communication and vaccine hesitancy. *Vaccine, 33*(34), 4212−4214.

Gust, D. A., Brown, C. J., Sheedy, K., Hibbs, B., Weaver, D., & Nowak, G. (2005). Immunization attitudes and beliefs among parents: Beyond a dichotomous perspective. *American Journal of Health Behavior, 29*(1), 81−92.

Gust, D. A., Strine, T. W., Maurice, E., Smith, P., Yusuf, H., Wilkinson, M., Battaglia, M., Wright, R., & Schwartz, B. (2004). Underimmunization among children: Effects of vaccine safety concerns on immunization status. *Pediatrics, 114*(1), e16−e22.

Headley, S. A., Jones, T., Kanekar, A., & Vogelzang, J. (2022). Using memes to increase health literacy in vulnerable populations. *American Journal of Health Education, 53*(1), 11−15.

Health Resources and Services Administration (HRSA). (2019). *Health literacy.* https://www.hrsa.gov/about/organization/bureaus/ohe/health-literacy/index.html.

Hilgartner, S. (1990). The dominant view of popularization: Conceptual problems, political uses. *Social Studies of Science, 20*, 519−539.

Jang, S. M., Mckeever, B. W., Mckeever, R., & Kim, J. K. (2019). From social media to mainstream news: The information flow of the vaccine-autism controversy in the US, Canada, and the UK. *Health Communication, 34*(1), 110−117. https://doi.org/10.1080/10410236.2017.1384433

Kennedy, A., Basket, M., & Sheedy, K. (2011). Vaccine attitudes, concerns, and information sources reported by parents of young children: Results from the 2009 HealthStyles survey. *Pediatrics, 127*(Suppl. 1), S92−S99.

Larson, H. J., Jarrett, C., Eckersberger, E., Smith, D., & Paterson, P. (2014). Understanding vaccine hesitancy around vaccines and vaccination from a global perspective: A systematic review of published literature. *Vaccine, 32*(19), 2150−2159.

Lee, D. K. L., & Ramazan, O. (2021). Fact-checking of health information: The effect of media literacy, metacognition and health information exposure. *Journal of Health Communication, 26*(7), 491−500. https://doi.org/10.1080/10810730.2021.1955312

Leonardi, P. M., Huysman, M., & Steinfield, C. (2013). Enterprise social media: Definition, history, and prospects for the study of social technologies in organizations. *Journal of Computer-Mediated Communication, 19*(1), 1−19. https://doi.org/10.1111/jcc4.12029

Livingstone, S. (2004). What is media literacy? *Intermedia, 32*(3), 18−20.

Office of Disease Prevention and Health Promotion. (n.d.). Health Literacy. Healthy People 2030. U.S. Department of Health and Human Services. https://health.gov/our-work/national-health-initiatives/healthy-people/healthy-people-2030/health-literacy-healthy-people-2030.

Millar, B. C., & Lim, M. (2022). The role of visual abstracts in the dissemination of medical research. *Ulster Medical Journal, 91*(2), 67−78.

Pew Research Center. (2021). *Demographics of social media users and adoption in the United States. Social Media Fact Sheet*. https://www.pewresearch.org/internet/fact-sheet/social-media/#:~:text=When%20Pew%20Research%20Center%20began%20tracking%20social%20media,-some%20type%20of%20social%20media.%20CHART%20TABLE%20SHARE.

Plum Analytics. (2022). *About PlumX metrics*. https://plumanalytics.com/learn/about-metrics/.

Purtle, J., Dodson, E. A., Nelson, K., Meisel, Z. F., & Brownson, R. C. (2018). Legislators' sources of behavioral health research and preferences for dissemination: Variations by political party. *Psychiatric Services, 69*(10), 1105−1108.

Purtle, J., Peters, R., & Brownson, R. C. (2015). A review of policy dissemination and implementation research funded by the National Institutes of Health, 2007−2014. *Implementation Science, 11*(1), 1−8.

Rescue Agency. (2022). *About us*. https://rescueagency.com/about.

Schafer, M. S. (2012). Online communication on climate change and climate politics: A literature review. *Wiley Interdisciplinary Reviews: Climate Change, 3*, 527−543.

Schillinger, D., Chittamuru, D., & Ramírez, A. S. (2020). From "infodemics" to health promotion: A novel framework for the role of social media in public health. *American Journal of Public Health, 110*(9), 1393−1396.

Smailbegovic, M. S., Laing, G. J., & Bedford, H. (2003). Why do parents decide against immunization? The effect of health beliefs and health professionals. *Child: Care, Health and Development, 29*(4), 303−311.

Sørensen, K., Van den Broucke, S., Fullam, J., Doyle, G., Pelikan, J., Slonska, Z., & Brand, H. (2012). Health literacy and public health: A systematic review and integration of definitions and models. *BMC Public Health, 12*(1), 1−13.

Tambuscio, M., Ruffo, G., Flammini, A., & Menczer, F. (2015). Factchecking effect on viral hoaxes: A model of misinformation spread in social networks. *Proceedings of the 24th International Conference on World Wide Web*, 977−982.

Tomeny, T. S., Vargo, C. J., & El-Toukhy, S. (2017). Geographic and demographic correlates of autism-related anti-vaccine beliefs on Twitter, 2009-15. *Social Science and Medicine, 191*, 168−175. https://doi.org/10.1016/j.socscimed.2017.08.041

U.S. Department of Health and Human Services (DHHS). (2008). *America's health literacy: Why we need accessible health information.* https://www.ahrq.gov/sites/default/files/wysiwyg/health-literacy/dhhs-2008-issue-brief.pdf.

Wilson, P. M., Petticrew, M., Calnan, M. W., & Nazareth, I. (2010). Disseminating research findings: What should researchers do? A systematic scoping review of conceptual frameworks. *Implementation Science, 5*(1), 1−16.

Wollmann, K., van der Keylen, P., Tomandl, J., Meerpohl, J. J., Sofroniou, M., Maun, A., & Voigt-Radloff, S. (2021). The information needs of internet users and their requirements for online health information—a scoping review of qualitative and quantitative studies. *Patient Education and Counseling, 104*(8), 1904−1932.

Zhu, H., Wang, Y., Yan, X., & Jin, Z. (2022). Research on knowledge dissemination model in the multiple-x network with enterprise social media and offline transmission routes. *Physica A: Statistical Mechanics and Its Applications, 587*, 126468. https://doi.org/10.1016/j.physa.2021.1264

Further reading

Altmetrics. What are Altmetrics? Retrieved from https://www.altmetric.com/about-altmetrics/what-are-altmetrics/.

Jang, S. M., Mckeever, B. W., Mckeever, R., & Kim, J. K. (2019).Despite increasing warnings about inaccurate information online, little is known about how social media contribute to the widespread diffusion of unverified health information.

Longo, A., & Hand, B. N. (2022). Brief report: The impact of social and news media coverage on the dissemination of autism research. *Journal of Autism and Developmental Disorders.* https://doi.org/10.1007/s10803-022-05464-8

U.S. Department of Health and Human Services, Office of Disease Prevention and Health Promotion. (2010). *National action plan to improve health literacy.* Washington, DC: Author. National Action Plan to Improve Health Literacy | health.gov.

Section IV

Social media, public health communication, and pedagogy

Chapter 11

Social media and policy campaigns

Gayle Walter
Department of Health and Human Physiology, University of Iowa, Iowa City, IA, United States

Objectives:

At the end of the chapter, the reader will be able to:

- Describe the role of social media in societal change.
- Analyze social media use by policymakers.
- Explain the challenges and threats in the use of social media in policy development.
- Evaluate best practices of social media in policy campaigns.
- Examine the effect of media interventions in shaping healthcare policies.

Introduction

As mentioned throughout the textbook, communication in many forms has shifted away from traditional telephone calls and written letters to social media channels such as Twitter, Facebook, and Instagram. Social media can reach a large audience across the nation, so it makes sense to use this platform for advocacy efforts and policy change. For the purpose of this chapter, media interventions are defined as "organized and purposive activities that utilize a variety of medial channels to inform, persuade, or motivate populations" (Bou-Karroum et al., 2017, p. 2). Media interventions relay health-related information to the public, policymakers, and health professionals. Many media outlets also post recent research findings and newly released statistics related to many health issues. As a result, individuals may change their health behaviors or align with political candidates who share the same health and healthcare values. In addition, there are numerous examples where policy shifts have had important constructive effects on how healthcare services are delivered and public programs are offered (Walley & Wright, 2010).

Overview of social media in advocacy and policy change

For policymaking, the media can promote agenda-setting for the press by underscoring salient issues that are newsworthy at a given time. That was evident at the beginning of the COVID-19 pandemic when most news sources were reporting on the outbreak and the number of individuals impacted. According to the Kaiser Family Foundation (KFF) (2021), approximately 78% of US adults either considered it as true or were not sure about at least one among eight false statements stated about COVID-19 including its vaccines. The Republicans and unvaccinated adults were most likely to hold such misperceptions (KFF, 2021). These eight false statements are listed in Fig. 11.1. The results underscore a crucial challenge regarding correct communication about the pandemic in the face of misinformation and disinformation, whether unintentionally or intentionally, being rapidly spread through social media (KFF, 2021).

Social media also provides an unparalleled chance to recognize values and prospects about health and health policy (Yeung, 2018). When users post about healthy eating, physical activity, mental health, and stress management, those analytics can be tracked. Social media check-ins at public places such as parks, restaurants, and fitness centers can reveal how often people visit healthy places. Trends can be evaluated over time to see an increase or a decrease in discussions or check-ins. This is a form of infoveillance which is described in more detail later in the chapter. Those trends and behaviors offer insight into how social media can reveal what people consider valuable, or in other words, the topics they consider worth their participation (Yeung, 2018). As a result, policymakers can then use this acumen to advise either concerted interventions or lengthier projects.

Social media use by members of Congress

According to Charalambous (2019), the use of social media in the political area has increased dramatically over the past several years. A recent report by the Pew Research Center (2020) found that Democrat policymakers posted more content on Twitter as opposed to their Republican counterparts. Twitter and Facebook have now become ingrained in both popular and political culture. A Pew Research Center (2020) report, that included an analysis of every

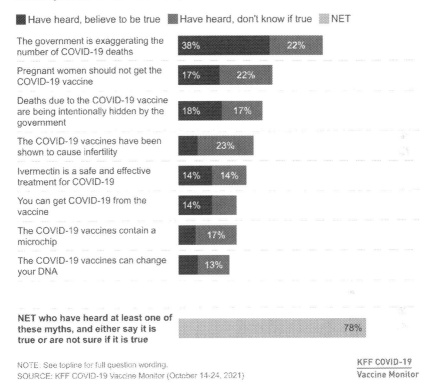

FIGURE 11.1 Common falsehoods about COVID-19 or the vaccine.

Facebook post and Tweet from Congressional members since 2015, discovered drastic changes in social media use. On Facebook, the average Congress produced 48% more posts between 2015 and 2016 and increased their total number of followers by twice (Pew Research Center, 2020). In a comparison between 2015 and 2016, the average Congress member tweeted 81% more and had approximately nearly thrice as many followers, and received greater than six times retweets on their average posts (Pew Research Center, 2020) (Please see Fig. 11.2).

In the present-day context, Democrats are inclined to have more followers and post more regularly on Twitter. When compared to a usual Republican

236 SECTION | IV Social media, public health communication, and pedagogy

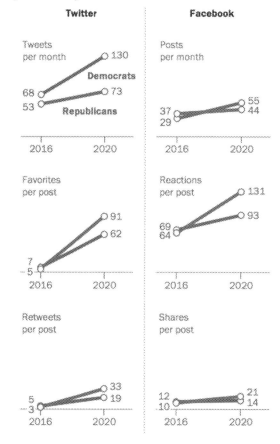

FIGURE 11.2 Members of Congress average use of Facebook and Twitter.

Congress member, the usual Democrat has more than 17,000 followers on Twitter. Accordingly, Democrats have close to twice as many tweets as compared to Republicans in a given month (Pew Research Center, 2020).

These variations tend to reveal variations in the demographic compositions of these two platforms. For example, 62% of US adult Twitter users identify themselves as Democrats or independents, while 50% of US adults use Facebook identify as Democrats (Pew Research Center, 2020).

Another interesting finding is that a small group of policymakers with relatively large followings dominate the social media among Congress members. For example, Nancy Pelosi had a large group of followers on both Facebook and Twitter. As another example in the 116th Congress, approximately 10% of Congressional members with the most followers on social media obtained in excess of 75% of all likings, other reactions, retweets, and shares on these platforms (Pew Research Center, 2020) (Please see Fig. 11.3).

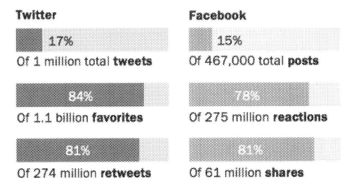

FIGURE 11.3 Audience engagement through Facebook and Twitter.

It is worth noting that Congress generates a colossal magnitude of social media matter each month. Even though the use of Twitter and Facebook serves as the means to foster communication, governmental openness, and engagement with the public, the public is bombarded by conflicting messages.

Benefits of using social media in policy development

A positive to this abundant use of social media posts is the increased ability to accumulate and broadcast real-time communication from politicians to voters that could be significant for policymaking or influencing the voting outcomes (Charalambous, 2019). As a result of social media, not only are constituents more informed but social media can also influence policymakers (Charalambous, 2019). Public opinion on a specific topic, for example, gun control, will force lawmakers to act. The number of recent mass shootings in the past month in the United States has caught the attention of both the American public and policymakers. Healthcare professionals, grassroots organizations, and nonprofit organizations are able to inform public policy through the use of social media.

Douglas Yeung from the RAND Corporation provided a viewpoint using social media as a catalyzing agent for policy activity and social transformation (Yeung, 2018). In addition to the information being disseminated, analysis techniques are useful ways to offer a vision for decision-making and policymaking. Social media is extensively used, but it is also especially used by specific sections of the population that work on issues pertaining to health equity. For example, racial and ethnic minorities tend to be more likely compared to other groups to gain access to the Internet and social media mostly through their cell phones. Hence, such actions are likely to be tracked to comprehend health and well-being behaviors. Since social media postings are often unplanned and repeated, the substance of these postings are opportune and can provide synchronous or real-time information. If the user has their location turned on in their mobile device, location information is also known. This proves crucial since sound policy choices often are contingent on obtaining an accurate and recent image of where citizens are situated. One thing to remember about social media and the abundance of content that is shared, care needs to be shown when describing inferences from what is told and shared (Yeung, 2018).

As mentioned earlier, the political backdrop has changed dramatically over the past 20 years due to the increasing use of the Internet. Government officials are more engaged in social media through different websites, social networking sites, and blogs. The work of Giacomini and Simonetto examined the influence of social media in the policy cycle in local governments. One research question looked at the extent of social media influence on the policy cycle, and the second question explored whether that influence changed depending on the certain characteristics of the mayors in local governments.

Mayors were asked about three putative effects of social media on the policy cycle changes namely (a) agenda setting, (b) policy formulation along with adoption, and (c) implementation along with evaluation (Giacomini & Simonetto, 2020). After analyzing the data, the authors found that the current level of the apparent influence of social media was relatively low which was surprising based on the increasing importance of social media. The mayors also saw a constrained role for the ideas conveyed by citizens on social networks in the setting of the agenda. The continued popularity of use of social media may lead to more stakeholder engagement (Giacomini & Simonetto, 2020).

Examples of use

Epidemiology and surveillance

There are several examples in health policy areas to indicate where social media may suggest insights or propose specific policy connotations. One example involves the field of infoveillance and infodemiology exploring the use of web-based data and social media for public health to assist in predicting disease epidemics (Yeung, 2018). The benefit of infodemiology is that information is collected in "real-time" rather than waiting for reports of confirmed cases. Analytics can monitor status updates mentioning COVID-19 and influenza (Eysenbach, 2009). This was demonstrated using Google Flu Trends to predict early outbreaks of influenza. This service was introduced in 2008 to monitor trends in the number of online search requests related to influenza-like symptoms (Pervaiz et al., 2012). Results of this monitoring have revealed that there is a coherent relationship between actual number of influenza cases data compiled by the Centers for Disease Control and Prevention (CDC) and this service. However, it needs to be clarified that Google Flu Trends is not an initial epidemic recognition system for influenza. Instead, it is created as a reference indicator of the trends in the number of influenza cases (Pervaiz et al., 2012). Another example looks at repetitive words to garner public opinion on health issues such as smoking and gun violence. Debates continue over the true prognostic power and effectiveness of using search data for health and disease surveillance.

Health promotion messaging

The following section will provide examples of social media use in health promotion and public health campaign messaging. Herrera-Peco and colleagues examined the part that health professionals have in applying Twitter to augment the campaign of a public institution on promoting COVID-19 vaccination (Herrera-Peco et al., 2021). When the COVID-19 outbreak began, people were searching for information and answers to the many questions people had about transmission, susceptibility, prevention, and treatment. One of the swiftest ways that health-associated information can be found is on the Internet, especially social media. The authors reported that Twitter had about 271 million users who accounted for over half a billion tweets every day (Herrera-Peco et al., 2021). Based on that information, Twitter would be the most effective social media channel for disseminating information during pandemics like the COVID-19 pandemic. At the beginning of the COVID-19 pandemic, there was an abundance of information on social media sites, especially inaccurate information, which makes it essential for healthcare professionals to take an active role in providing accurate, scientific, and reliable information. The specific purpose of the study by Herrera-Peco et al. was to evaluate the role of healthcare professionals during the initiation of the campaign begun on Twitter by the Ministry of Spain, in support of vaccination against COVID-19. The data from the tweets were mined through an application programming interface (API) search tool. The keyword that was selected was "yomevacuno" and the hashtag selected was #yomevacuno. This was the very hashtag employed by Spanish health authorities to initiate a support campaign for debunking anti-COVID-19 vaccines and promoting the COVID-19 vaccination to stop the spread (Herrera-Peco et al., 2021).

The results within the #yomevacuno network found 3038 users, of which 346 self-reported as healthcare professionals in their descriptions (Herrera-Peco et al., 2021). Those hashtags generated an amount of 58,177 impressions with users identified as pharmacists having the highest number of those impressions. The keyword "yomevacuno" generated 397 messages. One thing to mention is that participation by healthcare professionals in social media campaigns is voluntary, so there were different levels of engagement and participation among them. Their role tends to be more personal in nature rather than professional nature. The results demonstrated that individual users, healthcare professionals, and noncare healthcare professionals had a lower weight than institutions within all the traffic in the analyzed network. There may be specific reasons for the absence of posts regarding the importance of vaccinations, but healthcare professionals are considered an essential element by the population who are expecting to see that information. The lack of posts about health-related issues may lead to skepticism and distrust toward useful health information. The authors concluded that it is of supreme importance that healthcare professionals comprehend the need to be present on social media platforms from a professional perspective, rather than a personal point of view, so they can become central elements in the propagation and establishment of trustworthy information with scientific merit (Herrera-Peco et al., 2021).

As mentioned above, social media platforms can strengthen messages by healthcare professionals. In 2017, the Center for Nutrition and Health Research (CNHR) initiated a methodical social media communication campaign focused on the broadcasting of science-based and dietetics-based policies (Mendoza-Herrera et al., 2020). The intent of those messages was to educate the public on policies related to obesity, taxes, laws for food advertising, and information on food package labels. That information was not likely to reach the public through other channels. A group composed of nutritionists, graduate scholars, graphic designers, and a community-based manager composed a monthly plan to disseminate information in six or most posts per week on Facebook. All of the information provided was evidence-based visual content including an infographic, key messages, recommendations, and additional resources to get more information. The number of Facebook fans has consistently grown since this campaign began in addition to engagements and shares. This social media experience underscores the possibility of social media to be utilized for health promotion, even when the resources are limited. The Center did not use any type of paid advertising for this campaign (Mendoza-Herrera et al., 2020).

Social media offers the capacity for people not only pursue information about health and disease-related issues but to also engage with others. For example, a person posting about their struggle with substance abuse may connect with a follower that also has an issue with substance abuse. These are where communities start to build to address a disease or condition that is common among many. Public health professionals can also use social media to disseminate tailored messages (Meghan Mahoney & Tang, 2016). Users with similar conditions are led to a site where there is information on the illness, risks, advantages of treatment, and solutions to apparent barriers. Mahoney et al. also suggest that incentives are provided for users to act.

Many messages posted on social media may also convey misleading information. This was evident in a study by McCausland et al. examining the messages related to e-cigarettes (McCausland et al., 2019). They looked at various data sources and examined content captured from vaping-related social media discussions or promotions. E-cigarettes have been mistakenly perceived by the general public as a safer option to tobacco and a viable option as a smoking cessation aid. Retailers and manufacturers recognize this and capitalize on those perceptions to market e-cigarettes on social media. The authors found that there was limited content on the dangers of e-cigarettes on social media from public health and the government sector to counteract those mixed messages. This is another instance where those messages discussing e-cigarettes and vaping should be monitored more closely. Government agencies, including public health departments, could also bridge the gap in mixed messaging by providing accurate information in their own social media sites, and disseminating research, and educational programs (McCausland et al., 2019).

The information presented thus far has shown the influence of social media on health promotion messages and behavior change, but limited information is available on how self-presentation on social media (the one posting) influences an individual's (the reader's) subjective well-being. A study by Xiaojun and colleagues examined whether there was a meaningful influence on subjective well-being and if there was, how was it affected. Self-presentation was defined as the expression of thoughts, feelings, and life experiences. Subjective well-being is an evaluation of one's life according to self-determined criteria. The criteria may include the person's emotional responses, domain satisfaction, and global judgments of life satisfaction (Fan et al., 2019). The mobile communication tool, WeChat, was used in data collection. Participants included undergraduates from a university in Shanghai, China, and were divided into two groups. One group submitted posts focusing on positive experiences such as travel, time with friends and family, scholarships, awards, etc. The other group, the nonself-presentation group strictly limited their posts to news, knowledge, and other items not related to their selves. Participants were then asked a series of questions after reading the various posts. An interesting finding was that others' self-demonstration on social media increased individuals' feelings of relative deficiencies and decreased their subjective well-being. Many of the respondents felt they could not compete or compare with self-presenters, even though the information presented may not be entirely true (Fan et al., 2019). From a public health perspective, health messages that are posted on social media with good intentions may further alienate the very populations they are hoping to reach.

Public health policy

A study by Hatchard et al. (2019) also used content on Twitter to evaluate public opinion on the UK Government's announcement on standardized tobacco packaging. In the United Kingdom, standardized packaging is an argumentative policy attracting public health, political, and media attention. To clarify the term, standardized packaging requires the "mandatory removal of brand images, colors, and messages from tobacco product packs" (Hatchard et al., 2019, p. 2). All tobacco packs are mandated to be the same size, color, style, shape, print, and font size. The purpose of this mandate, which includes the color being a very drab color, is to deter youth from smoking. Youth are attracted by colorful packaging with imaging specific to them. Standardized packaging is in place in Australia and evidence shows that reduced pack exhibit and appeal boosts quit attempts and the effectiveness of health warnings. In addition, it helps to correct the misperceptions of harm and does not increase illicit tobacco purchases (Hatchard et al., 2019). Hatchard and colleagues (2019) collected a random sample of 1038 tweets over two 4-week periods to determine whether the content of those tweets supported, or did not support, the proposed policy.

For the purpose of the study, they explored global Twitter communication relating to a public health policy change by examining the case study of standardized policy related to tobacco packaging in the United Kingdom. During the first 4-week period, the United Kingdom completed its discussion on the guidelines and proposed them to the European Union (EU) for their approval. The second 4-week period began when the UK Government announced that there would be a vote in the Parliament on standardized packaging before the general elections in May 2015. Based on content analysis, support for the policy changed from one period to the next. Before the announcement of a Parliamentary vote, two-thirds of the tweets expressed a positive sentiment toward standardized packaging. After the announcement, negative tweets increased while positive tweets decreased with only one-third in favor. The authors found that after the initial time frame, more research and evidence were disseminated due to the nature of the policy, which may have influenced the public. They also posed that those against the standardized packaging were utilizing Twitter in a more effective manner compared to their health-related counterparts.

Three main themes emerged from the study. The first finding demonstrated that Twitter messages validly reflect the political landscape which has been proven in similar research studies. The second finding alluded to the need for public health advocates to be actively involved in Twitter to promote peer-reviewed evidence on public policy options, share URLs of academic peer-reviewed journal articles, and translate that information into suitable formats for the public. Lastly, the paper's search terms centered on the policy rather than a brand, product, or specific company. As such, the paper provided limited evidence that adversaries to standardized packaging were using computerized accounts or bots to distort perceptions of public opinion and were more likely to manipulate this policy issue to market tobacco-related products (Hatchard et al., 2019).

Public opinion

With the abundance of users engaging in social media, how do we know for certain if social media platforms are an effective and efficient tool for raising awareness and creating sustainable public health movements. Survey data were collected between July 13 and 19, 2020, and gathered information on five goals that included (1) raising public awareness about social and political issues; (2) creating lasting social actions; (3) getting elected representatives to pay consideration to issues; (4) influencing policy outcomes; and (5) changing people's opinions about political and social issues. Findings from a survey by the Pew Research Study in July 2020 revealed regarding social media platforms that approximately 80% of Americans opined these were very (31%) or somewhat (49%) effective for boosting public awareness about social or political issues. Similarly, approximately 77% believed that these social media platforms were at least fairly effective for creating sustained social changes (Auxier & McClain, 2020). Those percentages were further broken down by political party affiliation. Across both part politicians, more of them described these social media platforms as being effective as opposed to being ineffective as related to the achievement of goals (Please see Fig. 11.4).

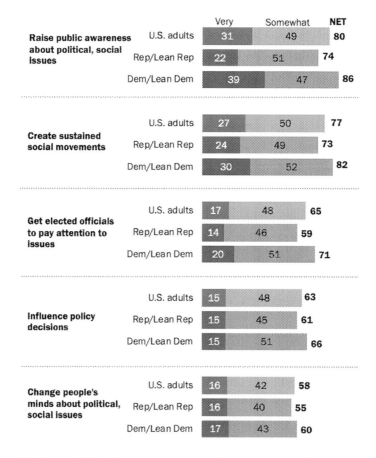

FIGURE 11.4 Effectiveness of social media platforms.

Younger Americans were found to be more likely than older adults to use social media platforms such as TikTok, Twitter, Instagram, YouTube, Snapchat, or Facebook. Kent and Taylor (2021) noted that millennials are more likely to use image-based social media, such as Instagram or Snapchat, on an everyday basis rather than Facebook or Twitter. There are also age-connected variations in interpretations of these sites' efficacy for political action and party variations continue more so among younger adults for many of the goals. As an example, "87% of Democrats ages 18 to 29 say social media sites are at least somewhat effective for raising awareness, compared with 76% of Republicans in the same age group. Democrats ages 18 to 29 are also more likely than their Republican counterparts to say these sites are at least somewhat effective at creating sustained social movements (84% vs. 74%) and getting elected officials to pay attention to issues (72% vs. 60%)" (Auxier & McClain, 2020). A bigger share of the public believe that users are simply engaging in "slacktivism" to discuss political issues easily and quickly without knowing its true meaning or purpose.

Social media influence in crises

The content of the chapter thus far included data on social media use and examples of social media in policy development and societal change. Another important issue to consider is how times of crises influence social media messages. Zhao, Zhan, and Liu (2018) proposed a new paradigm that theorizes how distinct aspects of social media influence publics' communicative actions during crises. Their study examined different dimensions of social media influence in crises by conducting a multigroup confirmatory factor analysis of Twitter data. The Twitter data stemmed from organization accounts and government accounts. The organizations vie for prominence during crises with the traditional media outlets and the leading social media influencers and users. After a crisis, the abundance of clutter is often hard to penetrate through—especially separating fact from fiction. Practitioners in public relations, or perhaps a public health information officer, must be able to evaluate whether or not their messages are getting through to users in times of crisis, meeting their organizational goals, getting through the online chaos, and vying with other media on social messages. In public health, getting through the clutter to transmit information on a current public health emergency is critical to reducing the misinformation that is getting through instead. An interesting finding from the study showed that "mentioning" was a firmer sign of active engagement compared to "replying" (Zhao et al., 2018).

The dialogic theory and engagement framework

The article by Kent and Taylor (2021) described how dialogic theory and engagement offer conceptual paradigms for ruminating about ethical interactions among societies, especially relative to social media. When conversations occur between two parties on social media, certain features of dialogic theory need to be present before a meaningful exchange of information on the issue can commence. People are also not prepared for discussions on societal issues that are very sensitive and likely to cause disagreements. The authors continue to explain that using social media to promote social change is challenging for at least five reasons. The first reason is that the preferences of the users vary by demographics such as gender, age, education, and platform used. The second reason is that the metrics collected tend to focus on what was written rather than how that information was perceived by others. The third reason is that the technical designs of several social media platforms limit the capacity to engage in effective and efficient discussions. The fourth reason is that the users have a wide range of experiences in using social media and thus all may not be able to communicate effectively. Finally, some people tend to be more of the "listeners" of what is posted rather than posting their own thoughts and views on societal issues (Kent & Taylor, 2021). These reasons are summarized in Table 11.1.

In the communication framework, there are three levels of engagement based on the level of interaction between parties. In Tier 1 engagement, there is a deliberate activity of having initial engagement interactions. Examples of Tier 1 engagement are having interactions on social media or meeting in public with an organized group or committee. In Tier 2 engagement, there is a higher level of behavioral interaction where people begin to form relationships among other people, groups, and organizations. The highest level of engagement is Tier 3 where groups such as activists or nongovernmental organizations (NGOs) coordinate action to create social capital. Examples of social capital is forming coalitions, creating social movements, and joining people together to promote a cause. If there is dialogic engagement, the outcomes of Tiers 2 and 3 are boosted (Taylor & Kent, 2021). Simply having a conversation on social media is unlikely to produce change unless a high level of engagement is demonstrated.

TABLE 11.1 Challenges in using social media for social change.

Number	Reason
1.	Variations in user preferences by demographic characteristics such as gender, age, education, platform, etc.
2.	Metrics collected focus on what is written rather than how the information is perceived
3.	Variations in technical designs
4.	Wide range of user experiences
5.	Some people tend to be more of the "listeners" of what is posted rather than posting their own thoughts

Future directions

The use of social media to inform policy is also a type of advocacy. Advocacy is broadly understood as any type of activity, process, or strategy to speak out on behalf of a group of people, idea, cause, or course of action (Jackson et al., 2021). An example of policy advocacy is disseminating research findings on Facebook to draw the attention of the public. Social media platforms provide avenues to involve stakeholders and affect effectual change. Numerous examples were provided in this chapter but only represent a small fragment of successful health policy campaigns. Jackson and colleagues performed a scoping review of academic literature published between 2011 and 2020. Posts related to COVID-19 bombarded social media when the first case was discovered in the United States on January 20, 2020 (CDC, 2022). The objective was to investigate the key attributes of academic scholarship and research on the public health community's use of social media for the purpose of advocacy in policy and provide insight into directions for future research.

The inclusion criteria only produced 22 studies for the use which implies the public health community's use of social media for advocacy in policy is an evolving topic. Research into the use of social media use for advocacy in policy and policy change presents an opportunity for inquiry, one that could health public health regulate its process. The authors also discovered that in the articles they examine, very few identified challenges such as health issues, health and climate change issues, conflict and crises issues, malaria, and water hygiene. Even though these are global challenges, they received very little attention in the United States. Of the 22 articles selected, the lack of authors from institutions in Africa, Asia, Central America, and South America was noteworthy. Jackson et al. (2021) hypothesized that authors from these geographies discussed these topics in languages other than English. Their research supports calls for a more transdisciplinary approach in social media and health advocacy research, such as the intersection of public health and political science.

Conclusions

Social media as a channel for policy development and societal change holds great potential. Many of these conversations on social media are initiated by federal agencies, politicians, nonprofit organizations, and grassroots organizations. These conversations also begin with a post from a general user and often spark disagreement from friends, family, and followers (Kent & Tyler, 2021). The use of social media for societal change comes with limitations such as age preferences for social media channels, limited discussions based on the maximum number of words allowed, and metrics measuring the spoken word rather than how that word is perceived and interpreted by others (Kent & Tyler, 2021). Even though there are limitations in place, social media provides the opportunity to deepen public discussions and public action around timely and sensitive issues.

Questions

- Do you think that political leaders are investing too much time in the use of social media and potentially isolating themselves from other constituents who may not use social media?
- How can the general public "fact check" the information that is posted on social media?
- Infodemiology and infoveillance are becoming increasingly popular in identifying early outbreaks. How reliable are these tools in accurately forecasting an epidemic?
- What are promising areas of research in social media, policy development, and health advocacy?

Websites to explore

Centers for Disease Control and Prevention—CDC social media tools, guidelines, and best practices

https://www.cdc.gov/socialmedia/tools/guidelines/

Explore this website to learn more about the social media policy that governs the official use of social at the Centers for Disease Control and Prevention for the organization and use by employees. *After you read the social media policy, do you feel that employees' personal use of social media may conflict with CDC information? Explore best practices for the use of social media for personal and professional use.*

The Community Toolbox—using social media for digital advocacy

https://ctb.ku.edu/en/table-of-contents/advocacy/direct-action/electronic-advocacy/main

The Community Toolbox has an abundance of information and step-by-step guidance in community-building skills that are relevant to nonprofit organizations, academia, and the private sector. Section 19 of Chapter 33 examines the reasons for using social media for digital advocacy. *As you are exploring this section, what social media sites grab the users' attention over others? How do you use social media for digital advocacy?*

U.S. Department of Health and Human Services—social media policies

https://www.hhs.gov/web/social-media/policies/index.html

Similar to the CDC, the U.S. Department of Health and Human Services has a social media checklist to make sure that relevant standards and policies related to social media use are followed. One of the policies provides guidance on comment moderation for social media posts. The policy requires that all comments must be reviewed and cleared before they are posted. Comments may not be posted if they contain blatantly partisan political views or commercial endorsements. *Does this policy infringe on Freedom of Speech?*

Definitions of key terms

Infodemiology: Defined as the science of distribution and determinants of both information and misinformation provided on the Internet which guides the general public to recognize quality information to make informed decisions.

Infoveillance: A type of syndromic surveillance that specifically utilizes information found online in order to detect infectious disease outbreaks or gather other information on human behavior.

Media intervention: Organized and purposive activities that utilize a variety of medial channels to inform, persuade, or motivate populations.

Slacktivism: The practice of supporting a political or social cause by means such as social media or online petitions, characterized as involving very little effort or commitment.

References

Auxier, B., & McClain, C. (September 9, 2020). *Americans think social media can help build movements, but can also be a distraction*. Pew Research Center. https://www.pewresearch.org/fact-tank/2020/09/09/americans-think-social-media-can-help-build-movements-but-can-also-be-a-distraction/.

Bou-Karroum, L., El-Jardali, F., Hemadi, N., Faraj, Y., Ojha, U., Shahrour, M., & Akl, E. A. (2017). Using media to impact health policy-making: An integrative systematic review. *Implementation Science, 12*(1), 52. https://doi.org/10.1186/s13012-017-0581-0

Centers for Disease Control and Prevention (CDC). (2022). *CDC Museum COVID-19 timeline*. https://www.cdc.gov/museum/timeline/covid19.html#Early-2020.

Charalambous, A. (2019). Social media and health policy. *Asia-Pacific Journal of Oncology Nursing, 6*(1), 24–27.

Eysenbach, G. (2009). Infodemiology and Infoveillance: Framework for an emerging set of public health informatics methods to analyze search, communication and publication behavior on the Internet. *Journal of Medical Internet Research, 11*(1). https://doi.org/10.2196/jmir.1157

Fan, X., Deng, N., Dong, X., Lin, Y., & Wang, J. (2019). Do others' self-presentation on social media influence individual's subjective well-being? A moderated mediation model. *Telematics and Informatics, 41*, 86–102. https://doi.org/10.1016/j.tele.2019.04.001

Giacomini, D., & Simonetto, A. (2020). How mayors perceive the influence of social media on the policy cycle. *Public Organization Review, 20*(4), 735−752. https://doi.org/10.1007/s11115-020-00466-5

Hatchard, J., Quariguasi Frota Neto, J., Vasilakis, C., & Evans-Reeves, K. (2019). Tweeting about public health policy: Social media response to the UK Government's announcement of a Parliamentary vote on draft standardised packaging regulations. *PloS One, 14*(2), e0211758. https://doi.org/10.1371/journal.pone.0211758

Herrera-Peco, I., Jimenez-Gomez, B., Jose Pena Deudero, J., Benitez De Gracia, E., & Ruiz-Nunez. (2021). Healthcare professionals' role in social media public health campaigns: Analysis of Spanish pro vaccination campaign on Twitter. *Healthcare, 9*(662), 1−12. https://doi.org/10.3390/healthcare9060662

Jackson, M., Brennan, L., & Parker, L. (2021). The public health community's use of social medial for policy advocacy: A scoping review and suggestions to advance the field. *Public Health, 198*, 146−155. https://doiorg/10.1016/j.puhe.2021.07.015.

Kaiser Family Foundation. (November 8, 2021). COVID-19 misinformation is ubiquitous: 78% of the public believes or is unsure about at least one false statement, and nearly a third believe at least four of eight false statements tested. https://www.kff.org/coronavirus-covid-19/press-release/covid-19-misinformation-is-ubiquitous-78-of-the-public-believes-or-is-unsure-about-at-least-one-false-statement-and-nearly-at-third-believe-at-least-four-of-eight-false-statements-tested/.

Kent, M., & Taylor, M. (2021). Fostering dialogic engagement: Toward an architecture of social media for social change. *Social Media and Society, 7*, 1. https://doi.org/10.1177/2056305120984462

McCausland, K., Maycock, B., Leaver, T., & Jancey, J. (2019). The messages presented in electronic cigarette-related social media promotions and discussion: Scoping review. *Journal of Medical Internet Research, 21*(2), 1−16. https://doi.org/10.2196/11953

Meghan Mahoney, L., & Tang, T. (2016). *Strategic social media: From marketing to social change.* John Wiley & Sons, Inc.

Mendoza-Herrera, K., Valero-Morales, I., Ocampo-Granados, M., Reyes-Morales, H., Arce-Amare, F., & Barquera, S. (2020). An overview of social media use in the field of public health nutrition: Benefits, scope, limitations, and a Latin American experience. *Preventing Chronic Disease, 17*, 200047. https://doi.org/10.5888/pcd17.200047

Pervaiz, F., Pervaiz, M., Abdur Rehman, N., & Saif, U. (2012). FluBreaks: Early epidemic detection from Google flu trends. Journal of Medical Internet Research, 14(5): e125. https://doi.org/10.2196/jmir.2102. PMID: 23037553; PMCID: PMC3510767.

Pew Research Center. (July 16, 2020). *Congress soars to new heights on social media.* https://www.pewresearch.org/internet/2020/07/16/congress-soars-to-new-heights-on-social-media/.

Walley, J., & Wright, J. (2010). *Public health − An action guide to improving health.* Oxford University Press.

Yeung, D. (2018). Social media as a catalyst for policy action and social change for health and well-being: Viewpoint. *Journal of Medical Internet Research, 20*(3), e94. https://doi.org/10.2196/jmir.8508

Zhao, M. Z. & Liu, B. F. (2018). Disentangling social media influence in crises: Testing a four-factor model of social media influence with large data. Public Relations Review, 44 (4): 549−561. ISSN 0363-8111. https://doi.org/10.1016/j.pubrev.2018.08.002.

Further reading

Naslund, J. A., Kim, S. J., Aschbrenner, K. A., McCulloch, L. J., Brunette, M. F., Dallery, J., & Marsch, L. A. (2017). Systematic review of social media interventions for smoking cessation. *Addictive Behaviors, 73*, 81−93. https://doi.org/10.1016/j.addbeh.2017.05.002

Rathore, A. K., Maurya, D., & Srivastava, A. K. (2021). Do policymakers use social media for policy design? A Twitter analytics approach. Australasian Journal of Information Systems, 25. https://doi.org/10.3127/ajis.v25i0.2965.

Solnick, R. E., Chao, G., Ross, R. D., Kraft-Todd, G. T., Kocher, K. E., & Yarris, L. M. (2021). Emergency physicians and personal narratives improve the perceived effectiveness of COVID-19 public health recommendations on social media: A randomized experiment. *Academic Emergency Medicine, 28*(2), 172−183. https://doi.org/10.1111/acem.14188

Stabile, B., Purohit, H., & Hattery, A. (2020). Social media campaigns addressing gender-based violence: Policy entrepreneurship and advocacy networks. *Sexuality, Gender & Policy, 3*(2), 122−133. https://doi.org/10.1002/sgp2.12021

Sweet-Cushman, J. (2019). Social media learning as a pedagogical tool: Twitter and engagement in civic dialogue and public policy. *PS, Political Science & Politics, 52*(4), 763−770. https://doi.org/10.1017/S1049096519000933

Chapter 12

Social media and Infodemiology—use of social media monitoring in emergency preparedness

Kavita Batra[1,2], Ravi Batra[3,4] and Manoj Sharma[5,6]

[1]*Department of Medical Education, Kirk Kerkorian School of Medicine at UNLV, University of Nevada, Las Vegas, NV, United States;* [2]*Office of Research, Kirk Kerkorian School of Medicine at UNLV, University of Nevada, Las Vegas, NV, United States;* [3]*Department of Information Technology, Coforge Ltd., Atlanta, GA, United States;* [4]*Social and Behavioral Health, School of Public Health, University of Nevada, Las Vegas, NV, United States;* [5]*Department of Internal Medicine, Kirk Kerkorian School of Medicine at UNLV, University of Nevada, Las Vegas, NV, United States;* [6]*Department of Social and Behavioral Health, School of Public Health, University of Nevada, Las Vegas, NV, United States*

Learning objectives

- Delineate the newly emerging field of Infodemiology and its research landscape.
- Describe the use of Infodemiology and social media analytics in the historical context as well as the current era.
- Identify the characteristics of Infodemic.
- Define information disorder and its sources (misinformation, disinformation, and malinformation).
- Identify fake news and understand the fake news triangle.
- Discuss supply-side and demand-based features of Infodemiology for preventing the pollution of the health information systems.
- Explain the role of Infodemiology in surveillance and combating infodemic.
- Use a set of preestablished guidelines to evaluate the information presented on social media.
- Utilize legitimate resources to check/validate facts.

Introduction

Social media (SM) is defined as a set of Internet-based applications and user interfaces, which allow the creation and sharing of content in the computer-mediated environment (hereafter CMEs), which can be asynchronous or virtual synchronous (Newkirk et al., 2012). Compared to the traditional face-to-face communication model, SM offers the benefit of engaging in direct user interactions in emergency crisis management (Hossain et al., 2016; Yates & Paquette, 2011). For instance, wikis were created by the United States Departments of State, the U.S. military, and the United States Agency for International Development (USAID) to relay important and time-sensitive information related to the Haiti earthquake, which was followed by the long-term reconstruction assistance to Haiti (Yates & Paquette, 2011). During an emergency crisis, SM tools, such as Facebook, Google Trends, and Twitter, have been extensively used in increasing situational awareness and risk communication to inform public health response, strategic planning efforts, and health assessments (Gasper et al., 2014; Mavragani, 2020). The use of web-based sources in the identification, forecasting of diseases, and predicting health behaviors introduces a concept of Infodemiology or Information epidemiology (Mavragani, 2020). The importance of a better information network has already been felt historically during the West African Ebola outbreak in 2014 (Hossain et al., 2016). Proactive rather than reactive public health responses are integral to addressing global health emergencies via building effective surveillance systems (Hossain et al., 2016). Through the medium of this chapter, the authors aim to discuss the essential features of Infodemiology, including the effective use of SM to optimize public health response (along with some applications in the historical context), application of SM in research (aka social media analytics), and challenges associated with the improper use of SM as it relates to infodemic and characteristics associated with it.

Information Epidemiology or Infodemiology

Infodemiology refers to "the science of distribution and determinants of information in an electronic medium, especially the internet, or in a population, with the ultimate aim to inform public health and public policy" (pg. 1, paragraph 1, Eysenbach, 2009). The applications of Infodemiology include but are not limited to the analysis of the queries used in the search engine to predict outbreaks, examining people's status for syndromic surveillance, assessing the effectiveness of health marketing campaigns, and measuring the diffusion of information and knowledge translation (Eysenbach, 2009). Infodemiology can also provide insights into the health-related behaviors of the population and how this information was communicated or propagated through SM analytics (Eysenbach, 2009).

Uses of Infodemiology on historical context

Previous studies investigated the relationship between search behaviors, information available on the Internet, and public health events, which were instrumental in historic outbreaks (Eysenbach, 2009). In 2006, the first Infodemiology study was conducted by Eysenbach (2006), which investigated the correlation between influenza-related queries on search engines and influenza cases that were announced in the following week in Canada. This method was replicated by other researchers to predict outbreaks. For instance, a study by Wilson and Brownstein (2009) analyzed the texts on the Internet before the official announcement of the *Listeriosis* outbreak.

SM analytics played a critical role in managing all types of crisis events, including natural as well as environmental disasters (Gasper et al., 2011). During the *Escherichia coli* outbreak in Spain (2014), data from Twitter tweets were used for monitoring population response to guide risk communication strategies (Gasper et al., 2011). The SM was extensively used for early identification and monitoring of some outbreaks of zoonotic origin, including H5N1, Severe Acute Respiratory Syndrome (SARS), and Ebola (Bernardo et al., 2013; Eysenbach, 2006; Madoff, 2004). During these historical times, an Internet-based reporting system—the Program for Monitoring Emerging Diseases (ProMED-mail) was utilized to rapidly disseminate the information related to the trajectory of the outbreaks (Hossain et al., 2016; Madoff, 2004). In addition to serving as an alert system, ProMED-mail was also used as a collaboration tool between laboratories globally to understand new challenges, which arrived in the form of outbreaks and emerging diseases (Madoff, 2004). And, SM analytics seems to play a vital role in outbreak risk communication, which utilizes an expanded framework of Infodemiology in the evolving research landscape (Rahmanti et al., 2021).

Research landscape of Infodemiology

Infodemiology is relatively a new research discipline, which helps in filling out the knowledge translation gaps between the best evidence and practice (Eysenbach, 2002). In 1996−97, the first Infodemiological study was published and the research discipline of Infodemiology gained prominence through another study being published in reputed journals (Davison, 1997; Impicciatore et al., 1997). Previous infodemiological studies were largely descriptive in nature, which reported the percentage of websites with inaccurate information and helped to identify conflicting evidence (Davison, 1997). Interestingly, about 90% of diet and nutrition-related information available on the websites was unreliable as opposed to 5% of information related to cancer (Davison, 1997). Given the inherent limitation of the descriptive infodemiological designs in not revealing the relationships between quality indicators and websites' characteristics, analytical infodemiological studies were deemed more useful (Davison, 1997). The uses of analytical infodemiological studies include but are not limited to determining the design, content, and features of the websites, which would in turn influence accessibility, readability, and usability (Davison, 1997). Moreover, it also helps in investigating the association between the type of source and the presentation/content of the information (Davison, 1997). For instance, one analytical study found that academic websites, sites with editorial boards, and government or nonprofit websites provide more accurate information and satisfy accountability criteria (Chen et al., 2000; Griffiths & Christensen, 2000). One study provided "CREDIBLE" guidelines published in the *American Journal of Medicine* to assess the health information on the Internet (Eysenbach, 2002). Fig. 12.1 explains all components of "CREDIBLE" guidelines. Infodemiological metrics can be measured through an analytic approach, which serves as the basis of an expanded framework of Infodemiology.

The expanded framework of Infodemiology

Information (concept) prevalence, information (concept) occurrence ratio, and information incidence are the basic metrics of Infodemiology, which distinguishes supply-based methods from demand-based methods (Eysenbach, 2009). The supply-based method relies on the content which is available on the Internet, websites, blogs, and SM, whereas demand-based methods analyze the search and navigation behavior, which can utilize active and passive strategies

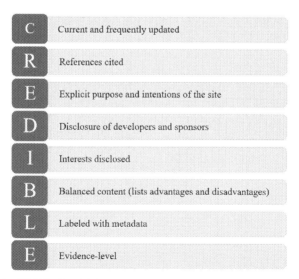

FIGURE 12.1 CREDIBLE guidelines. *Note: This image has been created by authors of this chapter and was based on the information provided in the textual form by Eysenbach in 2002.*

to follow and predict public health events (Eysenbach, 2009). As indicated above, information prevalence and information occurrence ratio are some basic Infodemiology indicators, which measure the absolute or relative occurrences of a specific keyword or concept (also account for synonyms) in a big pool of information (Eysenbach, 2009). The pool of information can be Internet postings, blogs, webpages, tweets, status pages, etc. From this, we calculate the prevalence of concept or information, which is a useful tool for longitudinal follow-up to evaluate outcomes related to certain external events, such as media campaigns or disease outbreaks (Eysenbach, 2009). Next, the information or concept ratio, which is expressed as a proportion of information units about a specific concept compared to a controlled concept (Eysenbach, 2009). Additionally, the information or concept incidence rate (analogous to the disease incidence rate), which measures the number of new information units created in a specified period of time. The concept incidence rate can point to the emerging public health threats (Eysenbach, 2009). However, inaccurate information about any concept may be more dangerous as compared to the event itself due to inappropriate use of information environment, such as SM.

Social media and the COVID-19 pandemic

Undoubtedly, SM helps in shaping public views during health crisis events, with one being the COVID-19 pandemic (Rahmant et al., 2021). This allows public health officials to quickly broadcast important and time-sensitive information to a large group of people (Rahmant et al., 2021). Additionally, SM platforms, such as Twitter, also help us to understand public response/perspectives to the government messages and alerts related to the associated risks and protective strategies (Hossain et al., 2016). For instance, one Indonesian study observed that public attention to the implementation of the "New Normal" was affected by the government actions and transmission of the disease (Rahmant et al., 2021). In addition, there was a significant correlation between the number of Twitter posts and issues of public interest (Rahmant et al., 2021). As of results of the sentiments' analysis, a shift from "negative" to "positive" perceptions was observed in the initial 2 months, with a majority of the posts expressing emotions of "trust," "anticipation," and "joy" (Rahmant et al., 2021). This highlights the effective use of SM to disseminate information to minimize the spread of the contagion; however, if not done properly, this can pose obstacles in the road to recovery in the form of "infodemic."

Infodemic: A new front of challenge in the COVID-19 battle (mis vs. dis vs. mal)

Infodemic as characterized by the abundance of information reduces one's ability to distinguish between facts, opinions, and biases (Naeem & Bhatti,

2020). With such an extensive amount of information available, it is important to delineate false from true information (Naeem & Bhatti, 2020). Infodemic is primarily characterized by the volume and velocity of the information; however, secondary characterization depends upon various forms of wrong information (Zielinski, 2021). There may be an instance when a large volume of some helpful information disseminates quickly, and its characterization as "infodemic" may depend upon other factors. It may depend on the manner which the information is going to be managed and translated to the appropriate audience to close the knowledge gaps, which might be useful for the decision-makers to devise new policies (Zielinski, 2021). The characteristics of an "infodemic" depending upon two primary characteristics of volume and velocity with its associated effects are shown in Table 12.1 given below.

The sudden increase in the volume of the information may be associated with immediate consequences on several aspects, including location, quality, visibility, validity, and capacity (Zielinski, 2021). At the onset of infodemic,

TABLE 12.1 Components and effects of volume and velocity of the infodemic.

Characteristic	Component	Effect
The volume of information	Location	Difficult to locate as the information has originated from multiple sources
	Capacity	Difficult to collect, store, and publish due to the overabundance of information
	Quality	Difficult to delineate high quality from low quality
	Visibility	Hard to identify new evidence
	Validity	Hard to debunk the false news
The velocity of the information	Assessment	Lack of time to assess or judge the information due to quick turnaround
	Gatekeeping	Little time to fact-check
	Application	Delays in identifying/acting on correct information
	History	Hard to keep track of the chronology of the information events
	Waste	Unnecessary replication of research and findings

Source: Adapted from Zielinski, C. (2021). Infodemics and infodemiology: A short history, a long future. *Revista Panamericana de Salud Pública =Pan American Journal of Public Health*, 45, 1. https://doi.org/10.26633/RPSP.2021.40.

when the information is outpoured through broad dissemination, it becomes challenging to locate and collect the information for researchers, particularly those belonging to the developing countries, and further exacerbates by the digital divide (Zielinski, 2021). With a plethora of information, it becomes difficult to rank the quality of the evidence being presented as a part of information appearing in large volumes. Quality assessment is deemed necessary in an infodemic; however, it takes time to prove the validity of the information. For this, we will need more effective tools (Zielinski, 2021). Next, the velocity which is another characteristic of an infodemic may pose problems in assessment, gatekeeping, application, and our ability to track the information. Both volume and velocity characteristics of the infodemic lead to inferior coordination on the research front (Zielinski, 2021). For instance, facts or factoids appearing at a fast speed on the Internet and other media would not allow sufficient time to perform the quality assessment. This may lead to the suppression of the right information by the wrong information. Moreover, it becomes challenging to identify true historical facts with such quickly prevailing information (Zielinski, 2021). Finally, the explosive nature of the emergency situation (e.g., COVID-19 pandemic in this case) may yield two byproducts of velocity and volume of the information. At present, we are not equipped with efficient tools to assess the quality of the information; this would need artificial intelligence or robotic solutions for quality scanning and sorting. The third characteristic of infodemic is the "spread of bad information or misinformation" (Zielinski, 2021). Three types of information can be used to examine information disorder: first, misinformation when the false information is shared with no resulting harm; second, disinformation, which is the intentional sharing of the information to cause harm; and third, malinformation, which entails moving the private yet original information to the public sphere to cause harm (Wardle & Derakhshan, 2017). As suggested by Wardle and Derakhshan, mis- and disinformation (fake news [FN]) can come in different forms (misleading content, satire or parody, fabricated content, imposter content, manipulated content, false context, and false connection) as shown in Table 12.2.

Fake news triangle and meta-framework of fake news impact

FN mimics the characteristics of true news (TN) to mislead the audience (Olan et al., 2022). With the advancement of new technology and SM, dissemination of FN amplified using one-to-many or many-to many transactions (Olan et al., 2022). Despite the presence of some innovative tools to identify factual discrepancies, the spread of FN remains a big challenge (Pierri et al., 2020). Moreover, a gap between fact-checking and fundamental beliefs still exists and makes the public more likely to accept the dangers of FN instead of performing fact-checking (Lukyanenko et al., 2014). To reduce the spread of FN,

TABLE 12.2 Types of fake news with definitions.

Type	Definition (s)
Misleading content	Misleading use of information to frame an issue or individual
Satire or parody	It has the potential to fool with no intention to cause harm
Fabricated content	Absolutely false information with an intention to cause harm and deceive the readers. In other words, fabricated content completely lacks a factual basis
Imposter content	Misinformation that is created and presented under the branding of an established informational source
Manipulated content	Genuine information that is edited or manipulated to deceive the readers
False context	Genuine information that is shared with false contextual information
False connection	Pieces of misinformation that are packaged as headlines, visuals, or captions that do not support the content

Source: Wardle and Derakhshan (2017).

it is important to understand the FN triangle, which adopted the disease triangle based on the epidemiological model (Olan et al., 2022). Similar to the disease triangle having three fundamental components of environment, host, and agent, the FN triangle also has three components as described by Olan and colleagues. In the FN triangle (as shown in Fig. 12.2), FN is analogous to the infectious agent; SM is conceptualized as the environment through which infection can be controlled; and hosts will include individual readers or society. The host interacts with others having similar interests/agendas and grows as a social network (Olan et al., 2022). SM as an environment includes multiple communication channels, such as websites, mobile apps, and other platforms that will help facilitate relationships among users (host) of the same interest (Olan et al., 2022).

The COVID-19 pandemic has been accompanied by a large amount of misleading and false information, particularly on SM. The rampant flow of misinformation is impacting the risk perception of people toward the virus and the effectiveness of containment strategies (Van der Linden et al., 2020). With the sudden arrival of the COVID-19 pandemic, there was a need to publish the data related to epidemiology and the impact of pandemics to quickly implement preventive strategies (Toth et al., 2020). However, this came with several pitfalls, especially the fast-track publications skipping a traditional peer-review process (Toth et al., 2020).

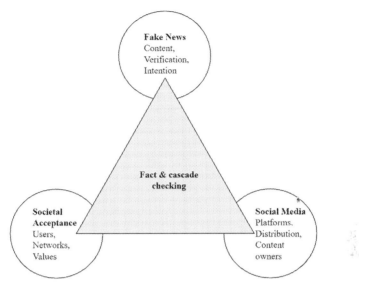

FIGURE 12.2 Fake news (FN) triangle. *Note: This figure was adapted based on the information provided by Olan, F., Jayawickrama, U., Arakpogun, E. O., Suklan, J., & Liu, S. (2022). Fake news on social media: The impact on society. Information Systems Frontiers: A Journal of Research and Innovation, 1–16. Advance online publication. https://doi.org/10.1007/s10796-022-10242-z.*

COVID-19: A pandemic of publications

With the onset of the COVID-19 pandemic, the volume and output of the information raised exponentially with a new manuscript getting published every 3 min (Yan, 2020; Zielinski, 2021). According to the latest estimates, the highest number of articles were published in the context of the COVID-19 pandemic than any other disease in the past (Tentolouris et al., 2021). Posting of preprints in MedRxiv has increased by 400%, and views/downloads have increased by 100 times (Latthe et al., 2000; Yan, 2020; Zielinski, 2021). Preprint articles are not peer-reviewed but can be cited by reviews, which may lead to the perpetuation of invalid information (Martins et al., 2020). During the pandemic, peer review has been virtually and extensively replaced by preprints thereby increasing the speed of publication (Elmore, 2018; Majumder & Mandl, 2020; Powell, 2016). While speed is important in an emergency situation, particularly when the time from review to publication is prolonged (Elmore, 2018; Majumder & Mandl, 2020; Powell, 2016). For instance, the median time for review at *Nature* Publishing group journals has increased to twice in this pandemic situation. However, no methods exist to date to evaluate the preprint articles (Martins et al., 2020). Also, lay audiences cannot understand the difference between preprint and peer-reviewed articles, unlike scientists. Therefore, it is critical to developing a mechanism to assess the quality of the preprint articles, which will allow relatively quick dissemination of factual and valid information. Even the peer-reviewed articles can have inferior quality to the preprints, especially when it relates to the fast-tracked review process offered by the open access journals (Martins et al., 2020). The critical evaluation of the articles might be missing in the process of fast-track review (Martins et al., 2020). Interestingly, upon analyzing the COVID-19 articles indexed in the PubMed database, it was found that around 8% of the articles were reviewed and published in a day, which might have compromised the robust review process (Besancon et al., 2021). The fast-tracked review process has the potential to propagate poor quality work, which may lead to retractions (Strojil & Suchánková, 2020). One worrisome example in this context can be the retrieval of a virology preprint article, which hypothesized the presence of HIV "insertions" in the SARS-CoV-2 virus (Tentolouris et al., 2021). Another typical example was the "hydroxy-chloroquine hypothesis" (being effective to treat COVID-19 illness) which gained popularity among researchers and political leaders in the early periods of the pandemic (Strojil & Suchánková, 2020). However, later this hypothesis was dismissed as follow-up trials fail to demonstrate efficacy (Elavarasi et al., 2020). Upon analyzing the retrieval rates, it was found that articles related to the COVID-19 pandemic were 0.097, which was the highest rate compared to previous outbreaks, such as Ebola, H1NI, HIV, and Middle East Respiratory Syndrome (Yeo-The & Tang, 2021). Among total of 28 COVID-19 articles retrieved, eight articles had errors in the results and conclusions (Soltani & Patini, 2020). The retrieval rate of the articles is anticipated to rise in the future as the median time from publication to retraction is 28 months (Trikalinos et al., 2008).

"Infodemic knowledge" and public health at risk

Information overload related to the disease is called "infodemic knowledge," which stems from conspiracy theories (Hua & Shaw, 2020). This is often accompanied by fabricated news or FN, which quickly multiplies and alters the factual information (Hua & Shaw, 2020; Naeem et al., 2021; Rocha et al., 2021). Some worrisome examples of "infodemic knowledge" include but are not limited to the stories of biological weapons produced in China, the effectiveness of water and lemon to treat COVID disease, and false information related to the drug shortage, supplies, and fuel (Hua & Shaw, 2020; Naeem et al., 2021; Rocha et al., 2021; Tentolouris et al., 2021). Islam and colleagues investigated 2311 reports from multiple sources, including television networks, and newspapers published during January—April 2020, and found that 89%, 8%, and 3.5% of these reports included rumors, conspiracy theories, and stigmatization comments, respectively (Islam et al., 2020). Exposure to false information forms false beliefs, and the etiology of the false belief is multifactorial (Ecker et al., 2022). It includes cognitive as well as socio-affective drivers, which facilitate the formation of false beliefs. Both types of drivers and their components are described in Table 12.3.

At the time of emergency situations such as the COVID-19 pandemic, it is commonplace for the public to engage in the opportunistic adoption of unverified information, which creates challenges in achieving good compliance of the public toward containment efforts (Badell-Grau et al., 2020). Given the rapid flow of digital information, real-time data collection and analysis using SM analytics can help shape public reactions, behaviors, and beliefs (Badell-Grau et al., 2020). Previous studies explored the positive association between high reliance on emotion and susceptibility to believing FN (Martel et al., 2020). The content of the information which appeals to the emotions of recipients increases the persuasion (Ecker et al., 2022). Given the positive association between the use of emotional language and persuasion, the likelihood of increased susceptibility toward misinformation and deception increases (Ecker et al., 2022). This may attribute to the psychological consequences (e.g., fear, panic, anxiety, etc.) among the public besides those being originated from the emergency event itself.

Misinformation over SM and mental health outcomes share a long-standing association (Daniels et al., 2021; Wang et al., 2020). Reportedly, the flow of exaggerated and false information can cause psychological morbidities, including fear, anxiety, stress, depression, and sleep problems (Wang et al., 2020). Misinformation worsens an already complex emotional situation, which arose through the COVID-19 pandemic (Lai et al., 2020; Nelson et al., 2020). Previous studies revealed a strong association between intense exposure to negative information about the crisis and public anxiety (Xu & Liu, 2021). Reports about media vicarious traumatization were also discussed according to which stories of traumatic experiences of others can elicit aversive emotions

TABLE 12.3 Etiology of the false beliefs.

	Cognitive drivers
1. Intuitive thinking	Lack of analytical thinking
2. Cognitive failures	Not paying attention to sources and counter-evidence
3. Illusory truth	Familiarity, cohesion, and fluency
	Socio-affective drivers
1. Source cues	Attractive, eelite, in-group
2. Emotion	Emotional state and emotional information
3. World-view	Personal views and partisanship

and psychological distress (Thompson et al., 2019). The relationship between trauma-related media exposure and distress has been reported as cyclical in nature, which means that distress can increase trauma-related media consumption, which in turn increases the distress to follow-up events (Thompson et al., 2019).

Certain demographic groups, such as older adults, adolescents, those with less educational attainment, and people with preexisting biases, are more susceptible to the misinformation and associated psychological consequences (Lai et al., 2020; Nelson et al., 2020). This susceptibility increases with repeated exposures (Lai et al., 2020; Nelson et al., 2020). Like other groups, healthcare workers are also vulnerable to the psychological impact of misinformation (Lai et al., 2020; Nelson et al., 2020). In a China-based study, it was found that the psychological burden among healthcare workers was increased by misinformation, confusion, and conspiracy theories prevailing over SM (Lai et al., 2020). This underscores the need to reinforce guidelines to use SM effectively, particularly during this period, when SM is ubiquitous.

Social media: A double-edged sword

Nowadays, media is becoming a permanent fixture of our lives and may have pro-social or antisocial effects, in other words, a "double-edged sword" (Srivastava et al., 2018). On the pro-social front, mass media can serve as a socializing vehicle that helps reflect public attitudes or opinions, shape behaviors, propagate learned behaviors, and demystify misinformation (Srivastava et al., 2018). Additionally, it allows the creation of new avenues for research collaborations, which will also help build resilient communities (Venegas-Vera et al., 2020). SM platforms also help in continuous learning efforts through live or recorded webinars (González-Padilla & Tortolero-Blanco, 2020). With this bright side, also comes the darker side of the SM. On the antisocial front, an abundance of information shared through multiple sources can be overwhelming, increasing cognitive overload, and maximizing the probability of error (Venegas-Vera et al., 2020). Perpetuation of misinformation and panic through SM puts public health at risk (Himelein-Wachowiak et al., 2021; Venegas-Vera et al., 2020). SM and their search engines are set up with algorithms (called bubble filters) to identify the "personalized ecosystem" of the user and help to predict users' preferences based on the previous search made by the users (Holone, 2016). Facebook and Twitter use these algorithms, which prevent users to see contrasting information to make informed decisions (Holone, 2016). In other words, bubble filters do not provoke curiosity to discover different points of view (Holone, 2016).

The misinformation can be directly spread by humans as well as bots (Himelein-Wachowiak et al., 2021; Venegas-Vera et al., 2020). Bots (software robots) pose as real human users, who use behaviors, such as excessive posting about political or public health matters, tweeting the emerging news early and frequently, and tagging influential figures in the anticipation of wider dissemination (Himelein-Wachowiak et al., 2021). Bots are categorized into subclasses, including content polluters, traditional spambots, social bots, and cyborgs (hybrid human bot) (Himelein-Wachowiak et al., 2021). A coordinated network of bots is called botnets and are found on SM, including Reddit, Facebook, YouTube, Twitter, and several other platforms (Himelein-Wachowiak et al., 2021). Bots are reported to be super-spreaders of misinformation as they can quickly diffuse misinformation and has a disproportionate role in disseminating the information from low credibility sources (Shao et al., 2018). Reportedly, around 9%—15% of Twitter active accounts were bots with over 60 million bots infested on Facebook during the 2016 election campaign that included a substantial portion related to political content (Committee on the Judiciary, 2017). Notably, the threat of misinformation has been a long-standing issue, particularly in the political domain. For instance, fake stories favoring Trump were shared 30 million times on Facebook before the presidential election in 2016. It was also noted that the velocity of FN was significantly higher than the velocity of true stories (Allcott & Gentzkow, 2017; Martel et al., 2020; Vosoughi et al., 2018). This highlights the need of curbing the diffusion of FN or misinformation to prevent the "infodemic."

Efforts to combat "infodemic"

The World Health Organization (WHO) is leading efforts to curtail the COVID-19 infodemic to ensure that people receive reliable information to act appropriately (Child et al., 2020; Naeem et al., 2020; WHO, 2020). It is a well-established fact that people tend to trust the information which appeals to their sentiments as opposed to factual or objective information (Maoret, 2017). A previous study by Wardle and Derakhshan (2017) proposed a helpful framework related to the information disorder. In this "posttruth" era, it is critical to educate people, particularly younger adults, about the dangers associated with sharing the disinformation (Naeem et al., 2020). Several agencies, such as UNESCO, are taking efforts to combat this infodemic and promote the facts-sharing of the COVID-19-related information (Naeem et al., 2020; UN News, 2020). These efforts include the use of hashtags #ThinkBeforeClicking, #ShareKnowledge, and #ThinkBeforeSharing (Naeem et al., 2020; UN News, 2020). The World Economic Forum suggested three important guidelines to curb COVID-19 "infodemic." These guidelines include: (1) embracing uncertainty in a responsible manner; (2) checking the credibility of the information source; and (3) seeing who is backing up the claim (Word Economic Forum, 2020). Lately, at the global level, the WHO launched a "Myth buster" Website to counter misinformation associated with COVID-19 and to track the authenticity of the information (Naeem & Bhatti, 2020). This underscores the need of developing information literacy programs to evaluate facts. Previous studies indicated that health science librarians (HSLs) have been playing a critical role in delivering information literacy programs since 1980 and can offer usable frameworks to identify FN (Dempsey, 2017; Zarocostas, 2020). The International Federation for the Library Association created eight-step guidelines, including (1) consider the source, (2) check the author, (3) check the date, (4) check your biases, (5) read beyond, (6) look for supporting sources, (7) ask "is it a joke?", and (8) seek an opinion from experts (Chan et al., 2020; Naeem et al., 2020). Recent studies suggested some guidelines to promote the responsible use of SM for disseminating information (Chan et al., 2020; González-Padilla & Tortolero-Blanco, 2020; Naeem et al., 2020), which are described in Table 12.4.

Strategies to flatten the "Infodemic curve"

First, the interventions should be planned to overcome cognitive and social-affective drivers of misinformation (Ecker et al., 2022). There can be two broad areas of interventions, such as prebunking and debunking (reactive) interventions. Prebunking is preemptive in nature, which helps people to recognize and resist misinformation. Prebunking interventions can include simple warning systems or more advanced literacy programs. Debunking interventions focus on responding to the specific piece of misinformation after an individual gets exposed to it and allows a reality check with the help of other more reputed sources. In other words, it involves strategies to counter

TABLE 12.4 Guidelines suggested by previous studies to promote the responsible use of social media.

1. Using established professional platforms or communication groups for disseminating the information.
2. Avoid sharing the information without a trusted source. Always supplement the information with a source.
3. Avoid sharing information which may lead to panic or anxiety among the public.
4. Quality takes precedence over quantity. Be wary of sharing the results of low-quality and in vitro studies.
5. Declare conflicts of interest wherever possible.
6. Avoid providing medical advice and recommendations on social media which are not backed up with sufficient evidence, as this may lead to confusion among the public.
7. Using transparent methods for peer-review and feedback, like platforms for postpublication peer-review process or preprint.

misinformation. This can be done by fact-based correction and addressing logical fallacies and undermining the plausibility of the misinformation (Ecker et al., 2022). Debunking strives to unveil factual information (Ecker et al., 2022).

The approach to designing prebunking and debunking interventions is multipronged, which will involve the media consumer (user e.g., public), the scientific community (e.g., practitioners), and policymakers (e.g., government organizations) (Tentolouris et al., 2021). In addition, the efforts will target supply as well as consumption of misinformation (Ecker et al., 2022). At the individual level, strategies such as checking the credibility of authors, sources, and fact-checking, and diversification of sources to validate information would be essential (Tentolouris et al., 2021). At the government level, investments should be directed toward increasing the amount of fact-checking and verification tools and educating the public about the role of FN and information verification. The government agencies can negotiate with SM providers to block the diffusion of misinformation (Olan et al., 2022). Organizations and providers can allocate sufficient resources to perform continuous monitoring of online activities and to develop automated solutions to delete FN contents to ensure a safer communication environment. This will increase the confidence of the society in delineating FN from TN and will prevent FN cascading (Olan et al., 2022). The scientific community can help the lay audience to discriminate between facts and opinions.

Future research should be performed to reveal a comprehensive theoretical view, and complex analysis can be performed using the big data in infodemiological studies (Tran et al., 2021). The robust infodemiological studies will provide better insights into the public perceptions and sentiments. New technology such as artificial intelligence algorithms can be utilized to address the infodemic. Government should take prompt actions to effective communication and should take advantage of the SM tools. For instance, promoting the use of official accounts of government organizations through SM platforms could be enhanced so the public will be exposed to the official sources of information (Hadi & Fleshler, 2016). This will help build trust and will prevent the diffusion of misinformation.

Conclusions and future directions

Infodemic is increasingly affecting society and undermining trust. It is critical to get prepared for this disinformation war. As suggested by Calleja et al., in 2021, the management of infodemic has five work streams, including the following: (1) continuous monitoring of the impact of infodemic during health emergencies; (2) identifying signals and understanding spread and risk associated with infodemic; (3) remediation of the impact associated with infodemic; (4) promoting and developing and implementing toolkit for infodemic management; and (5) evaluating infodemic interventions and building resilience in the community. Infodemic is not new; however, addressing it during the existing and future pandemics is centrally important. Infodemic when intertwined with the COVID-19 pandemic comes as a double burden, which definitely needs actions. Long-term capacity building through multi-disciplinary coordination will be essential. Research studies informed through the perspectives of media practitioners, policymakers, and media consumers will be needed to address the infodemic. Analytical Infodemiology can be a useful tool to fathom the dynamic of infodemic and its associated corollaries. The field of Infodemiology is still under evolution, which would need concerted efforts from the scientific community as well as governing bodies to fight against infodemic, which spread faster than the virus. The WHO is developing tools to provide an evidence-based response to the infodemic to strengthen epidemic and pandemic response activities and is fostering the growth of the field of Infodemiology. In the future, all emergencies and pandemics will be accompanied by infodemic that will be better addressed with the tools and insights developed today. Infodemic is viewed as a threat by health authorities, and measuring its true burden will be necessary to develop scalable social inoculation interventions. The efforts across countries should not be limited to emergency preparedness but should also extend to knowledge preparedness.

Questions for discussion

Through the medium of this chapter, the authors would like to set a stage to discuss the following:

- Can we address the challenge of fake use with a newly evolving field of artificial intelligence?
- What is the effect of rumors, stigma, and conspiracy theories on community trust?
- What lessons have we learned from the COVID-19 pandemic?
- Can we leverage our education system to counter the information disorder?
- Lastly, how people can be immunized against misinformation?

Websites to explore and a list of credible sources

Several countries developed fact-checking websites. Given below are some helpful links to assess these websites.

 I. Websites
   a. MythBusters as a part of COVID-19 advice for the public developed by the World Health Organization (WHO). This site can be accessed at https://www.who.int/emergencies/diseases/novel-coronavirus-2019/advice-for-public/myth-busters

 b. Misinformation on NewsGuard: This releases a periodic newsletter covering the landscape of digital misinformation. This can be accessed at https://www.newsguardtech.com/misinformation-monitor/
 c. Media bias/Fact Check (MBFC): The overarching objective of this independent website is to educate people about media bias and deceptive news practices. This also provides the tools necessary to check the credibility of the sources. More information can be accessed at https://mediabiasfactcheck.com/about/
 d. COVID19MISINFO.ORG can be accessed at https://covid19misinfo.org/misinfowatch/. This site has public-facing dashboards to track false information. This uses a simple fact-checking rating system to categorize information as true, false, misleading, and unproven.
 e. BBC News Reality Check available at https://www.bbc.com/news/reality_check
 f. Boom (https://www.boomlive.in/) is a fact-checking website.
 g. Fact-check through artificial intelligence, which can be accessed at http://www.fakenewschallenge.org/
 h. International fact-checking network available at https://twitter.com/factchecknet
 i. CREDIBLE Guidelines available at https://www.amjmed.com/article/S0002-9343(02)01473-0/fulltext

II. List of Credible Sources
 a. Center for Disease control and Prevention (https://www.cdc.gov/coronavirus/2019-ncov/index.html/)
 b. WHO COVID-19 database (https://www.who.int/emergencies/diseases/novel-coronavirus-2019/global-research-on-novel-coronavirus-2019-ncov)
 c. Official site of UK Government: https://www.gov.uk/coronavirus
 d. Coronavirus Resource Center at John Hopkins University: https://coronavirus.jhu.edu/map.html
 e. How to sort facts from fiction available at https://sites.google.com/umich.edu/library-fake-news/home
 f. UNESCO and Athabasca University's Information literacy course available at https://www.athabascau.ca/
 g. A toolkit developed by the Pan American Health Organization and World Health Organization, available at https://iris.paho.org/bitstream/handle/10665.2/52052/Factsheet-infodemic_eng.pdf

Key terms with definitions

Confirmation bias: The tendency to interpret information that confirms one's preexisting beliefs or hypotheses
 Debunking information: An act of detecting false information

Diffusion of information: The process through which information flows from one source to other through interactions

Disinformation: Disinformation is a type of misinformation that is intentionally false and intended to deceive or mislead

Fact: Reality, truth, or actual experience

Factoid: A statement based on the assumption

Infodemic: An excessive amount of information that is typically unreliable and spreads rapidly

Infodemiological study: Area of scientific research focused on scanning the Internet for user-contributed health-related content, with the ultimate goal of improving public health

Media-bias: The perceived bias of the journalists

Misinformation: Misinformation refers to false or out-of-context information that is presented as fact regardless of an intent to deceive

Post-truth: Adjectives that relate to or denote circumstances in which objective facts are less influential in shaping public opinion than appeals to emotion and personal belief

Propaganda: Deliberate spread of information or rumors to help or harm a person

Social bots: Programs simulating human behaviors that can be used on social media platforms to do online activities

Social media analytics: The approach of collecting and analyzing information available on social media networks

Social media (SM): Interactive digital channels that allow the exchange of information and other forms of expression through virtual communities and networks

References

Allcott, H., & Gentzkow, M. (2017). Social media and fake news in the 2016 election. *Journal of Economic Perspectives, 31*, 211−236.

Badell-Grau, R. A., Cuff, J. P., Kelly, B. P., Waller-Evans, H., & Lloyd-Evans, E. (2020). Investigating the prevalence of reactive online searching in the COVID-19 pandemic: Infoveillance study. *Journal of Medical Internet Research, 22*(10), e19791. https://doi.org/10.2196/19791

Bernardo, T. M., Rajic, A., Young, I., Robiadek, K., Pham, M. T., & Funk, J. A. (2013). Scoping review on search queries and social media for disease surveillance: A chronology of innovation. *Journal of Medical Internet Research, 15*(7), e147. https://doi.org/10.2196/jmir.2740

Besançon, L., Peiffer-Smadja, N., Segalas, C., Jiang, H., Masuzzo, P., Smout, C., Billy, E., Deforet, M., & Leyrat, C. (2021). Open science saves lives: Lessons from the COVID-19 pandemic. *BMC Medical Research Methodology, 21*(1), 117. https://doi.org/10.1186/s12874-021-01304-y

Chan, A., Nickson, C. P., Rudolph, J. W., Lee, A., & Joynt, G. M. (2020). Social media for rapid knowledge dissemination: Early experience from the COVID-19 pandemic. *Anaesthesia, 75*(12), 1579−1582. https://doi.org/10.1111/anae.15057

Chen, L. E., Minkes, R. K., & Langer, J. C. (2000). Pediatric surgery on the internet: Is the truth out there? *Journal of Pediatric Surgery, 35*(8), 1179−1182. https://doi.org/10.1053/jpsu.2000.8723

Child, D. (2020). *Fighting fake news: The new front in the coronavirus battle: Bogus stories and half-backed conspiracy theories are surging through the internet.* https://www.aljazeera.com/news/2020/04/fighting-fake-news-front-coronavirus-battle-200413164832300.html.

Committee on the Judiciary. (2017). *Extremist content and Russian disinformation online: Working with tech to find solutions.* https://www.judiciary.senate.gov/meetings/extremist-content-and-russian-disinformation-online-working-with-tech-to-find-solutions.

Daniels, M., Sharma, M., & Batra, K. (2021). Social media, stress, and sleep deprivation: A triple "S" among adolescents. *Journal of Health and Social Sciences, 6*(2), 159−166.

Davison, K. (1997). The quality of dietary information on the World Wide Web. *Clinical Performance and Quality Health Care, 5*(2), 64−66.

Dempsey, K. (2017). What's behind fake news and what you can do about it? Information Today, 34, 6. http://www.infotoday.com/it/may17/Dempsey−Whats-Behind-Fake-News-and-What-You-Can-Do-About-It.shtml.

Ecker, U. K., Lewandowsky, S., Cook, J., Schmid, P., Fazio, L. K., Brashier, N., Kendeou, P., Vraga, E. K., & Amazeen, M. A. (2022). The psychological drivers of misinformation belief and its resistance to correction. *Nature Reviews Psychology, 1*(1), 13−29. https://www.nature.com/articles/s44159-021-00006-y.

Elavarasi, A., Prasad, M., Seth, T., Sahoo, R. K., Madan, K., Nischal, N., Soneja, M., Sharma, A., Maulik, S. K., Shalimar, & Garg, P. (2020). Chloroquine and hydroxychloroquine for the treatment of COVID-19: A systematic review and meta-analysis. *Journal of General Internal Medicine, 35*(11), 3308−3314. https://doi.org/10.1007/s11606-020-06146-w

Elmore, S. A. (2018). Preprints: What role do these have in communicating scientific results? *Toxicologic Pathology, 46*(4), 364−365. https://doi.org/10.1177/0192623318767322

Eysenbach, G. (2002). Infodemiology: The epidemiology of (mis)information. *The American Journal of Medicine, 113*(9), 763−765. https://doi.org/10.1016/s0002-9343(02)01473-0

Eysenbach, G. (2006). Infodemiology: Tracking flu-related searches on the web for syndromic surveillance. AMIA Annual symposium proceedings (pp. 244−248). AMIA Symposium.

Eysenbach, G. (2009). Infodemiology and infoveillance: Framework for an emerging set of public health informatics methods to analyze search, communication and publication behavior on the Internet. *Journal of Medical Internet Research, 11*(1), e11. https://doi.org/10.2196/jmir.1157

Gaspar, R., Gorjão, S., Seibt, B., Lima, L., Barnett, J., Moss, A., & Wills, J. (2014). Tweeting during food crises: A psychosocial analysis of threat coping expressions in Spain, during the 2011 European EHEC outbreak. *International Journal of Human-Computer Studies, 72*(2), 239−254. https://doi.org/10.1016/j.ijhcs.2013.10.001

González-Padilla, D. A., & Tortolero-Blanco, L. (2020). Social media influence in the COVID-19 pandemic. *International Brazilian Journal of Urology: Official Journal of the Brazilian Society of Urology, 46*(Suppl. 1), 120−124. https://doi.org/10.1590/S1677-5538.IBJU.2020.S121

Griffiths, K. M., & Christensen, H. (2000). Quality of web-based information on treatment of depression: Cross sectional survey. *BMJ (Clinical Research ed.), 321*(7275), 1511−1515. https://doi.org/10.1136/bmj.321.7275.1511

Hadi, T. A., & Fleshler, K. (2016). Integrating social media monitoring into public health emergency response operations. *Disaster Medicine and Public Health Preparedness, 10*(5), 775−780. https://doi.org/10.1017/dmp.2016.39

Himelein-Wachowiak, M., Giorgi, S., Devoto, A., Rahman, M., Ungar, L., Schwartz, H. A., Epstein, D. H., Leggio, L., & Curtis, B. (2021). Bots and misinformation spread on social media: Implications for COVID-19. *Journal of Medical Internet Research, 23*(5), e26933. https://doi.org/10.2196/26933

Holone, H. (2016). The filter bubble and its effect on online personal health information. *Croatian Medical Journal, 57*(3), 298−301. https://doi.org/10.3325/cmj.2016.57.298

Hossain, L., Kam, D., Kong, F., Wigand, R. T., & Bossomaier, T. (2016). Social media in Ebola outbreak. *Epidemiology and Infection, 144*(10), 2136−2143. https://doi.org/10.1017/S095026881600039X

Hua, J., & Shaw, R. (2020). Corona Virus (COVID-19) "Infodemic" and emerging issues through a data lens: The case of China. *International Journal of Environmental Research and Public Health, 17*(7), 2309. https://doi.org/10.3390/ijerph17072309

Impicciatore, P., Pandolfini, C., Casella, N., & Bonati, M. (1997). Reliability of health information for the public on the World Wide Web: Systematic survey of advice on managing fever in children at home. *BMJ (Clinical Research ed.), 314*(7098), 1875−1879. https://doi.org/10.1136/bmj.314.7098.1875

Islam, M. S., Sarkar, T., Khan, S. H., Mostofa Kamal, A. H., Hasan, S., Kabir, A., Yeasmin, D., Islam, M. A., Amin Chowdhury, K. I., Anwar, K. S., Chughtai, A. A., & Seale, H. (2020). COVID-19-Related infodemic and its impact on public health: A global social media analysis. *The American Journal of Tropical Medicine and Hygiene, 103*(4), 1621−1629. https://doi.org/10.4269/ajtmh.20-0812

Lai, J., Ma, S., Wang, Y., Cai, Z., Hu, J., Wei, N., Wu, J., Du, H., Chen, T., Li, R., Tan, H., Kang, L., Yao, L., Huang, M., Wang, H., Wang, G., Liu, Z., & Hu, S. (2020). Factors associated with mental health outcomes among health care workers exposed to coronavirus disease 2019. *JAMA Network Open, 3*(3), e203976. https://doi.org/10.1001/jamanetworkopen.2020.397

Latthe, M., Latthe, P. M., & Charlton, R. (2000). Quality of information on emergency contraception on the Internet. *The British Journal of Family Planning, 26*(1), 39−43.

Lukyanenko, R., Parsons, J., & Wiersma, Y. F. (2014). The IQ of the crowd: Understanding and improving information quality in structured user-generated content. *Information Systems Research, 25*(4), 669−689. https://doi.org/10.1287/isre.2014.0537

Madoff, L. C. (2004). ProMED-mail: An early warning system for emerging diseases. *Clinical Infectious Diseases, 39*(2), 227−232. https://doi.org/10.1086/422003

Majumder, M. S., & Mandl, K. D. (2020). Early in the epidemic: Impact of preprints on global discourse about COVID-19 transmissibility. *The Lancet. Global Health, 8*(5), e627−e630. https://doi.org/10.1016/S2214-109X(20)30113-3

Maoret, M. (2017). *The social construction of fats: Surviving a post-truth world [Video File]* [TEDx Talks] https://www.youtube.com/watch?v=7tHbSasnvno.

Martel, C., Pennycook, G., & Rand, D. G. (2020). Reliance on emotion promotes belief in fake news. *Cognitive Research: Principles and Implications, 5*(1), 47. https://doi.org/10.1186/s41235-020-00252-3

Martins, R. S., Cheema, D. A., & Sohail, M. R. (2020). The pandemic of publications: Are we sacrificing quality for quantity? *Mayo Clinic Proceedings, 95*(10), 2288−2290. https://doi.org/10.1016/j.mayocp.2020.07.026

Mavragani, A. (2020). Infodemiology and infoveillance: Scoping review. *Journal of Medical Internet Research, 22*(4), e16206. https://doi.org/10.2196/16206

Naeem, S. B., & Bhatti, R. (2020). The Covid-19 'infodemic': A new front for information professionals. *Health Information and Libraries Journal, 37*(3), 233–239. https://doi.org/10.1111/hir.12311

Naeem, S. B., Bhatti, R., & Khan, A. (2021). An exploration of how fake news is taking over social media and putting public health at risk. *Health Information and Libraries Journal, 38*(2), 143–149. https://doi.org/10.1111/hir.12320

Nelson, T., Kagan, N., Critchlow, C., Hillard, A., & Hsu, A. (2020). The danger of misinformation in the COVID-19 Crisis. *Missouri Medicine, 117*(6), 510–512.

Newkirk, R. W., Bender, J. B., & Hedberg, C. W. (2012). The potential capability of social media as a component of food safety and food terrorism surveillance systems. *Foodborne Pathogens and Disease, 9*(2), 120–124. https://doi.org/10.1089/fpd.2011.0990

Olan, F., Jayawickrama, U., Arakpogun, E. O., Suklan, J., & Liu, S. (2022). Fake news on social media: The impact on society. Information Systems Frontiers: A Journal of Research and Innovation, 1–16. Advance online publication. https://doi.org/10.1007/s10796-022-10242-z.

Pierri, F., Artoni, A., & Ceri, S. (2020). Investigating Italian disinformation spreading on Twitter in the context of 2019 European elections. *PloS One, 15*(1), e0227821. https://doi.org/10.1371/journal.pone.0227821

Powell, K. (2016). Does it take too long to publish research? *Nature, 530*(7589), 148–151. https://doi.org/10.1038/530148a

Rahmanti, A. R., Ningrum, D., Lazuardi, L., Yang, H. C., & Li, Y. J. (2021). Social media data analytics for outbreak risk communication: Public attention on the "new normal" during the COVID-19 Pandemic in Indonesia. *Computer Methods and Programs in Biomedicine, 205*, 106083. https://doi.org/10.1016/j.cmpb.2021.106083

Rocha, Y. M., de Moura, G. A., Desidério, G. A., de Oliveira, C. H., Lourenço, F. D., & de Figueiredo Nicolete, L. D. (2021). The impact of fake news on social media and its influence on health during the COVID-19 pandemic: A systematic review. Zeitschrift fur Gesundheitswissenschaften = Journal of Public Health, 1–10. Advance online publication. https://doi.org/10.1007/s10389-021-01658-z.

Shao, C., Ciampaglia, G. L., Varol, O., Yang, K. C., Flammini, A., & Menczer, F. (2018). The spread of low-credibility content by social bots. *Nature Communications, 9*(1), 4787. https://doi.org/10.1038/s41467-018-06930-7

Soltani, P., & Patini, R. (2020). Retracted COVID-19 articles: A side-effect of the hot race to publication. *Scientometrics, 125*(1), 819–822. https://doi.org/10.1007/s11192-020-03661-9

Srivastava, K., Chaudhury, S., Bhat, P. S., & Mujawar, S. (2018). Media and mental health. *Industrial Psychiatry Journal, 27*(1), 1–5. https://doi.org/10.4103/ipj.ipj_73_18

Strojil, J., & Suchánková, H. (2020). Lessons for teaching from the pandemic. British Journal of Clinical Pharmacology. Advance online publication. https://doi.org/10.1111/bcp.14529.

Tentolouris, A., Ntanasis-Stathopoulos, I., Vlachakis, P. K., Tsilimigras, D. I., Gavriatopoulou, M., & Dimopoulos, M. A. (2021). COVID-19: Time to flatten the infodemic curve. *Clinical and Experimental Medicine, 21*(2), 161–165. https://doi.org/10.1007/s10238-020-00680-x

Thompson, R. R., Jones, N. M., Holman, E. A., & Silver, R. C. (2019). Media exposure to mass violence events can fuel a cycle of distress. *Science Advances, 5*(4), eaav3502. https://doi.org/10.1126/sciadv.aav3502

Toth, G., Spiotta, A. M., Hirsch, J. A., & Fiorella, D. (2020). Misinformation in the COVID-19 era. *Journal of NeuroInterventional Surgery, 12*(9), 829–830. https://doi.org/10.1136/neurintsurg-2020-016683

Tran, H., Lu, S. H., Tran, H., & Nguyen, B. V. (2021). Social media insights during the COVID-19 pandemic: Infodemiology study using big data. *JMIR Medical Informatics, 9*(7), e27116. https://doi.org/10.2196/27116

Trikalinos, N. A., Evangelou, E., & Ioannidis, J. P. (2008). Falsified papers in high-impact journals were slow to retract and indistinguishable from nonfraudulent papers. *Journal of Clinical Epidemiology, 61*(5), 464–470. https://doi.org/10.1016/j.jclinepi.2007.11.019

UN News. (2020). During this coronavirus pandemic, 'fake news' is putting lives at risk. UNESCO. https://news.un.org/en/story/2020/04/1061592.

Van der Linden, S., Roozenbeek, J., & Compton, J. (2020). Inoculating against fake news about COVID-19. *Frontiers in Psychology, 11*. https://doi.org/10.3389/fpsyg.2020.566790

Venegas-Vera, A. V., Colbert, G. B., & Lerma, E. V. (2020). Positive and negative impact of social media in the COVID-19 era. *Reviews in Cardiovascular Medicine, 21*(4), 561–564. https://doi.org/10.31083/j.rcm.2020.04.195

Vosoughi, S., Roy, D., & Aral, S. (2018). The spread of true and false news online. *Science, 359*, 1146–1151.

Wang, C., Pan, R., Wan, X., Tan, Y., Xu, L., Ho, C. S., & Ho, R. C. (2020). Immediate psychological responses and associated factors during the initial stage of the 2019 coronavirus disease (COVID-19) epidemic among the general population in China. *International Journal of Environmental Research and Public Health, 17*(5), 1729. https://doi.org/10.3390/ijerph17051729

Wardle, C., & Derakhshan, H. (2017). Information disorder: Toward an interdisciplinary framework for research and policy making (pp. 1–108). Council of Europe Report.

Wilson, K., & Brownstein, J. S. (2009). Early detection of disease outbreaks using the Internet. *CMAJ: Canadian Medical Association Journal = journal de l'Association medicale canadienne, 180*(8), 829–831. https://doi.org/10.1503/cmaj.090215

World Economic Forum. (2020). *How to read the news like a scientist and avoid the COVID-19 'infodemic'*. https://www.weforum.org/agenda/2020/03/how-to-avoid-covid-19-fake-news-coronavirus/.

World Health Organization. (2020a). *Coronavirus disease (COVID-19) advice for the public: Myth busters*. https://www.who.int/emergencies/diseases/novel-coronavirus-2019/advice-for-public/myth-busters.

Xu, J., & Liu, C. (2021). Infodemic vs. pandemic factors associated to public anxiety in the early stage of the COVID-19 outbreak: A cross-sectional study in China. *Frontiers in Public Health, 9*, 723648. https://doi.org/10.3389/fpubh.2021.723648

Yan, W. (April 21, 2020). Coronavirus tests science's need for speed limits. New York Times. www.nytimes.com/2020/04/14/science/coronavirus-disinformation.html.

Yates, D., & Paquette, S. (2011). Emergency knowledge management and social media technologies: A case study of the 2010 Haitian earthquake. *International Journal of Information Management, 31*(1), 6–13. https://doi.org/10.1016/j.ijinfomgt.2010.10.001

Yeo-Teh, N., & Tang, B. L. (2021). An alarming retraction rate for scientific publications on coronavirus disease 2019 (COVID-19). *Accountability in Research, 28*(1), 47–53. https://doi.org/10.1080/08989621.2020.1782203

Zarocostas, J. (2020). How to fight an infodemic. *Lancet (London, England), 395*(10225), 676. https://doi.org/10.1016/S0140-6736(20)30461-X

Zielinski, C. (2021). Infodemics and infodemiology: A short history, a long future. *Revista Panamericana de Salud Pública =Pan American Journal of Public Health, 45*, 1. https://doi.org/10.26633/RPSP.2021.40

Further reading

Calleja, N., AbdAllah, A., Abad, N., Ahmed, N., Albarracin, D., Altieri, E., Anoko, J. N., Arcos, R., Azlan, A. A., Bayer, J., Bechmann, A., Bezbaruah, S., Briand, S. C., Brooks, I., Bucci, L. M., Burzo, S., Czerniak, C., De Domenico, M., Dunn, A. G., ... Purnat, T. D. (2021). A public health research agenda for managing Infodemics: Methods and results of the first WHO Infodemiology conference (Preprint). https://doi.org/10.2196/preprints.30979.

Chapter 13

Social media, online learning, and its application in public health

Janea Snyder
Health Education and Promotion, University of Arkansas at Little Rock, Little Rock, AR, United States

Objectives

- To define social media, health communication, psychosocial health, social marketing, public health, and online learning.
- To examine the key constructs of the health promotion, prevention, and psychosocial health innovative action model.
- To analyze how social media and online learning are being utilized to promote public health.
- To identify implications and risk factors of social media usage on mental health.
- To discuss different health communication theories.
- To discuss health equity lens and the guiding principles for inclusive communication.

Social media

The popularity of social media has impacted the way information is shared given that a large number of people globally are active users of this technology. Its multifaceted abilities made it the dominant channel of communication with the potential of continuous growth of users. This growth has stemmed from the multitude of advancements in technology and Internet usage, which inadvertently has impacted the various modes of communication and how social media is being utilized. Decades ago, the Internet and social media were not as popular as they are today. These days social media platforms are used on a daily basis by many globally. The social media authors (Dobre et al., 2021) have referenced that the term *social media* has taken on multiple meanings and remains open to interpretation, given the dynamic nature of these online environments. However, in a broad sense, *social media* refers to online platforms that facilitate interaction among users, including the creation and distribution of information, content, and ideas as cited in Kaplan & Haenlein, 2010. Appel et al. (2020) concur that social media can be thought of in a few different ways. In a practical sense, it is a collection of software-based digital technologies—usually presented as apps and websites—that provide users with digital environments, in which they can send and receive digital content or information over some type of online social networks. In this regard, we can think of social media as the major platforms and their features, such as Facebook, Instagram, and Twitter. We can also in practical terms use social media as another type of digital marketing channel that marketers can use to communicate with consumers through advertising. But we can also think of social media more broadly, seeing it less as digital media and specific technology services, and more as digital places where people conduct significant parts of their lives. From this perspective, it means that social media becomes less about the specific technologies or platforms and more about what people do in these environments (Appel et al., 2020).

Appel et al. (2020) have deemed social media as culturally significant, since lately it became the primary domain in which people often receive vast amounts of information, share content and aspects of their lives with others, as well as receive local and or global news. Most would agree that social media is always changing. Social media as we know it today is different than even a year ago, and social media a year from now will likely be different than now. This is due to constant innovation, which is taking place on both the technology side (e.g., by the major platforms constantly adding new features and services) and the user/consumer side (e.g., people finding new uses for social media) of social media (Appel et al., 2020). For instance, combining social media usage with online learning may be beneficial to some and not preferred by others. Researchers Salmon et al. in 2015 conclude that it may be useful to outline in detail to students the contributions that social learning can bring to any online learning environment. Those who believe that conversations on social media are a waste of time may view things differently, if they understand how conversations and knowledge sharing with their peers can support their learning experience. However, we must be fully knowledgeable of two important concepts when it comes to online learning. First, we must have a clear understanding of what online learning fully entails, and second, we must understand how to best serve online learners.

Online learning is an emerging approach to learn at students' own premise through advanced information and communication technologies (such as Blackboard, Moodle, YouTube, and Virtual Reality) either asynchronously or synchronously (Yang & Kang, 2020). Therefore, when it comes to the various methods for online education course delivery, embracing the various learning styles of learners must be considered and also prioritized. Lee et al. (2011) recommends when preparing a course, faculty and instructional designers need to address how to support students in various ways. Creating learning environments where appropriate support for student learning is designed and provided becomes critical, particularly in online courses. Online learning should be designed around multiple learner preferences, if possible, to tailor the learning experience to each participant. Some learners prefer to separate their social and professional identities, while others are unconcerned about merging their social and professional lives (Salmon et al., 2015). With advancements in the ways that social media can be readily accessed and utilized, it is continuing to grow in diverse groups of users as well as those who engage in online learning. The continuous expansion in technology, emerging teaching modalities, and social networking resources will continue to challenge us to be open to adapting to change to improve online learning. This includes embracing the ever-growing diversity of users and adapting to their learning styles and preferences.

Chou et al., (2009) concluded that recent growth of social media is not equally distributed across all age groups and also recommended that communication programs utilizing social media must first consider the age of the targeted population to help ensure that messages reach to the intended audience. Given the range of possible disparities in Internet access that exist as it relates to race and health status, these factors do not have an effect on social media usage. Findings suggest that a new wave of technology advancements may be changing the various modes of communication throughout the United States (Chou et al., 2009). The increasing popularity of social networking sites has drawn scholarly attention in recent years, and a large amount of efforts have been made in applying social networking sites to health behavior change interventions (Yang, 2017). A meta-analysis of 21 studies examining the effects of health interventions using social networking sites was conducted. Results indicated that health behavior change interventions using social networking sites are effective in general (Yang, 2017). Korda and Itani (2013) noted that social media offers opportunities for modifying health behavior. It allows users to choose to be either anonymous or identified. People of all demographics are adopting these technologies on their computers or through mobile devices, and they are increasingly using these social media for health-related issues. By increasing interaction and engagement, social media may complement traditional health promotion by raising awareness, spreading influence, and contributing to health behavior change. It is essential that health organizations incorporate social media in their tailored communication strategies, to modernize their approaches and increase the likelihood of reaching different age groups (Levac & O'Sullivan, 2010). Korda and Itani (2013) although social media have considerable potential as tools for health promotion and education, these media, like traditional health promotion media, require careful application and may not always achieve their desired outcomes.

Health groups such as the World Health Organization (WHO) and the Centers for Disease Control and Prevention (CDC) find themselves fighting online rumors spread through social media while they themselves are working to raise public awareness of health issues through public dissemination. While these organizations can work to produce information that is timely, accurate, and reliable, there is uncertainty when it comes to what will happen with public health information once it is published via social media (Robledo, 2012). Korda and Itani (2013) further conclude that there is a need for evaluating the effectiveness of various forms of social media and incorporating outcomes research and theory in the design of health promotion programs for social media. Ghenai in 2017 concurs because people often use web searches and social media to research health issues. However, findings might reveal incorrect information which may contradict accurate medical information, thereby risking individuals being influenced by the presented misinformation and increasing their likelihood of making poor health decisions (Ghenai, 2017; Hahn & Truman, 2015).

Health communication

Health communication is defined as the study and use of communication strategies to inform and influence individual and community decisions that enhance health (The Community Guide, 2022). Health communication encompasses components of verbal and written communication. Therefore, when it comes to using health communication strategies to help with encouraging individuals to modify behavior to improve health outcomes, it is vital that the importance and benefits of using theories and models as a foundation to help promote and encourage positive health behavior are fully understood. According to the CDC (2022a,b,c), health communication and social marketing may have some differences, but they share a common goal: creating social change by changing people's attitudes, external structures, and/or modify or eliminate certain behaviors. Often, health communications and social marketing mirror each other's purpose in helping to serve as a catalyst for promoting and encouraging positive behavior change but with the use of effective strategies.

Health communication is related to social marketing, which involves the development of activities and interventions designed to positively change behaviors. Kreuter and McClure (2004) affirm when it comes to planning and executing health communication campaigns, programs, or educational materials, certain operational decisions must be made. Thus, at the most basic level, operational decisions in planning and carrying out communication efforts include selecting credible sources, choosing message strategies, and determining optimal settings or channels for delivery of the communication. Behavior change communication is a core part of health promotion and one of the most significant strategies health professionals deploy to achieve the objectives of public health. Unfortunately, health awareness and availability of appropriate information does not necessarily translate to changes in health behavior or adoption of healthier lifestyles, despite the efforts and wishes of health professionals (Adewuyi & Adefemi, 2016).

The Rural Health Information Rural Health Hub (2022) identified some effective communication and social marketing strategies, including utilizing evidence-based strategies, acknowledging different culture and unique perspectives to foster diversity and inclusion, improving accessibility to the resources, and disseminating useful information through several avenues, such as brochures, newspaper articles, broadcasts, public service announcements, newsletters, and videos. Relaying useful and correct information through appropriate communication channels (e.g., radio/TV, social media/Internet, newspapers, brochures, and broadcast messages) can shape people behaviors (Rural Health Information Rural Health Hub, 2022). Appropriate health communication strategies can change people's knowledge and attitudes by reinforcing the positive behaviors and empowering people through positive influence and social norms (Rural Health Information Rural Health Hub, 2022).

Health equity guiding principles for inclusive communication

The CDC's *Health Equity Guiding Principles for Inclusive Communication* emphasizes the importance of inclusively. These principles are designed to help public health professionals and educators use effective communication strategies that can adapt to the specific cultural, linguistic, environmental, and diverse populations or audience of focus (CDC, 2022a,b,c). Martin McKay, founder and CEO of Texthelp, an organization whose core purpose is to help people understand and be understood defines *inclusive communication* as sharing information in a way that everyone can understand and that includes tailoring to all people and those who may have sensory, cognitive, literacy or language challenges (Texthelp, 2021).

When it comes to using the health equity guiding principles for inclusive communication using a *health equity lens* in communication planning, development, and dissemination encourages intentionally looking at the potential positive and negative impacts of proposed messages on everyone with the goal to be inclusive, avoid bias and stigmatization, and effectively reach intended audiences, ideally with input from those intended audiences (CDC, 2022a,b,c). This information is vital in embracing cultural competence when working with diverse populations of people. This is also an important and effective strategy that will prove beneficial in more ways than one. Review the following table: *CDC's Health Equity Guiding Principles: Using a Health Equity Lens* to increase your awareness of the guiding principles in using a health equity lens as it relates to culturally competent communication (Table 13.1).

TABLE 13.1 CDC's health equity guiding principles: Using a health equity lens.

Health equity concept	Action (s)
1. Long-standing systemic social and health inequities have put some population groups at increased risk of getting sick, having overall poor health, and having worse outcomes when they do get sick.	• Avoiding perpetuating health inequities in communication by considering how racism and other forms of discrimination unfairly disadvantage people. • Avoiding implying that a person, community, or population is responsible for increased risk of adverse outcomes.
2. Community engagement should be a foundational part of the process to develop culturally relevant, unbiased communication for health promotion, research, or policy-making.	• remembering that successful community engagement is a continuous process that builds trust and relationships through a two-way communication process. • Starting with mindfulness and listening and continuing with joint decision-making and shared responsibility for outcomes.
3. Health equity is intersectional; diversity exists within and across communities and can be defined by several factors.	• remembering that people belong to more than one group and, therefore, may have overlapping health and social inequities, as well as overlapping strengths and assets. • Understanding that there is diversity within communities and members of population groups are not all the same in their health and living circumstances. • Acknowledging that communities can vary in history, culture, norms, attitudes, behaviors, lived experience, and many other factors.
4. Public health programs, policies, and practices are more likely to succeed when they recognize and	• Using language that is accessible and meaningful

TABLE 13.1 CDC's health equity guiding principles: Using a health equity lens.—cont'd

Health equity concept	Action (s)
reflect the diversity of the community they are trying to reach.	and tailoring interventions based on the unique circumstances of different populations. • Emphasizing positive actions and highlighting community strengths and solutions. • recognizing that some members of your audience of focus may not be able to follow public health recommendations due to their cultural norms, beliefs, practices, or other reasons.
5. Not all members of your audience of focus may have the same level of literacy and, specifically, health literacy.	• recognizing both the ability to read and the ability to understand the content in the language presented. • Using active verbs, plain language, and accessible channels and formats so all your audience can access and understand the information.

Source: CDC's Health Equity Guiding Principles: Using a Health Equity Lens. Adapted from the Centers for Disease Control and Prevention.

Skill building discussion questions activity

1. Which, if any, of these health equity concepts have you identified as an area that you could improve in and why?
2. If you are preparing a presentation for a diverse group of people, what health equity concept action(s) might you consider and why?
3. Why is it important to consider using a health equity lens?
4. Which, if any, of these health equity concepts might you consider a strength of yours?

Social marketing and public health

One defining characteristic of all social media is the potential to facilitate engagement—the magnitude of synchronous communication and collaboration among a multitude of people via the use of interactive technology yet in different places, thereby enabling public health organizations to move from typical traditional mass media to interactive information sharing dialog (Heldman, 2013). Social media has become a utilized platform for public health and public health interventions due to the benefits of social marketing and proves very beneficial because *public health* promotes and protects the health of people and the communities where they live, learn, work, and play (APHA, 2022). While *social marketing* is used to influence an audience to change their behavior for the sake of social benefits such as improving health, preventing injuries, protecting the environment, or contributing to the community (Kotler & Lee, 2008). Maibach et al. (2006) make a compelling case for the important role that marketing has played in public health and clearly defines key constructs of marketing. The authors suggest that core marketing activities (i.e., conducting customer research, building sustainable distribution channels, and improving access to easily adopted programs) can enhance the adoption and implementation of health behaviors and practices, specifically, evidence-based prevention strategies. Furthermore, they advocate that the effective marketing of evidence-based health programs can help close the gap that exists between public health research and everyday practice (Maibach et al., 2006).

Marketing is not well known or understood in the public health community, in part because it is rarely taught in public health or medical schools. Nonetheless, the time has come to increase our awareness, training, and application of health marketing strategies and principles, because the field holds great promise for increasing the adoption of health promotion and protection information and interventions (Bernhardt, 2006). Bernhardt (2006) references that the CDC defines health marketing as "creating, communicating, and delivering health information and interventions using consumer-centered and science-based strategies to protect and promote the health of diverse populations." A study offers new and important implications for health communication in this digital age: among Internet users, social media is found to penetrate the population regardless of education, race/ethnicity, or healthcare access (Chou et al., 2009). Findings also prompt the consideration of the connection between technologies and health disparities since the data point to the fact that social media are penetrating individuals of different demographics at the same rate. It would prove beneficial with the effective use of social media as communication and health promotion platforms to help contribute to reducing the health disparities gap. These media will not enable targeted communication messages but may have the capacity to reach a wider audience than traditional media have been able to reach (Chou et al., 2009).

Health promotion, prevention, and psychosocial health

There has been limited research on the adverse mental health consequences associated with social media and its ever-growing popularity among users.

It has been concluded that as social media started gaining popularity, the severity of mental health of Americans in the United States worsen (Patel et al., 2007; Twenge et al., 2019). However, authors (Sadagheyani & Tatari, 2021) have recently investigated the role of social media on mental health, and the findings revealed that social media does have an impact on mental health. It may be surprising that the impact includes both negative and positive outcomes. The negative effects included anxiety, depression, loneliness, poor sleep quality, poor mental health indicators, thoughts of self-harm and suicide, increased levels of psychological distress, cyberbullying, body image dissatisfaction, fear of missing out, and decreased life satisfaction. The positive effects included accessing other people's health experiences and expert health information, managing depression, emotional support and community building, expanding and strengthening offline networks and interactions, self-expression and self-identity, and establishing and maintaining relationships (Sadagheyani & Tatari, 2021). The study results of Braghieri et al. (2022) were consistent with the hypothesis that social media might be partly responsible for the recent deterioration in mental health among teenagers and young adults. It is up to social media platforms, regulators, and future research to determine whether and how these effects can be alleviated. The study findings of De Choudhury (2013) also suggested that individuals' social environments contain vital information useful for understanding and intervening on mental health. These authors' conclusions affirm the need for diverse interventions and strategic measures be considered to help address these known concerns regarding social media usage and the adverse effects on mental health. The increasing recognition of culture as an important factor in public health and health communication has the potential to contribute to the development of new and more effective strategies to help eliminate health disparities. However, the evidence base supporting such a focus is currently underdeveloped (Kreuter & McClure, 2004). Researchers Helfer et al. (2020) have been innovative in developing an action model designed to support health promotion and prevention to help improve psychosocial health and reduce health disparities.

According to Helfer et al. (2020), *psychosocial health* is as a complex interaction between the psyche of an individual and the social environment in which they live. To determine how health promotion and prevention could be used to promote the psychosocial health of individuals, an action model was developed by Helfer et al. in 2020 (Fig. 13.1). Helfer et al. (2020) describes the health promotion, prevention, and psychosocial model as an innovative action model that will support health professionals in promoting psychosocial health among target populations. In this figure, health promotion and prevention are joined with a dotted line, indicating they are united. The two gray triangles between health promotion and prevention symbolize the interdependence that salutogenesis has with health promotion and prevention. The triangle with

286 SECTION | IV Social media, public health communication, and pedagogy

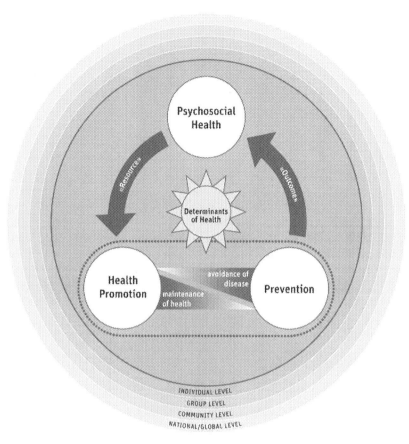

FIGURE 13.1 The innovative action model, health promotion, prevention, and psychosocial health. Bern university of applied Sciences, Switzerland, 2020. From *Helfer, T., Faeh, D., Luijckx, E., Frey, S., Berg, A., & Peter, K.A. (2020). The connection between health promotion, prevention, and psychosocial health: An innovative action model.* The Open Public Health Journal, 13(1).

health promotion illustrates what makes people healthy, and the triangle with prevention symbolizes avoiding what makes people ill.

Helfer et al., 2020 describes that the sun-like symbol depicts the interconnectedness that the Determinants of Health have with health promotion, prevention, and psychosocial health. The Determinants of Health impact the health of all individuals within their environments. Social Determinants of Health are not limited to one's physical environment, education, employment status, or access to healthcare resources. Improved statuses in these Social Determinants of Health are vital to health promotion and prevention strategies. Helfer et al., 2020 further describe that the circles surrounding the graphic portray how this action model can be used in all population health levels (individual, group, community, and global). The arrow labeled "outcome" written inside is the "action" part of the model. This displays how health promotion and prevention strategies can be used to promote psychosocial health. The improved psychosocial health of individuals is seen as an outcome. The arrow with "resource" represents that an individual who has adequate resources and sufficient psychosocial health, it is likely that she/he will be more motivated and able to maintain health and to avoid illness. Helfer et al., 2020 acknowledge limited resources may make it challenging to address preventive measures. This action model consists of established concepts, models, and frameworks in health promotion and prevention, making it an innovative model that is collectively constructed to be effective in its approach to address psychosocial health. See Fig. 13.1 and Table 13.2 below to learn more about each construct of this model.

Helfer et al., 2020 conclude that the use of health promotion and prevention is an innovative model approach for the promotion of psychosocial health in public health. The model is designed to guide health professionals in utilizing health promotion and prevention to promote psychosocial health. This innovative model would be a new approach to help promote psychosocial health among various populations (Helfer et al., 2020).

TABLE 13.2 Constructs of the innovative action model.

Name of the construct	What does it mean?
Psychosocial health	The psychosocial health is a dynamic relationship between the mental status and social dimension of a person psychosocially healthy and ill.
Health promotion	Health promotion is the process of enabling people to have more control to improve their health outcomes.
Prevention	Prevention refers to the activities that seek to reduce the predictable outcomes and to protect the current health status.
Relationship between health promotion and prevention	Health promotion and prevention differ when it comes to their approaches to health and illness. Health promotion often focuses on a positive and holistic view of health (Tengland, 2009). Preventative measures focus on diseases and aim to prevent them from occurring, reduce risk, detect them at an early stage, or to reduce further complications (Tengland, 2009). Health promotion and prevention commonality is to improve the health of the population.

Chapter summary

Because it is both an essential component and a major contributing cause to health, education achievement should be a legitimate arena for public health intervention. Thus, public health practitioners could legitimately promote educational programs to advance public health. Education should be essential for disrupting the cycle of poverty and inequities in health. The public health community should expand research better to understand the causal relationships between education and health and thereby identify evidence-based educational policies that have great potential to improve public health (Hahn & Truman, 2015). Technology-based approaches could make a significant contribution to the growth in expanding informative public health information. With social media platforms ever expanding in development and diversity, it proves beneficial for educators and health professionals to utilize these resources in hopes that such efforts might expand reaching more people. Thereby increasing awareness about critical public health issues in hopes such would inadvertently help increase access to preventive measures and improve the overall quality of healthy life for many.

Chapter key terms

Health communication: The study and use of communication strategies to inform and influence individual and community decisions that enhance health (The Community Guide, 2022).

Inclusive communication: As sharing information in a way that everyone can understand and that includes tailoring to all people and those who may have sensory, cognitive, literacy, or language challenges (Texthelp, 2021).

Public health: Promotes and protects the health of people and the communities where they live, learn, work, and play (APHA, 2022).

Social marketing: Is used to influence an audience to change their behavior for the sake of social benefits such as improving health, preventing injuries, protecting the environment, or contributing to the community (Kotler & Lee, 2008).

Social media: Online platforms which facilitate interaction among users, including the creation and distribution of information, content, and ideas (Kaplan & Haenlein, 2010).

Psychosocial health: Is defined as a complex interaction between the psyche of an individual and their social environment or community in which they live (Helfer et al., 2020).

Online learning: A term to describe an emerging approach to learn at students' own premise through advanced information and communication technologies (such as Blackboard, Moodle, YouTube, Virtual Reality) either asynchronously or synchronously (Yang & Kang, 2020).

Chapter review questions

1. Discuss what social media is.
2. Define health communication.
3. Differentiate between advocacy and public health.
4. Why is inclusive communication important?
5. Describe the constructs of the health promotion, prevention, and psychosocial health model and explain how they correlate.
6. How might you describe the implications of social media and online learning on public health?
7. What are some pros and cons of social media?
8. What factors should educators and health professional consider when it comes to the impact of social media on mental and psychosocial health?
9. How might you use social marketing effectively?
10. What are the constructs of the psychosocial health model?

Websites to explore to learn more

CDC social media tools, guidelines, and best practices

https://www.cdc.gov/SocialMedia/Tools/guidelines/

This website will provide you with tips for planning, development, and implementation of social media activities. It is equipped with various guidelines that will provide vital information on lessons learned and best practices.

Suggested activities to the readers

Explore to discover what social media activity you could implement. What did you find most interesting after visiting this website?

CDC public health media library

https://tools.cdc.gov/medialibrary/index.aspx#/learnmore

This website will provide you with access to credible, free health content for websites, apps, and social media.

Suggested activities to the readers

Visit this website and explore and select health education resources that you could use. What was most beneficial resource you found that you could use?

Quality matters

https://www.qualitymatters.org/

This website will provide you with resources that will increase your awareness of quality online course development.

Suggested activities to the readers

Visit this website to learn how you could earn an online teaching certification, attend free webinars, and learn about how to effectively design an online course. What free webinars might you consider participating in or viewing? What tips did you learn after exploring this website when it comes to effectively designing a course for online?

American public health association

https://www.apha.org/What-is-Public-Health

This website will provide you with helpful resources that will increase your awareness about the American Public Health Association (APHA).

Suggested activities to the readers

Visit this website to learn more about public health. Can you identify ways that you could contribute to increasing public health awareness within your community? What did you learn after exploring this website when it comes to public health issues? What public health issue is most important to you?

Texthelp

https://www.texthelp.com/resources/

This website will provide you with helpful resources that will increase your awareness about inclusive communication.

Suggested activities to the readers

Visit this website to learn more about the importance of inclusive communication resources. Can you identify what resources could be beneficial to you and/or others? What did you learn after exploring this website when it comes to communication and diversity and inclusion? What webinar resources could you recommend to others?

References

Adewuyi, E. O., & Adefemi, K. (2016). Behavior change communication using social media: A review. *International Journal of Communication and Health, 9*, 109–116.

American Public Health Association. (2022). *What is public health?* https://www.apha.org/what-is-public-health.

Appel, G., Grewal, L., Hadi, R., & Stephen, A. T. (2020). The future of social media in marketing. *Journal of the Academy of Marketing Science, 48*(1), 79–95.

Bernhardt, J. M. (2006). Improving health through health marketing. *Preventing Chronic Disease, 3*(3). https://www.cdc.gov/pcd/issues/2006/jul/05_0238.htm.

Braghieri, L., Levy, R. E., & Makarin, A. (2022). *Social media and mental health.* ECONSTOR. https://www.econstor.eu/bitstream/10419/256787/1/1801812535.pdf.

Centers for Disease Control and Prevention. (2022a). *Health equity guiding principles for inclusive communication.* https://www.cdc.gov/healthcommunication/Health_Equity.html.

Centers for Disease Control and Prevention. (2022b). *CDC media library.* https://tools.cdc.gov/medialibrary/index.aspx#/learnmore.

Centers for Disease Control and Prevention. (2022c). *Social media tools, guidelines and best practices.* https://www.cdc.gov/SocialMedia/Tools/guidelines/.

Chou, W. Y. S., Hunt, Y. M., Beckjord, E. B., Moser, R. P., & Hesse, B. W. (2009). Social media use in the United States: Implications for health communication. *Journal Of Medical Internet Research, 11*(4), e1249.

De Choudhury, M. (October 2013). Role of social media in tackling challenges in mental health. In *Proceedings of the 2nd international workshop on Socially-aware multimedia* (pp. 49–52).

Dobre, C., Milovan, A. M., Duţu, C., Preda, G., & Agapie, A. (2021). The common values of social media marketing and luxury brands. The millennials and generation Z perspective. *Journal of Theoretical and Applied Electronic Commerce Research, 16*(7), 2532–2553.

Ghenai, A. (July 2017). Health misinformation in search and social media. In *Proceedings of the 2017 international Conference on digital health* (pp. 235–236).

Hahn, R. A., & Truman, B. I. (2015). Education improves public health and promotes health equity. *International Journal of Health Services: Planning, Administration, Evaluation, 45*(4), 657–678. https://doi.org/10.1177/0020731415585986

Heldman, A. B., Schindelar, J., & Weaver, J. B. (2013). Social media engagement and public health communication: Implications for public health organizations being truly "social". *Public Health Reviews, 35*(1), 1–18.

Helfer, T., Faeh, D., Luijckx, E., Frey, S., Berg, A., & Peter, K. A. (2020). The connection between health promotion, prevention, and psychosocial health: An innovative action model. *The Open Public Health Journal, 13*(1).

Kaplan, A. M., & Haenlein, M. (2010). Users of the world, unite! The challenges and opportunities of Social Media. *Business Horizons, 53*(1), 59–68.

Korda, H., & Itani, Z. (2013). Harnessing social media for health promotion and behavior change. *Health Promotion Practice, 14*(1), 15–23. https://doi.org/10.1177/1524839911405850

Kotler, P., & Lee, N. (2008). *Social marketing: Influencing behaviors for good* (3rd ed.). Thousand Oaks, CA: Sage Publications.

Kreuter, M. W., & McClure, S. M. (2004). The role of culture in health communication. *Annual Reviews of Public Health, 25*, 439–455.

Lee, S. J., Srinivasan, S., Trail, T., Lewis, D., & Lopez, S. (2011). Examining the relationship among student perception of support, course satisfaction, and learning outcomes in online learning. *The Internet and Higher Education, 14*(3), 158–163.

Levac, J. J., & O'Sullivan, T. (2010). Social media and its use in health promotion. *Interdisciplinary Journal of Health Sciences, 1*(1), 47–53.

Maibach, E. W., Van Duyn, M. A. S., & Bloodgood, B. (2006). Peer reviewed: A marketing perspective on disseminating evidence-based approaches to disease prevention and health promotion. *Preventing Chronic Disease, 3*(3).

Patel, V., et al. (2007). Mental health of young people: A global public-health challenge. https://www.thelancet.com/journals/lancet/article/PIIS0140-6736(07)60368-7/fulltext.

Robledo, D. (2012). Integrative use of social media in health communication. *Online Journal Of Communication and Media Technologies, 2*(4), 77.
Rural Health Hub. (2022). Health communication. https://www.ruralhealthinfo.org/toolkits/health-promotion/2/strategies/health-communication.
Sadagheyani, H. E., & Tatari, F. (2021). Investigating the role of social media on mental health. *Mental Health and Social Inclusion*. https://www.proquest.com/docview/2533166242?fromopenview=true&pq-origsite=gscholar.
Salmon, G., Ross, B., Pechenkina, E., & Chase, A. M. (2015). The space for social media in structured online learning. *Research in Learning Technology*, 23.
Tengland, P. A. (2009). Power, empowerment and health. In *Presented at the European Conference on Philosophy of Medicine and Health Care*. Germany: Tübingen. http://urn.kb.se/resolve?urn=urn:nbn:se:mau:diva-10768.
Text Help. (2021). *Inclusive communication matters. for us, it's personal*. https://www.texthelp.com/resources/blog/inclusive-communication-matters-for-us-its-personal/.
The Community Guide. (2022). *Health communication and health information technology*. https://www.thecommunityguide.org/topic/health-communication-and-health-information-technology.
Twenge, J. M., Cooper, A. B., Joiner, T. E., Duffy, M. E., & Binau, S. G. (2019). Age, period, and cohort trends in mood disorder indicators and suicide-related outcomes in a nationally representative dataset, 2005–2017. *Journal of Abnormal Psychology, 128*(3), 185.
Yang, Q. (2017). Are social networking sites making health behavior change interventions more effective? A meta-analytic review. *Journal of Health Communication, 22*(3), 223–233.
Yang, K. C., & Kang, Y. (2020). What can college teachers learn from students' experiential narratives in hybrid courses?: A text mining method of longitudinal data. In L. Makewa (Ed.), *Theoretical and practical Approaches to Innovation in higher education* (pp. 91–112). IGI Global. https://doi.org/10.4018/978-1-7998-1662-1.ch006

Further reading

Kotler, P., & Zaltman, G. (1971). Social marketing: An approach to planned social change. *Journal of Marketing, 35*(3), 3–12. https://doi.org/10.2307/1249783
Mittelmark, M. B. (1999). The psychology of social influence and healthy public policy. *Preventive Medicine, 29*(6), S24–S29.
Quality Matters. (2022). *Quality matters*. https://www.qualitymatters.org/.
Thackeray, R., Neiger, B. L., & Keller, H. (2012). Integrating social media and social marketing: A four-step process. *Health Promotion Practice, 13*(2), 165–168. http://www.jstor.org/stable/26739546.

Section V

Social media in healthcare

Chapter 14

Application of social media in designing and implementing effective healthcare programs

Priyanka Saluja[1], Vishakha Grover[2], Suraj Arora[3], Kavita Batra[4,5] and Jashanpreet Kaur[6]

[1]*Department of Dentistry, University of Alberta, Edmonton, AB, Canada;* [2]*Department of Periodontology and Oral Implantology, Dr. H.S.J. Institute of Dental Sciences, Panjab University, Chandigarh, Punjab, India;* [3]*Department of Restorative Dental Sciences, College of Dentistry, King Khalid University, Abha, Saudi Arabia;* [4]*Department of Medical Education, Kirk Kerkorian School of Medicine at UNLV, University of Nevada, Las Vegas, NV, United States;* [5]*Office of Research, Kirk Kerkorian School of Medicine at UNLV, University of Nevada, Las Vegas, NV, United States;* [6]*Dr. H.S.J. Institute of Dental Sciences, Panjab University, Chandigarh, Punjab, India*

Learning objectives:
- To describe the benefits of social media for people to seek health-related information.
- To discuss the different social media platforms for healthcare providers.
- To explain the utility of social media for healthcare programs.
- To discuss the application of social media analytics in health program planning and evaluation.
- To describe the role of health education specialists in delivering social media interventions.
- To delineate the limitations and challenges of using social media for healthcare.

Introduction

Social media (Web 2.0) has a broad definition that is always changing. The term generally refers to the online tools that permit real-time information sharing, engagement, and communication between people and society (Chauhan et al., 2012; Lambert et al., 2012; Peck, 2014; von Muhlen & Ohno-Machado, 2012).

Social media use is ubiquitous today with over 4.5 billion users being reported in October 2021 (DataReportal, n.d.). Media sharing via Flickr or YouTube, professional networking via LinkedIn, blogs or microblogs via Tumblr/Twitter, information aggregation via Wikipedia, and social networking through Facebook, TikTok, Twitter, etc., are some of the commonly used social media applications. Besides the social integration, social media plays an important role in patient care, patient education/motivation/communication, and to better healthcare professional networking, public health awareness programs, and research. This chapter discusses different types and uses of social media in designing and implementing effective healthcare and promotion programs and also the associated challenges.

Role of social media in healthcare

The healthcare practitioners (HCPs) utilize social media for sharing knowledge and skills in addition to their personal and practical experiences (Alanzi & Al-Habib, 2020; Alanzi & Al-Yami, 2019). Additionally, there is available literature citing the use of social media by the HCPs in the health promotion by raising awareness about a particular health phenomenon and also to explore research opportunities (Alanzi & Al-Habib, 2020; Alsobayel, 2016; Justinia et al., 2019). Social media can also be used by physicians to engage the patient and interact with them through effective strategies. The information about medical organization or private practice can be shared through Facebook, Instagram, or Twitter. Additionally, HCPs can use social media in facilitating patient-to-patient connections through their networks. By developing routine patient communication techniques and building online communities for patients to interact with one another, medical professionals can help increasing the patient compliance to the treatment and/or lifestyle recommendations (Lin & Zhu, 2012). This highlights the role of peer support online groups as a part of solution to improve patients' adherence to the treatment (Lin & Zhu, 2012). Hoedebecke and Colleagues (2017) noted the role of health literacy in improving the patient compliance and how the use of social media can be optimized to ensure effective communication between healthcare providers, patients, and also the general public.

The existence of a web is important for the general public as they look to social media and the internet for answers related to their medical issues (Hoedebecke et al., 2017). Therefore, it is critical to evaluate the accuracy of the information as the false information can cause more harm than good (Southwell et al., 2020). Here, HCP and health educators play a critical role in educating the general public as well. According to the U.S. Bureau of Labor Statistics, the primary responsibility of a health educator is to evaluate the health needs of the community (Health Education Specialists and Community Health Workers, n.d.). Social media is important for public health educators to create a forum for patients to discuss concerns pertaining to their health

(Southwell et al., 2020). By offering health-related education that balances out erroneous information that patients could receive online elsewhere, HCPs and public health educators play a critical role in society (Southwell et al., 2020). HCPs and health educators can establish themselves online as a reliable source of information in order to help reduce the spread of medical misinformation (Southwell et al., 2020). HCPs and public health educators may use several social media platforms to ensure the flow of accurate information in an effective manner.

Role of social media in health promotion

In health promotion, social media is commonly used for the community building purpose as well as to inform healthcare decisions through an internet-mediated dialog (Southwell et al., 2020). These days, social media is predominantly used for the public health education and promotion, especially in the United States (Southwell et al., 2020). The integration of eHealth with public health has an ability to reduce barriers associated with the healthcare accessibility and community outreach, thereby promoting health equity (Southwell et al., 2020). The research landscape related to the effective use of social media in health promotion is still growing, and there is a need to develop culturally inclusive public health programs leveraging the new technological advancements (Hunter et al., 2019). A collective evidence suggests that social network interventions are positively associated with a wide range of health behaviors and outcomes (Hunter et al., 2019). Social media holds opportunities for health education specialists being involved in the formative research and evaluation planning, for consumers to share strategic health messages, and for HCP and researchers to conduct the health impact analysis (Bardus et al., 2020; Zhao & Zhang, 2017).

Health education specialists play an integral role in developing, directing, and evaluating health promotion programs and need to be well-versed with the use of social media and computer-mediated contexts (Stellefson et al., 2020). The National Commission for Health Education Credentialing, Inc. (NCHEC) and the Society for Public Health Education (SOPHE) started a new health education initiative—Health Education Specialist Practice Analysis II (HESPA II 2020), which outlined competencies and knowledge areas for health education specialists (Greenhalgh et al., 2017; Stellefson et al., 2020). The HESPA II 2020 identified a new model consisting of eight areas of responsibility, 35 competencies, and 193 subcompetencies of health education profession (Greenhalgh et al., 2017; Stellefson et al., 2020). As indicated in Table 14.1, two areas of responsibility of the health education specialists include advocacy and communication, which are directly related to the effective use of social media (Greenhalgh et al., 2017; Stellefson et al., 2020). In the responsibility area of communication, evaluation is the central to most social media activities within the health promotion field (Stellefson et al.,

TABLE 14.1 HESPA II 2020 competencies and subcompetencies addressing the use of social media by the health education specialists.

Responsibility area	Competencies	Subcompetencies
Advocacy	1. Engage in the collaborative networks and stakeholder groups to address health issues and in planning advocacy efforts	Identify strategies for the proposed policy change leveraging social media, political leaders, and stakeholders
	2. Engage in advocacy	Using social media (e.g., press releases, public service announcements for advocacy)
Communications	1. Identify factors that affect communication with the target audience	Identify social media channels available and accessible to/by the target audience
	2. Deliver the message appropriately using effective social media strategies	Utilize new communication tools to engage target audience
	3. Evaluating communication	Investigate reach and dose of communication using website and social media analytics

Source: Stellefson et al. (2020).

2020). Fig. 14.1 describes how social media can be utilized in evaluating communication.

As indicated earlier, social media can be used by a wide array of audience, including health institutions, researchers and healthcare professionals, and the general public (Chan et al., 2020; Chen, 2017).

- *Use by health institutions*: It includes all sorts of health corporations, for example, Government, Semigovernment, and Private, which use social media for surveillance and dissemination of health information. Further it tends to prevent the propagation of any misleading health-related knowledge and information (Chen, 2017). For the purpose of supervising, monitoring, and exploring the unorganized information available on the web, infoveillance is used to inform public and policy makers (Borges do Nascimento et al., 2022; Sutton et al., 2020).

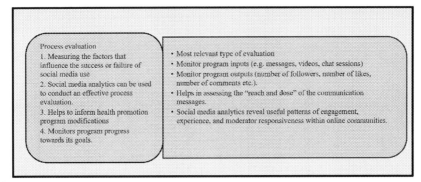

FIGURE 14.1 Process evaluation using social media analytics. *Note: This image has been created by the authors using the information from a source article by Stellefson et al. (2020) and Welch et al. (2016).*

- Next, the data available on social media can be investigated to predict the possible epidemic in a population. For example, in the recent COVID19 pandemic (Patrick et al., 2019), the statistics on social media helped us to know the burden and trajectory of disease, which helped with the resource planning (Grimes, 2020). Reportedly, Twitter was the maximally used social media for the surveillance for isolated disease cases and epidemics (Xie et al., 2020).
- Individual posts on social media help in revelation about people's perception and practices toward matter related to health. Studies have evaluated the available data on social media to know the people's insight toward the recent epidemics of COVID-19 (Steffens et al., 2019) and Zika virus (Mheidly & Fares, 2020). Some studies investigated people's viewpoint for the actual execution of the required actions to prevent the disease (Cuello-Garcia et al., 2020), kept a surveillance on people's opinion about the debatable topics on health, such as vaccination for human papilloma virus (Southwell et al., 2020), and to figure out the people's support for the policies on health (Gesser-Edelsburg et al., 2018). It was found that evaluating social media data for people's perception toward issues related to health provided consistent results with the national surveys (Southwell et al., 2020). Therefore, in comparison to time-consuming older survey techniques which required big budgets, social media helps save valuable time and money along with allowing timely implementation of the required action plans as suited for the needs of the public (Southwell et al., 2020).

Role of social media in circulating health facts and ceasing the wrong information

Health organizations can keep the people well informed about the day-to-day health-related or lifestyle-related topics, such as smoking and also about the consequences of epidemics, which immediately helps to adopt preventive practices to the public during epidemics (Ashton et al., 2017; Diddi & Lundy, 2017; Li et al., 2017). There are continuous experimental efforts to identify different modalities to produce health data on social media and to escalate retransmission (Chung, 2016; Park et al., 2016). It was also observed that there was a lot of false information circulated during the COVID-19 pandemic, and several studies raised the concern of fast misleading dissemination on social media and advocated modalities to curb the spread of such data. Scientists requested health organizations to observe and do quick data check on doubtful data/content on social media (Li et al., 2017; Mendoza et al., 2017; Yu et al., 2020) to refute false news and also used enhanced search engine modalities for redirecting the users looking data on health to authentic sources (Fang et al., 2019). This way of reducing the false content by the way of social media, the health organizations may also promote the development of personal engagements between the health officials and the patients and also asking the patients to come forward to speak about the false content they experience on social media (Pagkas-Bather et al., 2020). Several studies have investigated efficacy of various message layout traits for rectifying false content (Thornber et al., 2019). A study stated that theory-focused rectification content published on Facebook by health administration was more helpful in clarifying false information about vaccination than a normal message. So the health professionals may follow such modalities to design messages as put forward by these studies to make and publish false information rectifying notes on social media (Thornber et al., 2019).

In today's world, not only social media enhances the delivery of the health inventions but also it helps in collecting the feedback and critical analysis from the audience so that the information collected maybe utilized to improve the design of the social media-based health interventions (Chan et al., 2020). In general, suggestions from masses can be integrated very easily and cost effectively, without the physical meetings even, so as to make the future planning of public healthcare interventions more effective (Ronen et al., 2020). In fact, while building up the plans for public health, we may utilize social media to actively invite and involve populations with the advantage of maintaining their privacy and anonymity, in order to have minimally biased opinions (Steffens et al., 2019).

As social media has an advantage of reaching a wide variety of audience and stakeholders, thus it may further help private health organizations to mobilize social resources (Young & Scheinberg, 2017). Many times, it has been observed that such associations have utilized the media as a platform to emphasize and advocate the need for change in social health policies. In fact it has also served as a portal to raise funds for individual as well as group medical care and research (He et al., 2020). Social media may also help showcase various weaknesses and strength of the existing healthcare system including the availability and access to medical equipment and services (Fu & Zhang, 2019).

Role of social media in research

Social media impacts the health-related research, and broadly by two possible ways, it may help us know the patients' perspective better in terms of their understanding of disease concerns and various inhibitions to opt for many health-related behaviors (Jiang et al., 2020). Many times patient self-reported disease experience are not reported to a specific clinical environment rather on the social media in terms of reviews and feedbacks (Oser et al., 2019). Moreover, it may serve as an excellent platform to involve research participants from general population, studies have actually documented this advantage of social media over the much expensive traditional recruiting method and are cost effective to provide the information to "difficult-to-reach" population groups also (Guidry et al., 2016).

Role of social media in professional development

Social media may help the researchers and professionals to enhance their professional learning so that they have better chances of collaborations (Helm & Jones, 2016) and better careers options. Social media (Loeb et al., 2020) also makes it much easier to assess new career opportunities, explore potential funding agency, and stay updated regarding health events and conferences (Chen & Wang, 2020). Majority of the health professionals, approximately 2/3rd of population in a study, have indicated positively regarding the use of social media in health career development (Borgmann et al., 2015).

Role of social media in improving doctor—patient communication and health services

Social media may also be employed for better communication between the physician and the patient now a days offering teleconsultation or online consultation which has been much more in trend all the more so in the times of the COVID-19 pandemic (Benetoli et al., 2017). Social media has supported the health services to the people as it's also an important venue for pharmacists (Hermansyah et al., 2019). In fact, social media portals may be utilized for the intimation of the results of examination, compliance to medication, and regarding feedback of the health services. Social media can also be utilized to maintain the database in terms of patient appointments, queries, patient information, and the payment records to enhance the patient experience (Liu et al., 2020; Shen et al., 2019).

Popular social media platforms for doctors

Social networking sites: The establishment of medically focused professional communities is the result of social networking. These are the private networks for professionals only. Professional associations, advertising or data sales, research funding, and pharmaceutical firms are some of the sources of funding and financial assistance (Ventola, 2014).

"SERMO" is a "physician-only" program that demands credentials verification prior to registration. On this network site, medical professionals from 50 states and roughly 68 specializations come together to debate various case studies and solicit peer advice (Fogelson et al., 2013). Beginning in April 2014, SERMO expanded to include 260,000 positions held by United States citizens who choose to remain anonymous. SERMO is primarily a forum where medical professionals can debate treatment alternatives. This network also offers a rating system, whereby doctors score posts on the website according to perceived reliability (Chretien & Kind, 2013).

Another website for medical directors that offers a secure environment for interaction is The Medical Directors Forum (www.medicaldirectorsforum.skipta.com). This website's large library, group discussions, and notifications serve as its primary resources. Additionally, medical directors who work for hospitals, Medicare, veterans' affairs, group practices, behavioral healthcare management, and correctional facilities can benefit from it.

The intimacy of a "physician's lounge" is recreated online as necessary by these sites by doctors or other healthcare professionals submitting their credentials to a site gatekeeper.

Pharmacists can also use social networking sites, and some of them include ASHP Connect (www.connect.ashp.org), which is sponsored by the American Society of Health-System. Professional networking forums for nurses include the American Nurses Association and Nurse Space (www.ananursespace.org), Nursing Link (www.nursinglink.com), and SocialRN (www.twitter.com/socialRN).

Blogs: The "blog" a term formed from "web log" is the most known form of social media; it has been used widely in the medicine field since 2004. Blogs can reach worldwide; the content which becomes viral can establish interest in the public and the audience worldwide. A lot of information can be shared in an open forum in the form of text, video, audio, and animations. This blogging platform allows readers to submit their own comments, and the administrator can reply to those remarks. Examples of long-form blogging platforms are Blogger, WordPress, and Tumblr (www.wordpress.org) (www.blogger.com).

Blogs can be used for communication between patients and medical practitioners. For instance, the Clinical Case blog showcases case studies from a range of different specialties. Pharmacists use blogs very commonly. There is a unique section on admission note, procedure guidelines, and material related.

Microblogs: Microblogs provide most simplified form for the exchange of information via social media. Primary advantage of microblogs is that maximum number of messages and updates can be posted in shorter period. Twitter is the most used and most known platform of microblog. Users can post tweets with a maximum character count of 140. Links to other online media, including websites or movies, can be included in tweets. On Twitter, which is a kind of information indexing, hashtags can be used to help individuals find tweets that are relevant to their search. The most common hashtags are #HTSM, #MDChat, and #Health (Grajales et al., 2014).

Other networks used by professionals: The social media sites which allow medical professionals to participate online and listen to experts (zoom meetings) and communicate with other healthcare professionals regarding case studies. Beside the case studies and treatment discussion topics, these sites highlight diversity of subjects such as biostatistics, politics, ethics and management of the practice, and career strategies in medicine. Supportive environment for medical professionals who specialize can also be provided by these sites.

Another excellent method of professional networking is crowdsourcing, which involves using the expertise of a community to solve issues and exchange ideas with other practitioners. Additionally, third-world nations can communicate with medical professionals in more developed nations. For instance, surgical procedures can be streamed using networking, and questions can be submitted via Twitter. These networking websites enable the sharing and exchange of information.

Benefits of using social media by healthcare professionals

Medical education: Social media can be used as a dynamic platform to educate healthcare professionals in the medical field. Numerous researches have discussed how social media can help clinical students comprehend communication, professionalism, and ethics, such as the pharmacy undergraduate program, which makes extensive use of social media. For instance, Facebook is used as a platform for teaching by 38% of pharmacy professors (Ventola, 2014). As an example, a professor at the University of Rhode Island teaching geriatric pharmacology courses used Facebook to improve theory lectures and engage with their students and seniors who actively participate. It has also affected the education for the nurses, with one service reporting 53% of nursing schools are using Facebook as a social media tool. For example, Twitter has reported to improve decision-making skills of medical professional in emergency situations in a study in which the students were made to watch case discussions, clinical scenarios, and tweeted their observation on patient's condition (Ventola, 2014).

YouTube can also be used in the smart class to improve discussion and to make more clear concepts. Students can watch a video and then they are made to respond to question asked as a clinical exercise. The objectification of social media into health education has met with opposing reviews. Although the use of social media and various networking sites has generally been viewed favorably, some scholars have claimed that using Facebook to teach clinical education interferes with their personal lives. They have also reported that while social media has improved paternal communication, the increased distraction in the educational environment makes it challenging. Unlikely, rules guiding the suitable use of social media tools in education are in their immaturity (Grajales et al., 2014).

Health education and awareness: Information is shared via social media which helps medical professionals to discuss healthcare policy and emergency issues; it also enables them to interact with members of the public as well as the patient's caretakers and coworkers. Health-related outcomes can be improved, professional networks can be developed, news and new additions can be shared, patients can be encouraged, and the community can receive hygiene education. Various online communities are joined by healthcare professionals where they can explore articles, lend an ear to the experts, research medical development, and consult among the team regarding case discussion emergency. Ideas can be shared about the management of practice, make and advertise their research, market their practices, or indulge in healthy activism. Studies show that more than 90% of HCPs use social media for conditioning purposes but just 65% for raising public awareness of health issues.

Unlike medical healthcare professionals, pharmacists have adopted to social media slowly. Pharmacists use Facebook, where there are over 90 pages devoted to the industry, for example, pharmacists interest page, the American pharmacist Association. 9%–10% of pharmacist use Twitter, and if we search for pharmacist on LinkedIn, nearly 274,980 profiles are found (MacMillan, 2013).

The general public: People can use social media to gain knowledge about health conditions, preventive measures, and track and share health status or activities. By self-monitoring and sharing their health status, people can construct network with likeminded people and improve their health outcomes. Writing and sharing their health experiences can also additionally assist sufferers cope with their illness and attain fitness goals. Other people can learn from such sharing and can effortlessly locate friends who have similar experiences (Chen & Wang, 2020; Shaiket et al., 2019).

People use social media to supervene news related to medical conditions like COVID-19, to seek knowledge on daily basis such as choice of food, seek for health advice for his or her own medical conditions, request for alternate opinion after the appointment from medical professionals about the diagnosis, and to approach the laws related to fitness, they subscribe various channels and follow the real content pages on social media of medical professionals. It is the primary source of detailed data for people with different requirements of information and knowledge.

Professional recommendations for using social media

Social media in the healthcare industry may potentially be misused not only in terms of safety and security of patient information but also spreading misleading inaccurate information. Therefore, it is crucial that there be a set of rules regulating how employees should utilize social media in this regard. It has been documented that only information from the creditable sources should be shared, and any misleading or inaccurate information shall not be disseminated on social media platforms. Since the content, which is easily accessible, so it is very important to respect any copyright reservations and other privacy laws on the information which is being disseminated on the social media by the organization. It is not advisable to ask patients to join personal network probably; for such request, more secure means of communication need to be in place. Social media shall not be utilized to provide very pertinent specific medical advice to patients, and for all electronic communication, standard disclosure and disclaimer regarding the accuracy, timeliness, and privacy of the content need to be there. Patients' privacy shall be respected, and wherever possible, patient consent should be obtained as needed. For the healthcare person also adhering to most secure privacy settings available is advisable, and the personal and professional profiles shall be kept separate. Any kind of disclosure or financial compensation if have been received should be mentioned on the media platform, and no misleading claim such as credentials should be stated. It needs to be clearly mentioned on to the professional sites if an employer is being represented (Chretien & Kind, 2013; Dizon et al., 2012).

Limitations and challenges

Poor quality of data: The biggest drawback of health information acquired on social media and other online sources could be a lack of responsibility and quality (Reidy et al., 2019; Sagnia et al., 2020). Medical data found on social media networks typically have unidentified authors. Furthermore, the medical information might not include references, be incomplete, or be informal. Anecdotal reports are less important in evidence-based investigations, but social media tends to place more emphasis on them, relying on individual patient accounts for overall medical data. These issues are made worse by the participatory nature of social media because any user can upload anything to a website (Sagnia et al., 2020).

There are actions being made that will be beneficial in resolving this drawback. Wherever the information is amenable to internal control, HCPs should direct patients to reliable peer-reviewed sources. The World Health

Organization is spearheading the charge in requesting allocated names and numbers from the internet corporation in order to identify an alternative domain suffix that might be used just for accurate health data. The distribution of this domain suffix would be closely controlled, and the content of websites using these addresses would be checked to ensure that strict quality standards were being met. When search engines offer leads to respond to health-related queries, these domain addresses will be given priority (Grajales et al., 2014).

Damage to skilled image: The sharing of unprofessional content on social media, which could have a negative impact on HCPs, students, and affiliated establishments, is a significant danger associated with its use. Social media portrays information about people's personalities, values, and priorities, and the first impression created by this content will endure a lifetime. Any information included in an extremely social media profile, such as images, nicknames, posts, and comments likable or shared, as well as the friends, causes, organizations, games, and media that a person follows, may help reinforce perceptions (Ventola, 2014). Violating patient privacy, using derogatory language, posting images of oneself in a sexually provocative or intoxicated state, and making disparaging remarks about patients, associate leaders, or faculty members are all examples of conduct that could be viewed as unprofessional. Such well-publicized errors by HCPs are documented, including surgeons using a camera while operating, using guns or alcohol, and posting "tweets" that are detrimental to a person or their profession. The airing of frustrations, or "venting," concerning patients conjointly happens in online forums and isn't suggested.

Information gathered from social media can even effect choices concerning admission to medical or skilled programs, choice for residencies, or employment. Employers and residency programs currently search Facebook and alternative social networking sites before hiring candidates. It is known that HCPs have engaged in such public gaffes, such as snapping photos of patients during surgery, acting recklessly with firearms or alcohol, or sending "tweets" that are detrimental to a person or their profession.

The majority of social networking sites do, however, currently allow users to tailor each of their profile contents through privacy settings, and United Nations agencies will read it. Account and privacy settings should ideally be established in a way that allows one's network to grow while restricting the visibility of data to others outside of the network. Any options made available by the social media platform that allow users to categorize various relationships so that only appropriate information is shared with particular groups or persons should even be used. HCPs should regularly search for their own names or other personal information to ensure that their social media presence presents a professional image.

Breaches of patient privacy: Many times, worries about the use of social media by HCPs revolve on the possibility of adverse consequences resulting from the breach of patient confidentiality. Such transgressions may subject healthcare providers and healthcare organizations to liability under federal HIPAA and state privacy laws (Chretien & Kind, 2013; Dizon et al., 2012). The federal HIPAA Act and state privacy laws may hold HCPs and healthcare organizations liable for such violations. The HITECH legislation expands certain standards to include business associates and specifies the privacy breach reporting requirements. HIPAA/HITECH or state privacy laws may apply in a variety of ways when posting information about a patient to a social networking site, including comments, photographs, and videos. Legal action against an associate HCP and possibly his or her boss may emerge from patient information posted on social media with or without the patient's consent and breaches of patient confidentiality. It's important to note, though, that HIPAA does not forbid the dissemination of "de-identified" medical data (Ventola, 2014). Due to social media's widespread use, it is now more typical to employ it in the healthcare industry. Medical professionals must use these platforms to the best extent possible in their future care delivery strategies. While using social media in the healthcare sector has some disadvantages, there are also a lot of benefits for patients of all ages, nurses, doctors, public health professionals, and the sector as a whole. Careers in public health or marketing in the sector are available to students and professionals interested in the impact of social media on healthcare.

Conclusions

Social media is used by the healthcare industry to advance or better professional development, networking, patient education and care, organization promotion, public health initiatives, etc. It is employed to raise awareness, combat incorrect information, manage crises, address common inquiries, share decision-making, and organize the best possible patient care. Apps, for instance, can be used to track a patient's health behaviors and encourage healthy lifestyle choices. With enhanced health-related awareness, many common diseases may be prevented/diagnosed early and may be treated better to eventually cut down on the healthcare costs as a result.

It has been demonstrated that multidimensional healthcare, which combines medical assistance with social media and other types of communication, is quite effective. Along with the benefits, there will also be some challenges. Some of the challenges to be overcome include the privacy and confidentiality of patient health information, professionalism, a lack of time, the risks of broadcasting false health information, and cultural difficulties that may affect how prepared doctors are to communicate with their patients. The range and complexity of research on how patients and medical professionals utilize social media have significantly expanded. The coronavirus illness pandemic of 2019 has brought to light the different sources of false information and how they affect healthcare on many levels. To help medical providers create an online presence to counteract disinformation or address specific patients, best practices have been established.

Chapter summary

In this era, when digital advancements have become a part of almost all spheres of our daily life, it is vital and apt for healthcare sector to embrace the evolving span of social media. The way individuals connect, interact, and communicate in the modern world has truly altered as a result of social media interactions. Since the medical sciences have transitioned into the digital age, social media use in healthcare is no exception. It has integrated on various aspects of healthcare, including the conduct of scientific studies, improvized research opportunities and platforms, healthcare organization and access to medical facilities, communication between the public and healthcare professionals via health informatics systems, and healthcare organizations. In fact, social media has helped strategies built in for comprehensive public healthcare plans across nations. The use of social media in healthcare has produced significant benefits, including improved patient–doctor communication and widespread healthcare information flow. Healthcare professionals have a lot of chances, thanks to social media platforms to network, collaborate on research, learn about new discoveries, inspire their patients, and provide accurate health information. A smaller section of healthcare community also delves in providing medical consultation and care by telecommunication to patients in remote areas, where otherwise health consultation may not be possible. Even specific disease condition-related social media groups/websites are available including general public and professionals from variety of health sectors, which aid in the dissipation of health-related information for disease prevention and management. However, there are some possible problems and concerns with the use of social media by healthcare professionals in addition to this broad variety of advantages. There may appear some untoward incidences related to breach of privacy of healthcare workers or the patients and some criminal or malicious activities, which need certain regulations or guidelines for the use of social media in healthcare. Organizations and experts must constantly monitor the dissemination of accurate information about healthcare and prevent the spread of any false information or health myths. Use and adoption of social media in healthcare are bound to stay and increase in the times to come and should be exploited to best of its potential in enhancing the quantity and quality of healthcare administration to public at large.

Questions for discussion

Through the medium of this chapter, the authors would like to set a stage to discuss the following:

- Can we use social media to improve the healthcare experience of the people?
- What are the most important objectives on which healthcare social media should concentrate?

- Can social media aid in the development of detailed public healthcare plans?
- How can social media be a portal to disseminate health-rated information aiming prevention of most widely prevalent noncommunicable diseases?
- What are the best practices for healthcare professionals to utilize social media in healthcare?
- How can social media contribute toward healthcare education?
- How can we prepare healthcare social network that is helpful for emergency situations, such as the COVID-19 pandemic?
- How we can make sure that the health-related information shared on social media platforms is valid and from authentic source?
- Are the social networking interventions of health promotion effective?

Websites to explore and a list of credible sources

I. Websites

a. **Mayo Clinic:** A nonprofit institution, Mayo Clinic offers professional, all-encompassing care to anybody in need of healing and is dedicated to clinical practice, teaching, and research. This site can be accessed at: https://www.mayoclinic.org/

Suggested activity for the readers: Explore this website, and describe your experience. Discuss the advantages of this site.

b. **Online learning platforms for doctors**: Docvidya brings all the latest medical updates, webinars, courses, therapies blogs, videos, etc., across 50+ specialties for all healthcare professionals in India. This website and its contents are for the use of registered medical practitioner only. This site can be accessed at: www.docvidya.com

Suggested activity for readers: If you are registered medical professional practicing in India, create an account and explore the latest updates related to the monkeypox.

c. **Medical news:** Univadis provides a free access to essential medical news and clinical tools to facilitate decision. A must-have app for UK healthcare professionals (Doctor, Physician, Pharmacist, Nurses, Medical students). This website provides the information related to the world's top medical conferences. This website can be accessed at: www.univadis.com

Suggested activity for readers: Create your account and explore the latest updates on this website related to your research area of interest.

d. **Community for doctors**: Docplexus is 380,000+ Global Community of Doctors which enables peer-to-peer interaction and knowledge-sharing for better clinical decisions. The website is owned and operated by Docplexus Online Services Pt. Ltd., an India-based business with operations around the world. These website and materials are designed to comply with relevant laws and regulations. This site can be accessed at: www.docplexus.com

   Suggested activity for readers: If you are a healthcare professional, then you can explore groups depending upon your interest.
   e. **Online Health Information: Is It Reliable?**—An article by NIH exploring the reliability of online health information. This can be accessed at: https://www.nia.nih.gov/health/online-health-information-it-reliable

   Suggested activity for readers: Explore this article and discuss the reliability of health information available online.
   f. **Use the SHARE checklist** by the WHO to help you spot false information to make sure you don't contribute to the spread of harmful content. This can be accessed at: https://www.who.int/news/item/22-09-2021-be-careful-what-you-share.-things-aren-t-always-what-they-seem-online

   Suggested activity for readers: Use the information provided on the website and check the credibility of any news before you share it.
   g. **Explore the telemedicine guidelines by WHO** at https://www.who.int/news/item/10-11-2022-who-issues-new-guide-to-running-effective-telemedicine-services

   Suggested activity for readers: Explore the article and compare it with any telemedicine experience you had in the past.

II. List of credible sources
   a. Centers for Disease Control and Prevention (https://www.cdc.gov/)
   b. National Institute of Health (https://www.nih.gov/)
   c. National library of medicine (https://www.medlineplus.gov/)
   d. Official site of UK Government (https://www.gov.uk/coronavirus)
   e. WHO Health statistics and information systems (https://www.who.int/health-topics/universal-health-coverage/health-statistics-and-information-systems)

Definition of key terms

Health awareness

The fundamental goal of health awareness is to educate people about health issues so they can avoid and treat sickness. Since it varies from person to person, there are many different interpretations of what health is.

Health promotion

The practice of empowering people to exert more control over and improve their health is known as "health promotion."

Infodemiology

"Infodemiology can be defined as the science of distribution and determinants of information in an electronic medium, specifically the Internet, or in a population, with the ultimate aim to inform public health and public policy" (Eysenbach et al., 2009).

Professional networking

Networking is an intentional action to establish, strengthen, and maintain trusting relationships with others to accomplish your aims. Simply said, professional networking is networking with an emphasis on business objectives. Building and sustaining enduring connections with other professionals in your sector and adjacent professions based on mutual advantage. Having a strong network can give an opportunity to be in touch with a variety of people to call for seeking expert advice.

Social media

Websites and software programs that let users create, share, and engage in social networking.

Social Networking Site: A social networking site is an online community where users can connect with one another and build public profiles. Social networking sites typically let new users create a list of contacts they have a connection with, and those contacts can subsequently confirm or deny the link. Once connections have been made, a new user can look through existing connections to find new ones.

Social media analytics

Social media analytics is the process of gathering and analyzing data from social networks such as Facebook, Instagram, LinkedIn, or Twitter.

References

Alanzi, T., & Al-Habib, D. K. (2020). The use of social media by healthcare quality personnel in Saudi Arabia. *Journal of Environmental and Public Health, 2020*, e1417478. https://doi.org/10.1155/2020/1417478

Alanzi, T., & Al-Yami, S. (2019). Physicians' attitude towards the use of social media for professional purposes in Saudi Arabia. *International Journal of Telemedicine and Applications, 2019*, e6323962. https://doi.org/10.1155/2019/6323962

Alsobayel, H. (2016). Use of social media for professional development by health care professionals: A cross-sectional web-based survey. *JMIR Medical Education, 2*(2), e15. https://doi.org/10.2196/mededu.6232

Ashton, L. M., Morgan, P. J., Hutchesson, M. J., Rollo, M. E., & Collins, C. E. (2017). Feasibility and preliminary efficacy of the 'HEYMAN' healthy lifestyle program for young men: A pilot randomised controlled trial. *Nutrition Journal, 16*, 2. https://doi.org/10.1186/s12937-017-0227-8

Bardus, M., El Rassi, R., Chahrour, M., Akl, E. W., Raslan, A. S., Meho, L. I., & Akl, E. A. (2020). The use of social media to increase the impact of health research: Systematic review. *Journal of Medical Internet Research, 22*(7), e15607. https://doi.org/10.2196/15607

Benetoli, A., Chen, T. F., Schaefer, M., Chaar, B., & Aslani, P. (2017). Do pharmacists use social media for patient care? *International Journal of Clinical Pharmacy, 39*(2), 364–372. https://doi.org/10.1007/s11096-017-0444-4

Borges do Nascimento, I. J., Pizarro, A. B., Almeida, J. M., Azzopardi-Muscat, N., Gonçalves, M. A., Björklund, M., & Novillo-Ortiz, D. (2022). Infodemics and health misinformation: A systematic review of reviews. *Bulletin of the World Health Organization, 100*(9), 544–561. https://doi.org/10.2471/BLT.21.287654

Borgmann, H., DeWitt, S., Tsaur, I., Haferkamp, A., & Loeb, S. (2015). Novel survey disseminated through Twitter supports its utility for networking, disseminating research, advocacy, clinical practice and other professional goals. *Canadian Urological Association Journal = Journal De l'Association Des Urologues Du Canada, 9*(9–10), E713–E717. https://doi.org/10.5489/cuaj.3014

Chan, L., O'Hara, B., Phongsavan, P., Bauman, A., & Freeman, B. (2020). Review of evaluation metrics used in digital and traditional Tobacco control campaigns. *Journal of Medical Internet Research, 22*(8), e17432. https://doi.org/10.2196/17432

Chauhan, B., George, R., & Coffin, J. (2012). Social media and you: What every physician needs to know. *The Journal of Medical Practice Management: MPM, 28*(3), 206–209.

Chen, T. H. (2017). What is "stride" in convolutional neural network? Machine learning notes. https://medium.com/machine-learning-algorithms/what-is-stride-in-convolutional-neural-network-e3b4ae9baedb.

Chen, J. H., & Wang, Y. (2020). Social media usage for health purposes: Systematic review (preprint). *Journal of Medical Internet Research, 23.* https://doi.org/10.2196/17917

Chretien, K. C., & Kind, T. (2013). Social media and clinical care: Ethical, professional, and social implications. *Circulation, 127*(13), 1413–1421. https://doi.org/10.1161/CIRCULATIONAHA.112.128017

Chung, J. E. (2016). A smoking cessation campaign on Twitter: Understanding the use of Twitter and identifying major players in a health campaign. *Journal of Health Communication, 21*(5), 517–526. https://doi.org/10.1080/10810730.2015.1103332

Cuello-Garcia, C., Pérez-Gaxiola, G., & van Amelsvoort, L. (2020). Social media can have an impact on how we manage and investigate the COVID-19 pandemic. *Journal of Clinical Epidemiology, 127*, 198–201. https://doi.org/10.1016/j.jclinepi.2020.06.028

Diddi, P., & Lundy, L. K. (2017). Organizational Twitter use: Content analysis of tweets during breast cancer awareness month. *Journal of Health Communication, 22*(3), 243–253. https://doi.org/10.1080/10810730.2016.1266716

Dizon, D. S., Graham, D., Thompson, M. A., Johnson, L. J., Johnston, C., Fisch, M. J., & Miller, R. (2012). Practical guidance: The use of social media in oncology practice. *Journal of Oncology Practice, 8*(5), e114–e124. https://doi.org/10.1200/JOP.2012.000610

Eysenbach, G. (2009). Infodemiology and infoveillance: Framework for an emerging set of public health informatics methods to analyze search, communication and publication behavior on the internet. *Journal of Medical Internet Research, 11*(1), e11. https://doi.org/10.2196/jmir.1157

Fang, Y., Ma, Y., Mo, D., Zhang, S., Xiang, M., & Zhang, Z. (2019). Methodology of an exercise intervention program using social incentives and gamification for obese children. *BMC Public Health, 19*(1), 686. https://doi.org/10.1186/s12889-019-6992-x

Fogelson, N. S., Rubin, Z. A., & Ault, K. A. (2013). Beyond likes and tweets: An in-depth look at the physician social media landscape. *Clinical Obstetrics and Gynecology, 56*(3), 495−508. https://doi.org/10.1097/GRF.0b013e31829e7638

Fu, J. S., & Zhang, R. (2019). NGOs' HIV/AIDS discourse on social media and websites: Technology affordances and strategic communication across media platforms. *International Journal of Communication, 13*(0), Article 0.

Gesser-Edelsburg, A., Diamant, A., Hijazi, R., & Mesch, G. S. (2018). Correcting misinformation by health organizations during measles outbreaks: A controlled experiment. *PLOS One, 13*(12), e0209505. https://doi.org/10.1371/journal.pone.0209505

Grajales, F. J., Sheps, S., Ho, K., Novak-Lauscher, H., & Eysenbach, G. (2014). Social media: A review and tutorial of applications in medicine and health care. *Journal of Medical Internet Research, 16*(2), e13. https://doi.org/10.2196/jmir.2912

Greenhalgh, T., Wherton, J., Papoutsi, C., Lynch, J., Hughes, G., A'Court, C., Hinder, S., Fahy, N., Procter, R., & Shaw, S. (2017). Beyond adoption: A new framework for theorizing and evaluating nonadoption, abandonment, and challenges to the scale-up, spread, and sustainability of health and care technologies. *Journal of Medical Internet Research, 19*(11), e367. https://doi.org/10.2196/jmir.8775

Grimes, D. R. (2020). Health disinformation and social media: The crucial role of information hygiene in mitigating conspiracy theory and infodemics. *EMBO Reports, 21*(11), e51819. https://doi.org/10.15252/embr.202051819

Guidry, J., Zhang, Y., Jin, Y., & Parrish, C. (2016). Portrayals of depression on Pinterest and why public relations practitioners should care. *Public Relations Review, 42*(1), 232−236. https://doi.org/10.1016/j.pubrev.2015.09.002

Health Education Specialists and Community Health Workers. (n.d.). Occupational Outlook Handbook. U.S. Bureau of Labor Statistics. https://www.bls.gov/ooh/community-and-social-service/health-educators.htm.

Helm, J., & Jones, R. M. (2016). Practice paper of the academy of nutrition and dietetics: Social media and the dietetics practitioner: Opportunities, challenges, and best practices. *Journal of the Academy of Nutrition and Dietetics, 116*(11), 1825−1835. https://doi.org/10.1016/j.jand.2016.09.003

He, S., Ojo, A., Beckman, A. L., Gondi, S., Gondi, S., Betz, M., Faust, J. S., Choo, E., Kass, D., & Raja, A. S. (2020). The story of #GetMePPE and GetUsPPE.org to mobilize health care response to COVID-19: Rapidly deploying digital tools for better health care. *Journal of Medical Internet Research, 22*(7), e20469. https://doi.org/10.2196/20469

Hermansyah, A., Sukorini, A. I., Asmani, F., Suwito, K. A., & Rahayu, T. P. (2019). The contemporary role and potential of pharmacist contribution for community health using social media. *Journal of Basic and Clinical Physiology and Pharmacology, 30*(6). https://doi.org/10.1515/jbcpp-2019-0329. /j/jbcpp.2019.30.issue-6/jbcpp-2019-0329/jbcpp-2019-0329.xml.

Hoedebecke, K., Beaman, L., Mugambi, J., Shah, S., Mohasseb, M., Vetter, C., Yu, K., Gergianaki, I., & Couvillon, E. (2017). Health care and social media: What patients really understand. *F1000Research, 6*, 118. https://doi.org/10.12688/f1000research.10637.1

Hunter, R. F., de la Haye, K., Murray, J. M., Badham, J., Valente, T. W., Clarke, M., & Kee, F. (2019). Social network interventions for health behaviours and outcomes: A systematic review and meta-analysis. *PLoS Medicine, 16*(9), e1002890. https://doi.org/10.1371/journal.pmed.1002890

Jiang, T., Osadchiy, V., Mills, J. N., & Eleswarapu, S. V. (2020). Is it all in my head? Self-reported psychogenic erectile dysfunction and depression are common among young men seeking advice on social media. *Urology, 142*, 133−140. https://doi.org/10.1016/j.urology.2020.04.100

Justinia, T., Alyami, A., Al-Qahtani, S., Bashanfar, M., El-Khatib, M., Yahya, A., & Zagzoog, F. (2019). Social media and the Orthopaedic surgeon: A mixed methods study. *Acta Informatica Medica: AIM: Journal of the Society for Medical Informatics of Bosnia and Herzegovina: Casopis Drustva Za Medicinsku Informatiku BiH, 27*(1), 23−28. https://doi.org/10.5455/aim.2019.27.23-28

Lambert, K. M., Barry, P., & Stokes, G. (2012). Risk management and legal issues with the use of social media in the healthcare setting. *Journal of Healthcare Risk Management: The Journal of the American Society for Healthcare Risk Management, 31*(4), 41−47. https://doi.org/10.1002/jhrm.20103

Lin, R. J., & Zhu, X. (2012). Leveraging social media for preventive care-A gamification system and insights. *Studies in Health Technology and Informatics, 180*, 838−842.

Liu, Q. B., Liu, X., & Guo, X. (2020). The effects of participating in a physician-driven online health community in managing chronic disease: Evidence from two natural experiments. *MIS Quarterly, 44*(1), 391−419. https://doi.org/10.25300/misq/2020/15102

Li, H., Xue, L., Tucker, J. D., Wei, C., Durvasula, M., Hu, W., Kang, D., Liao, M., Tang, W., & Ma, W. (2017). Condom use peer norms and self-efficacy as mediators between community engagement and condom use among Chinese men who have sex with men. *BMC Public Health, 17*(1), 641. https://doi.org/10.1186/s12889-017-4662-4

Loeb, S., Carrick, T., Frey, C., & Titus, T. (2020). Increasing social media use in urology: 2017 American urological association survey. *European Urology Focus, 6*(3), 605−608. https://doi.org/10.1016/j.euf.2019.07.004

MacMillan, C. (2013). Social media revolution and blurring of professional boundaries. *Imprint, 60*(3), 44−46.

Mendoza, J. A., Baker, K. S., Moreno, M. A., Whitlock, K., Abbey-Lambertz, M., Waite, A., Colburn, T., & Chow, E. J. (2017). A Fitbit and Facebook mHealth intervention for promoting physical activity among adolescent and young adult childhood cancer survivors: A pilot study. *Pediatric Blood and Cancer, 64*(12). https://doi.org/10.1002/pbc.26660

Mheidly, N., & Fares, J. (2020). Leveraging media and health communication strategies to overcome the COVID-19 infodemic. *Journal of Public Health Policy, 41*(4), 410−420. https://doi.org/10.1057/s41271-020-00247-w

von Muhlen, M., & Ohno-Machado, L. (2012). Reviewing social media use by clinicians. *Journal of the American Medical Informatics Association : JAMIA, 19*(5), 777−781. https://doi.org/10.1136/amiajnl-2012-000990

Oser, T. K., Minnehan, K. A., Wong, G., Parascando, J., McGinley, E., Radico, J., & Oser, S. M. (2019). Using social media to broaden understanding of the barriers and facilitators to exercise in adults with type 1 diabetes. *Journal of Diabetes Science and Technology, 13*(3), 457−465. https://doi.org/10.1177/1932296819835787

Pagkas-Bather, J., Young, L. E., Chen, Y.-T., & Schneider, J. A. (2020). Social network interventions for HIV transmission elimination. *Current HIV/AIDS Reports, 17*(5), 450−457. https://doi.org/10.1007/s11904-020-00524-z

Park, H., Reber, B. H., & Chon, M.-G. (2016). Tweeting as health communication: Health organizations' use of Twitter for health promotion and public engagement. *Journal of Health Communication, 21*(2), 188−198. https://doi.org/10.1080/10810730.2015.1058345

Patrick, M. D., Stukus, D. R., & Nuss, K. E. (2019). Using podcasts to deliver pediatric educational content: Development and reach of PediaCast CME. *Digital Health, 5*. https://doi.org/10.1177/2055207619834842, 2055207619834842.

Peck, J. L. (2014). Social media in nursing education: Responsible integration for meaningful use. *The Journal of Nursing Education, 53*(3), 164−169. https://doi.org/10.3928/01484834-20140219-03

Reidy, C., Klonoff, D. C., & Barnard-Kelly, K. D. (2019). Supporting good intentions with good evidence: How to increase the benefits of diabetes social media. *Journal of Diabetes Science and Technology, 13*(5), 974−978. https://doi.org/10.1177/1932296819850187

Ronen, K., Grant, E., Copley, C., Batista, T., & Guthrie, B. L. (2020). Peer group focused eHealth strategies to promote HIV prevention, testing, and care engagement. *Current HIV/AIDS Reports, 17*(5), 557−576. https://doi.org/10.1007/s11904-020-00527-w

Sagnia, P. I. G., Gharoro, E. P., & Isara, A. R. (2020). Adolescent-parent communication on sexual and reproductive health issues amongst secondary school students in Western Region 1 of the Gambia. *African Journal of Primary Health Care and Family Medicine, 12*(1), e1−e7. https://doi.org/10.4102/phcfm.v12i1.2437

Shaiket, H. A. W., Anisuzzaman, D. M., & Saif, A. F. M. (2019). Data analysis and visualization of continental cancer situation by Twitter scraping. *International Journal of Modern Education and Computer Science, 11*, 23−31. https://doi.org/10.5815/ijmecs.2019.07.03

Shen, L., Wang, S., Chen, W., Fu, Q., Evans, R., Lan, F., Li, W., Xu, J., & Zhang, Z. (2019). Understanding the function constitution and influence factors on communication for the WeChat official account of top tertiary hospitals in China: Cross-sectional study. *Journal of Medical Internet Research, 21*(12), e13025. https://doi.org/10.2196/13025

Southwell, B. G., Wood, J. L., & Navar, A. M. (2020). Roles for health care professionals in addressing patient-held misinformation beyond fact correction. *American Journal of Public Health, 110*(S3), S288−S289. https://doi.org/10.2105/AJPH.2020.305729

Steffens, M. S., Dunn, A. G., Wiley, K. E., & Leask, J. (2019). How organisations promoting vaccination respond to misinformation on social media: A qualitative investigation. *BMC Public Health, 19*(1), 1348. https://doi.org/10.1186/s12889-019-7659-3

Stellefson, M., Paige, S. R., Chaney, B. H., & Chaney, J. D. (2020). Evolving role of social media in health promotion: Updated responsibilities for health education specialists. *International Journal of Environmental Research and Public Health, 17*(4), 1153. https://doi.org/10.3390/ijerph17041153

Sutton, J., Renshaw, S. L., & Butts, C. T. (2020). COVID-19: Retransmission of official communications in an emerging pandemic. *PLOS One, 15*(9), e0238491. https://doi.org/10.1371/journal.pone.0238491

Thornber, K., Huso, D., Rahman, M. M., Biswas, H., Rahman, M. H., Brum, E., & Tyler, C. R. (2019). Raising awareness of antimicrobial resistance in rural aquaculture practice in Bangladesh through digital communications: A pilot study. *Global Health Action, 12*(Suppl. 1), 1734735. https://doi.org/10.1080/16549716.2020.1734735

Ventola, C. L. (2014). Social media and health care professionals: Benefits, risks, and best practices. *Pharmacy and Therapeutics, 39*(7), 491−520.

Welch, V., Petkovic, J., Pardo Pardo, J., Rader, T., & Tugwell, P. (2016). Interactive social media interventions to promote health equity: An overview of reviews. *Health Promotion and Chronic Disease Prevention in Canada: Research, Policy and Practice, 36*(4), 63−75. https://doi.org/10.24095/hpcdp.36.4.01

Xie, T., Tan, T., & Li, J. (2020). An extensive search trends-based analysis of public attention on social media in the early outbreak of COVID-19 in China. *Risk Management and Healthcare Policy, 13*, 1353−1364. https://doi.org/10.2147/RMHP.S257473

Young, M. J., & Scheinberg, E. (2017). The rise of crowdfunding for medical care: Promises and perils. *JAMA, 317*(16), 1623–1624. https://doi.org/10.1001/jama.2017.3078

Yu, Y., Li, Y., Li, T., Xi, S., Xiao, X., Xiao, S., & Tebes, J. K. (2020). New path to recovery and well-being: Cross-sectional study on WeChat use and endorsement of WeChat-based mHealth among people living with schizophrenia in China. *Journal of Medical Internet Research, 22*(9), e18663. https://doi.org/10.2196/18663

Zhao, Y., & Zhang, J. (2017). Consumer health information seeking in social media: A literature review. *Health Information and Libraries Journal, 34*(4), 268–283. https://doi.org/10.1111/hir.12192

Chapter 15

Role of social media in telemedicine

Rasika Manori Jayasinghe and Ruwan Duminda Jayasinghe
Faculty of Dental Sciences, University of Peradeniya, Kandy, Sri Lanka

Learning objectives

- State the role of social media in telemedicine.
- Discuss different acts of social media in medical practice.
- Describe the advantages of social media in telemedicine.
- Analyze the challenges of using social media in medical practice.
- Design the steps for overcoming challenges when using social media in telemedicine.

Telemedicine

Telemedicine is the combination of telecommunication with healthcare systems to provide healthcare services. Even though it is getting popular in the recent past, especially with the COVID-19 pandemic, it was initially introduced in the 1950s. The term "telehealth" is used at times because certain disciplines of health such as nursing and mental health which do not supply direct clinical services also use the technology (Whitten et al., 2010). For the healthcare provided through Internet, terms like e-health and Cybermedicine are generally used. The term telemedicine is mostly and commonly used as a general term describing all of the above aspects. Telemedicine further can be defined as the provision of clinical information, education, and healthcare services in all aspects over a different technology including electronic medical records, mobile phones, and the Internet (Whitten et al., 2010). Since its inception, telemedicine is growing mostly due to the development of broadband infrastructures, increasing human resources with expertise, and the reduction of costs associated with the technology. As its importance and usefulness have been recognized by healthcare experts, administrators, receivers, and policymakers, healthcare providers increasingly use the technology in all aspects of management and delivery.

Telemedicine, the provision of healthcare via telecommunication technology, has evolved dramatically from its inception, especially over the past two decades. Although the field of telemedicine has undergone remarkable changes in the recent past, it has retained its main objective, that is, to improve access to care as it is from day one to date (Barbosa et al., 2021). Telemedicine is helpful in multiple situations, but it will be an ideal option to be considered when there is no alternative like when managing emergencies in remote environments, and when it is better than existing traditional services. The use of telemedicine under these circumstances will improve equity in the provision of healthcare irrespective of the geographical location, level of the healthcare provided, and the efficiency by which it is delivered (Craig & Petterson, 2005).

Evidence suggests that telemedicine has improved the provision of care in a wide area of clinical situations and specialties. Telemedicine has been proven to improve health outcomes too. As barriers to healthcare are significant and vary, the use of this technology can be considered to address some of these barriers, especially the geographical barrier, but its effect on overcoming the social barriers appears to be low.

There are different types or classifications for telemedicine. It can be classified according to the time of information is transmitted (Chellaiyan et al., 2019).

(i) Synchronous (real-time) telemedicine (sending and receiving ends are both online at the information transfer).
(ii) Asynchronous (store-and-forward) telemedicine (information is stored by the sender using databases and sends it to the receiving end later; the receiver can review the data at any time once he receives it).
(iii) Self-monitoring (remote monitoring) telemedicine (uses a wide range of technological devices).

Most telemedicine services use developed sophisticated systems, and only a few such systems which are tested and proven to be clinically and cost-effective are available. The lack of an adequate number of such systems is considered a limiting factor in the use of telemedicine by many healthcare professionals. Such telemedicine programs can be created suddenly without resources and expertise; healthcare professionals and patients are using other available cost-effective methods as alternatives. The most popular method is the use of instant messaging applications available with social media.

In order to get the maximum use of telemedicine in providing healthcare, more evidence on health outcomes and cost savings needs to be collected and appropriate changes in the legislation and policy need to be introduced (Barbosa et al., 2021). Most researchers have identified that telemedicine can provide a significant and positive effect on clinical services compared to traditional healthcare delivery or at least equal to it (Whitten, 2010). As the initial investment is high, the cost of the telemedicine service can be significant. However, research data for the cost-effectiveness of telemedicine are not conclusive, with mixed results.

Social media

There are multiple definitions for "social media." They are mostly broad and constantly evolving. The term social media is usually referred to as Internet-based methods that allow individuals to communicate; share information, opinions, messages, and photos; and, in some situations, collaborate with others in real-time. Social media is further referred to as "Web 2.0" or "social networking" (Ventola, 2014). Like any other technology-related application, social media is ever-evolving. It has evolved in multiple directions to fulfill the requirements of the public. Competition among different platforms has resulted in the introduction of new features. Not all social media sites are with same types of features that provide different purposes for the individual. Common social media activities with some examples are given below.

- Social networking (Facebook, Myspace, Google Plus, Twitter)
- Professional networking (LinkedIn)
- Media sharing (YouTube, Flickr)
- Content production (blogs [Tumblr, Blogger] and microblogs [Twitter])
- Knowledge/information aggregation (Wikipedia)
- Virtual reality and gaming (Second Life) (Ventola, 2014)
- Social communication/instant messaging (WhatsApp, Viber, WeChat)

Social media is getting very popular day by day. In 2012, there were 1.48 billion social media users, which have increased to 4.62 billion by the end of the year 2022. According to the data published, 424 million people started using social media for the first time in 2021, or more than 1 million new users a day in 2021 highlighting the ever-increasing popularity of social media among people. Figures indicate that 58.9% of the total population is into social media now (Digital 2022: Global Overview Report). Social media is used by people of all ages, genders, geographical areas, and socioeconomical and educational statuses. Healthcare professionals are no exception.

Most people use social media to acquire information on all aspects of health. Facebook is the most famous social media platform in this regard. People use social media to acquire information on the diagnosis of a condition that they are suffering from and to understand the test results better. They are interested in reading information posted on rare diseases by groups/organizations dealing with such diseases. Further, people also use social media to engage in discussions about their disease conditions and to find support groups (Rocha et al., 2018).

The use of social media in healthcare has a lot of advantages, but it comes with some serious threats as well. The abundance of fake news is the main issue. Fake news is inaccurate news that contains misleading or fabricated content. They can confuse care seekers and may result in the self-misdiagnosis of various types of health issues. These are without any base and are intentionally exaggerated or provide inaccurate information. Attention has been paid to identifying mechanisms to separate accurate information from fake news. Different types of modeling are too developed to increase the accuracy of fake news analysis (Jang et al., 2019).

The use of social media is popular among all healthcare personnel as well. Most of them use social media for personal as well as professional use. It provides an opportunity for healthcare professionals to share information, engage in discussions on various health-related issues like clinical case management, administrative issues, and healthcare policy, engage patients and the public, promote best health practices, and provide education to patients, professional students, and the public. Engaging in professional groups through social media allows healthcare professionals to communicate with others who are not in the vicinity. The main reason to use social media can vary from one professional group of healthcare providers to the other. Physicians use them mostly to listen to experts and to communicate with colleagues on patient issues, whereas pharmacists frequently focus on communication with colleagues (Ventola, 2014). As social media has high communication capabilities, it is widely used to improve clinical education. As the younger generation is more into social media and is very familiar with it, it has been used effectively to motivate the students on communication, professional practice, and ethics. Higher educational institutes also use social media to promote their institution, recruit students to follow courses, create virtual classrooms, and increase access to libraries and to provide a different learning experience in addition to traditional teaching. Even though social media is a very effective and helpful tool for healthcare professionals and in healthcare delivery, it also causes many potential risks. Careless use of social media can potentially affect patient security and safety/healthcare system information, patient consent, employment systems, and many other professionals and ethical issues. Therefore, it is essential to have proper guidelines established for healthcare professionals on the use of social media according to international, national, and institutional needs (Ventola, 2014).

Social media has been used in the field of medicine and other allied health disciplines over a long period of time. Its main use has been the dissemination of knowledge and patient education. A significant proportion of people, especially adolescents, use social networking sites as the initial resource to find information on health-related matters. Many individuals with medical questions use the web to find answers to their questions and concerns. Some health websites are maintained to provide answers to general health issues too. They rely on the Internet as it can virtually provide information anywhere at any time. Different social media platforms have been used by different healthcare professionals, institutions, and organizations at different levels. An array of social media platforms can support dialog between patients and healthcare providers and patients with patients (Colineau & Paris, 2010). Sites such as PatientsLikeMe allow patients to communicate and share information with each other related to disease conditions and advice such as information on treatment and medication (Frost & Massagli, 2008). YouTube has been used widely by the public to share information on health, medicines, and diagnoses (Fernandez-Luque et al., 2009) and by patients to share individual cancer experiences (Chou et al., 2011). Blog sites enable a space in which individuals can access personal resources (Adams, 2010) and allow healthcare professionals an opportunity to distribute information to patients and the general public. Facebook is being used by the general public, patients, caretakers, and health personnel to share their own experiences of disease management, investigations, and diagnosis (Farmer et al., 2009). Asthma groups use Myspace to share experiences, especially personal stories related to ill health (Versteeg et al., 2009). Social media can be used to acquire data on patient opinions and experiences on physicians' performance as well (Adams, 2010).

Selection of the type of social media depends on individual preference, availability and restrictions, cost, characteristics of the target group, rules and regulations governing the use of social media in that community, familiarity, and many others. Twitter is the most used platform in most developed regions, whereas Facebook is popular in many other countries (Rocha et al., 2018). Many patients are concerned about the privacy of information shared on social media. Most people do not feel comfortable sharing their personal/medical information or photos on social media groups that are open to the public due to privacy issues. Further, some are not comfortable sharing such information with groups that are private (Rocha et al., 2018).

Social media in healthcare

WhatsApp and Twitter Apps have been used frequently for medical education purposes, and Twitter is considered a useful platform for getting news and information. Some studies have identified WhatsApp and Twitter apps to be the two main social media platforms for the elderly, whereas others have reported YouTube and Facebook as the most used social media platforms. Different other studies have demonstrated different other results highlighting the diversity in selecting social media apps by different populations. Social media is popular among healthcare professionals. They believe that using social media helps to get connected with colleagues and network with the wider community (El Kheir et al., 2022). Data security, confidentiality, consent, record keeping, authentication, identification, privacy, and ethical issues are considered the major concerns in mHealth and telemedicine, and these concerns are equally important for using social media for telemedicine as well. Security of the data during the transmission of information from the patient to the healthcare provider, its subsequent storage on smartphones or with the service provider, and keeping records of chat messages are major concerns with social media usage. It is important to recognize the principles of information transfer happening in social media by the healthcare personnel in this read. For example, to send a WhatsApp message to the patient to the healthcare professional, an Internet connection is required, which can be a wireless one or a data connection. The message that is sent is then routed to a WhatsApp server, which may be in the same country as the sender. Then the server attempts to send the message to the recipient. To deliver the message, the recipient's smartphone must have an Internet connection. Until such time, a message will be stored in the server, and it will be delivered to the recipient's smartphone when it has an Internet connection followed by the deletion of the message from the server. The message will be stored in the server for a month if it was not delivered due to no Internet connection with the receiver, and the message is deleted from the server after this month's period.

Twitter is very popular among healthcare professionals, especially in the developed part of the world. Healthcare professionals can reach a broad audience including patients, trainees, other professionals, and organizations through Twitter. It is popular among healthcare researchers as well due to its ability to share and communicate. Research findings will be easily available for clinicians through Twitter (Pershad et al., 2018). Therefore, it will act as an excellent platform to connect patients, clinicians, and researchers. The availability of incorrect information and the lack of a mechanism to check the accuracy of the information are the major issues on Twitter. As it has a larger capacity to share information, it can lead to more and more discussions when used in groups resulting in the collection of inaccurate information, and opinions masking the important issues.

WhatsApp is a very popular social media app that is widely used for various clinical practices, mainly to support communication within intradisciplinary groups. It is mostly used in the developing part of the world. Healthcare professionals seem to have been able to incorporate WhatsApp into their day-to-day practice without the need for additional training. Like any other service, WhatsApp comes with advantages and disadvantages, but the practical advantages appear to outweigh the disadvantages. Data security with the transmission of the messages to the server and to the recipient has been considered a longtime problem with WhatsApp. Further, patient privacy, consent, confidentiality, data security, and record-keeping have been identified as other major issues, but they have been poorly reported. The issue of hacking into WhatsApp users' accounts seems to be resolved now with the recent versions of WhatsApp. Some of the issues are due to the misconception and a poor understanding of how WhatsApp encrypts, transmits, and stores messages and files (Mars & Scott, 2016). It has been popular with the new version as all messages are end-to-end encrypted giving more protection (Mars & Scott, 2016).

One of the major challenges in using WhatsApp in medical practice is the maintenance of autonomy and consent, which are the foundations of medical ethics. Available literature/guidance on how to obtain consent is limited.

Consent from the patient for WhatsApp use as a tool in telemedicine should cover multiple aspects including the use of the application, data attainments such as a photograph, its transmission, information about who will receive it, data storage on the sender's and receiver's phones, privacy laws and regulations such as the general data protection laws/regulation, and an explanation of how it will be used and by whom (Mars et al., 2019). There are no clear guidelines on the form in which consent should be obtained, and whether or how it should be recorded. This has created a situation where many doctors use/share information without proper consent. In instances where discussions are made with international participants, minimal attention is considered to liability or jurisdictional issues and legal concerns of transmission of data via servers located in another country. As there are no proper guidelines for the use of these social media platforms including WhatsApp, it is important for healthcare professionals to follow available guidelines for the ethical practice of telemedicine (Mars et al., 2019).

TikTok is the newest, very popular, video-based social media platform. Even though it was introduced only in 2017, it has more than 500 million active users. With TikTok, videos up to 60 s can be created and shared. It is used in all fields of the health profession, but research on its use is minimal. Its usefulness in healthcare especially in radiology has been reported (Lovett et al., 2021).

Role of smartphones in using social media in health

Over the last few decades, the mobile phone has played an integral part in the day-to-day activities of individuals replacing most of the traditional methods of communication. Advances in telecommunication have combined with advances in computer science resulting in smartphones. Today, smartphones are very popular replacing traditional mobile phones which were used only for limited functions like taking calls or sending simple text messages. The popularity of conventional voice calls is diminishing, and instant messaging is taking over, especially with the younger generation. Although instant messaging is designed for interpersonal communication, it is widely used for health-related needs as well. It assists in the provision of healthcare and patient-related information between patient/s and healthcare professionals through one-to-one communication through chats and/or one-to-many communications via chat groups. As there are multiple such services, healthcare professionals use their favorite or comfortable IM app to send messages to coworkers to seek or give clinical advice, supported by audio recordings, images, or videos made them use their phone (Mars et al., 2019). These different social media platforms include WhatsApp, Viber, Twitter, Facebook, VChat, TikTok, and many others. WhatsApp is one of the most widely used instant messaging apps with over 1.5 billion active users, sending 65 billion messages a day and more than 1 billion groups (Mars et al., 2019).

Use of social media in telemedicine

Governments, agencies, institutions, and professional associations are recognizing the value of social media as an effective tool in telemedicine and are in the process of relaxing some restrictions that they have imposed in the past. In the recent past, the Department of Health and Human Services of the United States of America has removed some restrictions placed on communication apps. This reduced the barriers to providing telemedicine service to individuals and allows the use of some popular video conferencing applications in telemedicine. Even though this will allow greater use of social media in telemedicine, overusage may result in overlooking information security and privacy concerns. The possible risk of cyberattacks on healthcare institutions and organizations needs to be seriously considered (Jalali et al., 2020).

Guidelines for social media use in telemedicine

There are a larger number of guidelines supporting, governing, and advising on the use of telemedicine not only in developed countries but also in developing regions. Six laws, six advisory guidelines, five policy statements, and two regulations are in place by different agencies, governments, or professional organizations in Southeast Asian countries. These are covering all aspects related to telemedicine. Most guidelines/regulations are for clinical governance, information and communication technology infrastructure, privacy issues, security and safety issues, data/information storage, record-keeping, legal and ethical aspects, applications of telemedicine, maintaining confidentiality, licensing procedures, and cost. Areas like mobile applications, reimbursement, and feedback were the least covered (Intan et al., 2021).

Even though social media is widely used in telemedicine, established or published obligatory guidelines regulating such activities are minimal if not nonexistent. Healthcare professionals are concerned and in agreement that the ethical and confidentially issues are significant and need to be well regulated but for WhatsApp which is one of the most popular social media platforms, there is no such guideline available. It is important to develop such guidelines to protect the proper use of social media, especially WhatsApp and other common instant messaging platforms, and to protect the interests of patients as well healthcare professionals. Existing guidelines on telemedicine could be used as a baseline in developing such obligatory generic guidelines for social media. As these platforms are immensely helpful, most health authorities have softened their approach toward their use of them by providing advice on the use and making attempts to find other IM solutions with the view of minimizing the shortcomings of WhatsApp. Development of guidelines for these platforms considering the international and national setting is an urgent requirement (Mars et al., 2019; Mars & Scott, 2016).

The use of social media in healthcare is with major implications on patient management, ethical aspects, profession, and to society. When a healthcare professional is directly communicating with a patient through social media, it is recommended to avoid third-party systems which are low in data security but to use closed social media platforms which are secure due to data encryption. It is essential to inform the patients about the privacy and protective measures that have been taken on privacy. Patients must provide prior approval for such terms before a healthcare professional engages in clinical care through social media. It is recommended to decide the accepted time taken for image and data transfer when to move from teleconsultation to face-to-face consultation and the procedure to follow in an emergency. These recommendations will allow healthcare workers to use social media ethically and professionally and thereby achieve public trust (Chretien & Kind, 2013). The use of social media in healthcare is becoming a common practice. In order to minimize technical, ethical, and professional issues, all healthcare professionals need to be formally trained. Training on appropriate and effective use of social media can minimize many issues. Incorporation of such training into the curricula of all health professional programs is necessary. Appropriate training programs as continuing professional development programs are needed to educate healthcare professionals in current practice (Pershad et al., 2018).

Use of social media by different specialties/branches in healthcare

The use of social media is not restricted to a single branch or specialty in healthcare. All medical and allied disciplines are using it alike. WhatsApp has been used extensively in the field of dentistry as well. Teledentistry is a discipline within telemedicine, which is involved with networking, sharing digital information, distant consultations, workup, and analysis related to dentistry. It has been used by all involved in the field of dentistry including dental specialists, general dentists, dental hygienists, and patients. General dentists, dental hygienists, and patients have used WhatsApp for transferring clinical information, images, and diagnostic questions for telemedicine services to specialists. Studies have demonstrated 82% acceptance of a clinicopathologic impression when used in telemedicine (Petruzzi & De Benedittis, 2016). Like in general radiology, WhatsApp has been used in oral radiology as well. It helps dental surgeons to get expert opinions from oral radiology specialists. In a study conducted among oral radiologists in India, it was reported that 95% of participants of the study were using teledentistry in their practice. WhatsApp was considered a useful tool for improving dental practice. Younger clinicians had a more positive perception of the use of WhatsApp in teledentistry (Ramdurg et al., 2016).

Different uses of social media in healthcare

Social media for telemedicine has been used by multiple groups including healthcare professionals, organizations, patients, and the public. The biggest advantage of social media is its availability and familiarity. Most people not only in the developed world but also in the developing region are having smartphones and Internet connectivity. Not like in the past, most geographical areas except areas with very a smaller number of people are having Internet connectivity. On the other hand, limited availability and accessibility for the most vulnerable groups including the elderly, poor socioeconomical groups, and people without adequate technical knowledge can be left out of the consultation.

The value of social media in health education and dissemination of knowledge was demonstrated well during the COVID-19 pandemic. Even though it has many advantages, there are several disadvantages as well. The biggest disadvantage of social media is the availability of incorrect information and therefore people get lost in selecting the correct from incorrect. Incorrect and misleading information available on social media has created a negative impact. For example, incorrect and misleading information on the adverse effects of the COVID-19 vaccine had a serious effect on the success of the vaccination campaign. Healthcare professionals will continue to be challenged by the misinformation freely available to patients on social media. They must be determined to follow evidence-based healthcare and be ready when challenged by affected patients. In addition, healthcare professionals must take a leading role in providing scientifically proven information more accessible to the public (Machado et al.).

mHealth and telemedicine

mHealth and telemedicine are not something new. It has been in practice for decades but reached its maximum popularity in the last few years because of the COVID-19 pandemic on the traditional healthcare systems. mHealth can be defined as "medical and public health practice supported by mobile devices, such as mobile phones, patient monitoring devices, personal digital assistants, and other wireless devices" (Kay et al., 2011). Telemedicine is a broad term that describes the delivery of healthcare services through communication technologies. Telemedicine can be defined as the provision of medical services from one site to another using electronic communication devices (Leung et al., 2018). These together have changed the way the traditional healthcare delivery systems are operating and provide greater opportunities for healthcare delivery, education, and research. Multiple modalities, devices, and telecommunication services are in use in this regard. Multiple opportunities are provided by telemedicine and mHealth technology, and most healthcare professionals believe that the new technology in telemedicine and mHealth will help in expanding the services that can be delivered to them. As there is a significant disparity in the availability of healthcare services and facilities between urban and rural communities, mHealth and telemedicine are evolved as mechanisms to provide better accessibility to specialist care. Increasing access to health services for patients and families from underserved areas via mHealth and telemedicine has been recognized by healthcare professionals (Kay et al., 2011).

With rapid advances in computer and allied sciences, mHealth and telemedicine have evolved dramatically, but their acceptance is still low especially the in the developing world where the use of telemedicine can assist in minimizing the inequalities in the distribution of healthcare facilities. Even though telemedicine and mHealth provide excellent opportunities for the healthcare system and caregivers as well as for the care receivers, there are multiple barriers associated with it in effective delivery. Multiple barriers have been identified which are responsible for this low acceptance in the developing world including high costs involved in developing required infrastructure and telecommunication, lack of expertise, lack of training, difficulties in maintenance, and the extra workload of healthcare professionals who are already overburdened with other activities (Mars & Scott, 2016).

Some healthcare professionals are eager to understand the incorporation of telemedicine with other existing technologies. Some healthcare professionals believe that the use of telemedicine and mHealth technology is limited to diagnosis and treatment within hospitals which is not the case (Kay et al., 2011). One of the major advantages of mHealth and telemedicine is the possibility of using it for patients in rural areas who are not able to obtain the services of specialist care, which is generally limited in selected areas. Some healthcare professionals are not sure how they can apply telemedicine or mHealth in real clinical practice. As there are existing systems in our healthcare systems already, another major concern is the interoperability among telemedicine, mHealth devices, and these already existing health information technologies (Kay et al., 2011). The cost involved and the attitudes of healthcare professionals in moving from these traditional systems to incorporate newer systems need to be considered. As there are multiple systems and devices used in telemedicine and mHealth, standardization is considered a concern. The cost involved, both physical and human resources required, data security, acceptability by both healthcare professionals and patients, institutional barriers, support from the authorities, legislations, professionalism, and ethics are also considered as factors in the use of telemedicine and mHealth.

Social media has been used extensively in the field of telemedicine in the recent past with some positive as well as negative implications. Social media platforms have made the transition from traditional healthcare delivery systems to telemedicine easier, more approachable, and with significant success. As social media platforms are universally available and people are used to them, their use in the field of telemedicine is considered relatively easy as the need for special preparation for apps or special training for both healthcare professionals and patients in the use is not necessary (O'Neill et al., 2020).

Ethical and professional considerations must be assessed carefully in the use of social media for telemedicine. Social media communications about or with patients can lead to a violation of patients' privacy and anonymity, which may end up with legal actions against healthcare professionals and/or healthcare institutions. To avoid legal outcomes, any post about patients, whether in the video, image, or text, should be deidentified and regulations governing such activities should be strictly adhered to. It is always advisable to take consent before sharing any information related to the patient, even if the content is anonymized (Farsi et al., 2022).

The use of social media in telemedicine can be on multiple levels depending on the requirement and preference of the healthcare professional or the healthcare institution or the organization. The level of operation can be decided by the available facilities, technical knowledge, stability of the Internet, regulations, and preferences. It can be effectively used as a communication tool between a specialist in a central hospital and a healthcare professional in a peripheral area. It can be used to communicate directly between the patient and healthcare professional as well, or it can be used in discussions of larger healthcare professionals' groups. Improper, unethical, and unprofessional use as well as safeguarding the patients as well as healthcare professionals' rights need to be seriously considered. A well-established system with preset guidelines and procedures will minimize the complications associated with the procedure. It is important to explain the procedure, its strengths, advantages, and limitations to the patients prior to the commencement of the teleconsultation. The patient must be given the opportunity to select any one of the options traditional face-to-face visit or teleconsultation via social media. As the initial diagnosis, making the management plan, and developing the confidence of the patient is critical, face-to-face physical meetings for the first visit unless it is near to impossible due to the situation like what we experienced during the COVID pandemic is appropriate. Depending on the patient's condition and wish, teleconsultations via social media can be effectively conducted from the second visit. Investigations and prescriptions can be shared via the same platform provided due to attention to confidentiality and patient safety. It may not be appropriate to provide prescriptions for certain drugs where there is possibility of misuse (Fig. 15.1).

Even though multiple apps dedicated to telemedicine have been developed and used worldwide, still they are not in wide usage due to the cost, availability, technical requirements, patient, and healthcare professional acceptability. Medical care has been presented remotely through these telemedicine apps, which have the best access to care for some populations, especially those in isolation, with limited access to specialist care and in rural areas. Monitoring patients in their homes and rural areas without easy access to healthcare services can improve the healthcare services of such individuals and communities. Good overall satisfaction by both healthcare professionals as well as

Role of social media in telemedicine **Chapter | 15** **329**

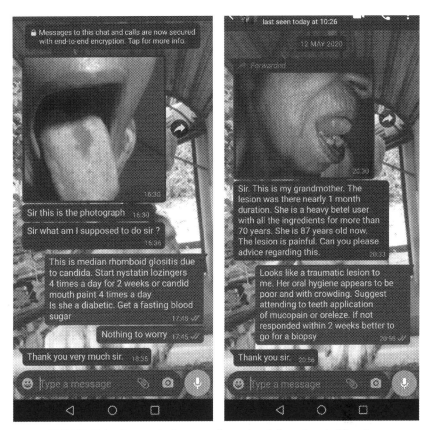

FIGURE 15.1 Communications regarding patient information via WhatsApp.

patients has been reported with these new telemedicine policies which shift care from one-to-one physical meetings to a more patient-centered one. Not only is telemedicine efficient but it is also time- and cost-saving. Healthcare professionals and patients are familiar with social media which is free and available; hence, they have been used widely for monitoring and communication in healthcare in recent years. Making appointments turned out to be web-based, health information became available on the Internet, and examinations and laboratory results appeared on the web-based portal of the facility.

Telemedicine, social media, and curriculum

Technology is advancing every day and at an extremely rapid pace. The incorporation of newer technology creates opportunities and challenges that will have an impact on healthcare education, delivery, and management. As telemedicine is considered as an important area by many medical schools, its inclusion in medical curricula is increasing (Pourmand et al., 2021). As the medical curricula are already overcrowded, introducing telemedicine as a separate area may not be easy. It is vital to identify alternative mechanisms to educate the future generation of healthcare professionals on the proper usage to serve the underprivileged populations through telemedicine as well as to use it in an appropriate manner to minimize ethical, professional, and technical issues. Incorporating telemedicine in the curriculums of healthcare professions needs to change the direction and the way that the objectives are looked at. Not only healthcare professionals and academics but even undergraduate as well as postgraduate students have also recognized the value of telemedicine as an educational tool. They are highly satisfied with the improvement in outcomes of the patient, care, practice-based learning, and knowledge (Boyers et al., 2015). It is important in improving technology and considers integrating telemedicine with available dedicated apps, social media, mobile health, and other newer technologies. This will be helpful in teaching/training, providing clinical care, training, development, and administration. The development of positive culture in this regard is an essential requirement together with leadership and participation by all levels of the organization (Hilty et al., 2019). Required competencies on social media and mobile health have been developed for some disciplines and are in practice. These competencies are useful in developing and accessing training, clinical care, development, and administration (Hilty et al., 2018a, 2018b).

Telepharmacy

Social media has been used not only to provide consultations by doctors to patients but has been used in many other healthcare services as well. The COVID-19 pandemic has created a unique opportunity to use and access such services. Telepharmacy service has been created and operated during the COVID pandemic. A remote pharmacy service platform named "Cloud Pharmacy Care" based on social software has been used to help patients solve medication-related problems (Li et al., 2021). This application has allowed patients to ask questions related to their health or medications, have online consultations, and allow patients to keep in touch with their pharmacists. It has been observed that the patients are primarily interested in getting information regarding drug efficacy, adverse effects, and chronic disease management through this service. A study conducted by Li et al. (2021) on "Cloud Pharmacy Care" has identified that it is a timely and interactive consultation platform that helps to carry out medication management for chronically ill patients, improve patients' medication adherence, and improve medical quality. This service which runs through a social media platform can play a positive role in improving the knowledge of safe medication. This service can be used beyond the pandemic as well (Li et al., 2021).

Positive and negative points of using social media in telemedicine

Even though social media has been identified as an excellent tool in patient education and management, still it has not gained the full trust of healthcare professionals or patients. Most of healthcare professionals identified social media as an excellent medium to offer and spread technological and scientific information among professionals through videos, articles, books, case studies, forums, and other materials that are helpful in enhancing the knowledge and professional skills of healthcare professionals. Social media can improve the skills and knowledge of healthcare professionals, a fact that most healthcare professionals agreed. In many instances, healthcare professionals do use social media for the dissemination of knowledge and patient education but are hesitant in interacting with patients using social media through online consultation because they believed that the use of social media can disturb the choice of the healthcare provider in such consultations. Further, some healthcare professionals are hesitant to interact with patients due to the issues with ethical and professional dilemmas related to the patient's privacy and the possibility of publishing unacceptable information (Alanzi & Al-Yami, 2019). Further, a recent review has identified social media as a bioethics issue that has serious implications for medical practice, research, and public health. Therefore, they have argued that bioethicists should focus their attention on the areas of consumer engagement, bias, and profit maximization to shape online content and related areas. They further offered a set of recommendations and suggested future directions for addressing ethical concerns in these domains (Terrasse et al., 2019).

Dedicated telemedicine programs consume a lot of time, money, and resources in development and maintenance. As we have experienced with the recent COVID pandemic, they cannot create suddenly. Social media is an ideal alternative in such situations and has already been adopted by different patients and healthcare professionals. Mobile social media applications based on text messaging and images can be very useful with instant communication with the patient and healthcare professionals allowing quick decisions. Even with some limitations, social media telemedicine alternatively helps people to rule out lesions, have an early diagnosis, and correct follow-up (Machado et al., 2020).

Even though the use of social media in telemedicine is well supported by evidence, the patient perspective on the transition from traditional healthcare delivery to telemedicine via social media has largely been ignored or less studied. Ignoring the patient-related factors can have a negative impact and may lead to reduced patient compliance and satisfaction unless the use of technology in healthcare delivery is aligned with patient preferences and wishes as well as being accessible and intuitive to use. It has been observed that most patients are having a positive attitude toward the use of social media-delivered telemedicine, and a significant proportion of patients are already using these platforms in their disease management. As patients desire individualized care like in-person care provided in traditional healthcare delivery systems, care must be taken to have a proper virtual environment, and technology to allow better communication between patients and healthcare professionals. Maintenance of patient privacy is having to be considered a top priority. Further research is required to investigate the optimal methods to integrate social media-delivered telemedicine into the healthcare delivery systems (O'Neill et al., 2020).

Recommendations for using social media in telemedicine

Social media usage is only a smaller fraction of telemedicine. Social media can have an impact on telemedicine as the same that telemedicine can influence social media. The relationship between social media and telemedicine has been investigated using multiple modalities. In a relatively newer method, social media data have been examined as part of health analytics. Using the above analysis, investigators have identified several interesting issues from social media platforms on telemedicine, mHealth, and related technology. In one such study using a selected social media platform, many participants have seen telemedicine as a system used primarily within hospitals and only a smaller portion of them have identified it as a method that can be used to receive consultation for patients at home by physicians from afar. If social media to develop as an easy and less costly alternative to telemedicine, it is important to convince patients and physicians that telemedicine and mHealth using social media can be used "outside" hospitals and that these newer technologies can provide a level of high-quality healthcare similar to traditional face-to-face consultations (Leung et al., 2018). Properly planned educational programs targeting the patients and potential users of social media as a modality in telemedicine will help in reducing these misconceptions and will allow better use of it as an alternative. Professional organizations can take a role in providing insights to the healthcare professionals on its use while providing proper guidelines in ethical and professional usage. It will be beneficial to use the full potential of social media as a tool in telemedicine till a universally acceptable, cost-effective, user-friendly telemedicine application is developed.

Social media platforms have proven to be very popular and very engaging with the public. Data analysis of social media content has been widely used in gathering information and predicting outcomes in political campaigns, media, and daily events. It has been used in healthcare as well to identify the perception and needs of people. A study, analyzing public data available on Twitter to investigate the rapid shift in telehealth adoption amidst the recent COVID-19 pandemic has highlighted the need for widespread implementation of digital health and the importance of favoring policy changes to increase the usage of this technology (Massaad & Cherfan, 2020). It appears that telemedicine and the use of social media in telemedicine have been considered as two separate issues without giving due recognition to the power of the combined approach. Considering the financial and technical difficulties faced by less developed countries in adopting telemedicine with advanced technology and resources, social media can be very effectively used to fill up the gap of such difficulties until such resources are fulfilled and telemedicine services are established. Developing proper guidelines and regulations and popularizing them with healthcare professionals as well as potential users to minimize improper, unprofessional, or unethical use of social media in telemedicine will allow this important tool to be used in a justifiable manner.

Telehealth services have rapidly and largely transformed healthcare delivery in areas of high infection rates during the COVID pandemic. In parallel, social media platforms became an immense source of information on day-to-day events and a reflection of social interactions and responses. Multiple studies separately and collectively have been conducted during the pandemic covering most aspects of telehealth and social media. Studies have shown that social media platforms can be used to assess the needs of communities and to embrace the healthcare response, resilience, and preparedness of the community during pandemics (Massaad & Cherfan, 2020).

As with any other systems involving technology, proper training in the use of social media in telemedicine is required to provide better healthcare delivery. One of the most important concerns in the use of some social media apps in telemedicine is the quality of images that are being transferred and whether these poor-quality images can have an impact on the final diagnosis. One reason for this reduction in the quality of images may be the device used to acquire images. Another reason is that the images are compressed when transferring from one device to another which leads to a loss of fine details in images resulting in poor-quality images. The third reason may be the poor image capturing by the user. Capturing images without showing the details of the lesion or even the small lesion can change the final diagnosis. Finally, this may be linked to the way that the receiver is looking at the images or the way that they interpreted them without giving due consideration to possible technical errors and artifacts. Proper training in capturing images and interpreting them will enhance their diagnostic accuracy of them. Even though it may not be practical to train the patients, such programs can be arranged for the healthcare professionals working in the periphery who require such services.

To provide protection to the healthcare professionals/institution who is using social media-delivered telemedicine, appropriate policy and legislation needs to be developed and implemented. More and more research studies are required to provide adequate evidence in this regard. Patient preferences and healthcare professionals' attitudes on the use of social media in telemedicine must be considered seriously in the development of policy/legislation to allow the best practice in providing safe and effective digital healthcare. Some countries have included appropriate regulations and policies for the use of telemedicine as a healthcare delivery method, and due attention has been given to social media as well. However, most countries do not have such regulations, and unfortunately, most of these countries are the ones that require social media as an effective alternative to telemedicine as they are not able to use sophisticated expensive specialized apps.

Summary

In summary, social media can be used as an effective method of telemedicine. It has many opportunities for improvement. More research is needed to increase the level of evidence to make proper decisions in the use of social media in telemedicine by reducing its disadvantages. The development of clear guidelines and policies is very important to minimize the risk associated with the use of social media in telemedicine.

Questions for discussion

1. What would be the steps taken in order to minimize ethical issues in relation to the use of social media in telemedicine?
2. What would be the role of professional medical organizations in maintaining ethical aspects of this practice?
3. What would be the alternatives to the use of social media in telemedicine?
4. What type/features of electronic communication mode would be able to replace the use of social media in the clinical practice?

A list of key terms used in the chapter with definitions:

1. **Bioethics:** The ethics of medical and biological research.
2. **Content production:** The process of developing and creating visual or written assets, such as videos, eBooks, blog posts, whitepapers, or infographics. The term might be useful in a broad sense, but the reality is that the details of content production vary wildly depending on the type of content.
3. **Knowledge information aggregation:** The problem of collecting information from various different sources and aggregating them into a merged knowledge base. One of the main challenges in this process has been handling with data incompleteness as data sources seldom contain complete answers to a user's inquiry.
4. **Media sharing:** Types of social media are used to find and share photographs, live videos, videos, and other kinds of media on the web. They are also going to help you in brand building, lead generation, targeting, and so on.
5. **Professional networking:** People who have connected with one another for business-related reasons or a career. Members can share the information which may include but are not limited to job leads.

6. **Social communication:** Allows sharing of personal experiences, emotions, and thoughts. Social communication skills are required for language expression and comprehension in nonverbal, spoken, written, and visual-gestural (sign language) modalities.
7. **Social media:** Websites and applications that enable users to create and share content or to participate in social networking.
8. **Social networking:** The practice of using a dedicated online platform to maintain contact, interact, and collaborate with like-minded individuals, peers, friends, and family.
9. **Telehealth:** The provision of healthcare remotely by means of telecommunications technology.
10. **Telemedicine:** The remote diagnosis and treatment of patients by means of telecommunications technology.
11. **Telepharmacy:** Considered to be a way of delivering pharmaceutical products and care by the means of telecommunication to different patients.
12. **Virtual reality and gaming environment:** The application of a three-dimensional (3-D) artificial *environment* to computer *games*.

Websites to explore

Facebook: A popular social networking site
https://www.facebook.com/
Explore this website if you have an account. If not, it is easy to create an account to attempt searching through this website. *Did you find it interesting? How can this be used in medicine? What might be the advantages and disadvantages of this in telemedicine?*

LinkedIn: A popular website among professionals in different fields
https://www.linkedin.com.
Visit this website and explore the pages of some healthcare professionals using telemedicine.
What did you find? What specialties were more represented? What were the motivations of these professionals that you could decipher?

YouTube: A popular website for watching and creating video clips
https://www.youtube.com/
Visit this website and explore videos on applications of telemedicine in healthcare.
What do you think might be the advantages or disadvantages of this social media in watching and creating video clips for telemedicine?

TikTok: One of the most popular websites nowadays among young adults and adolescents
www.tiktok.com
What do you think might be the advantages or disadvantages of this social media in telemedicine?

Patientslikeme: Patient empowering website
https://www.patientslikeme.com/
The purpose of this website is to create a trusted digital platform that empowers patients to navigate their health journeys together and with a smooth transition through peer support, personalized health insights, tailored digital health services, and patient-friendly clinical education. Its growing community of more than 850,000 members with over 2800 conditions share personal stories and information about their health. *Has any of your relatives or friends sought support through the site? Can you think of the pros and cons of this website for them?*

References

Adams, S. A. (2010). Blog-based applications and health information: Two case studies that illustrate important questions for Consumer Health Informatics (CHI) research. *International Journal of Medical Informatics, 79*(6), e89—e96. https://doi.org/10.1016/j.ijmedinf.2008.06.009

Alanzi, T., & Al-Yami, S. (2019). Physicians' attitude towards the use of social media for professional purposes in Saudi Arabia. *International Journal of Telemedicine and Applications.* https://doi.org/10.1155/2019/6323962. Article ID 6323962.

AlShaya, M., Farsi, D., Farsi, N., & Farsi, N. (2022). The accuracy of teledentistry in caries detection in children — a diagnostic study. *Digital Health, 8.* https://doi.org/10.1177/20552076221109075

Barbosa, W., Zhou, K., Waddell, E., Myers, T., & Dorsey, E.,R. (2021). Improving access to care: Telemedicine across medical domains. *Annual Review of Public Health, 42*(1), 463—481.

Boyers, L. N., Schultz, A., Baceviciene, R., Blaney, S., Marvi, N., Dellavalle, R. P., & Dunnick, C. A. (2015). Teledermatology as an educational tool for teaching dermatology to residents and medical students. *Telemedicine Journal and e-Health, 21*(4), 312—314. https://doi.org/10.1089/tmj.2014.0101

Chellaiyan, V. G., Nirupama, A. Y., & Taneja, N. (2019). Telemedicine in India: Where do we stand? *Journal of Family Medicine and Primary Care, 8*(6), 1872—1876. https://doi.org/10.4103/jfmpc.jfmpc_264_19

Chou, W. Y., Hunt, Y., Folkers, A., & Augustson, E. (2011). Cancer survivorship in the age of YouTube and social media: A narrative analysis. *Journal of Medical Internet Research, 13*(1), e7. https://doi.org/10.2196/jmir.1569

Chretien, K. C., & Kind, T. (2013). Social media and clinical care: Ethical, professional, and social implications. *Circulation, 127*(13), 1413—1421.

Colineau, N., & Paris, C. (2010). Talking about your health to strangers: Understanding the use of online social networks by patients. *New Review of Hypermedia and Multimedia, 16*(1—2), 141—160. https://doi.org/10.1080/13614568.2010.496131

Craig, J., & Petterson, V. (2005). Introduction to the practice of telemedicine. *Journal of Telemedicine and Telecare, 11*(1), 3—9. https://doi.org/10.1177/1357633X0501100102

El Kheir, D. Y. M., Alnufaili, S. S., Alsaffar, R. M., Assad, M. A, & Alkhalifah, Z. Z. (2022). Physicians' perspective of telemedicine regulating guidelines and ethical aspects: A Saudi experience. *International Journal of Telemedicine and Applications.* https://doi.org/10.1155/2022/5068998

Farmer, A. D., Bruckner Holt, C. E., Cook, M. J., & Hearing, S. D. (2009). Social networking sites: A novel portal for communication. *Postgraduate Medical Journal, 85*(1007), 455—459. https://doi.org/10.1136/pgmj.2008.074674

Fernandez-Luque, L., Elahi, N., & Grajales, F. J. (2009). An analysis of personal medical information disclosed in YouTube videos created by patients with multiple sclerosis. *Studies in Health Technology and Informatics, 150*, 292−296.

Frost, J. H., & Massagli, M. P. (2008). Social uses of personal health information within PatientsLikeMe, an online patient community: What can happen when patients have access to one another's data. *Journal of Medical Internet Research, 10*(3), e15. https://doi.org/10.2196/jmir.1053

Hilty, D. M., Chan, S., Torous, J., Luo, J., & Boland, R. J. (2018). A telehealth framework for mobile health, smartphones and apps: Competencies, training and faculty development. *Journal of Technology in Behavioral Science, 4*, 106−123. https://doi.org/10.1007/s41347-019-00091-0

Hilty, D. M., Maheu, M. M., Drude, K. P., & Hertlein, K. M. (2018). The need to implement and evaluate telehealth competency frameworks to ensure quality care across behavioral health professions. *Academic Psychiatry, 42*(6), 818−824. https://doi.org/10.1007/s40596-018-0992-5

Hilty, D. M., Unützer, J., Ko, D. G., Luo, J., Worley, L., & Yager, J. (2019). Approaches for departments, schools, and health systems to better implement technologies used for clinical care and education. *Academic Psychiatry, 43*(6), 611−616. https://doi.org/10.1007/s40596-019-01074-2

Intan, S. M., & Defi, I. R. (2021). Telemedicine guidelines in South East Asia—a scoping review. *Frontiers in Neurology, 11*, 581649. https://doi.org/10.3389/fneur.2020.581649

Jalali, M. S., Bruckes, M., Westmattelmann, D., & Schewe, G. (2020). Why employees (still) click on phishing links: investigation in hospitals. *Journal of Medical Internet Research, 22*(1), e16775.

Jang, Y., Park, C.-H., & Seo, Y.-S. (2019). Fake news analysis modeling using quote retweet. *Electronics, 8*, 1377. https://doi.org/10.3390/electronics8121377

Kay, M., Santos, J., & Takane, M. (2011). *mHealth: New horizons for health through mobile technologies*. Geneva: World Health Organization.

Leung, R., Guo, H., & Pan, X. (2018). Social media users' perception of telemedicine and mHealth in China: Exploratory study. *JMIR mHealth and uHealth, 6*(9), e181. https://doi.org/10.2196/mhealth.7623

Li, H., Zheng, S., Li, D., Jiang, D., Liu, F., Guo, W., Zhao, Z., Zhou, Y., Liu, J., & Zhao, R. (2021). The establishment and practice of pharmacy care service based on internet social media: Telemedicine in response to the COVID-19 pandemic. *Frontiers in Pharmacology, 12*, 707442. https://doi.org/10.3389/fphar.2021.707442

Lovett, J. T., Munawar, K., Mohammed, S., & Prabhu, V. (2021). Radiology content on TikTok: Current use of a novel video-based social media platform and opportunities for radiology. *Current Problems in Diagnostic Radiology, 50*(2), 126−131.

Machado, R. A., de Souza, N. L., Oliveira, R. M., Martelli Júnior, H., & Ferreti Bonan, P. R. (2020). Social media and telemedicine for oral diagnosis and counselling in the COVID-19 era. *Oral Oncology, 105*, 104685.

Mars, M., Morris, C., & Scott, R.,E. (2019). WhatsApp guidelines—what guidelines? A literature review. *Journal of Telemedicine and Telecare, 25*(9), 524−529. https://doi.org/10.1177/1357633X19873233

Mars, M., & Scott, R. E. (2016). WhatsApp in clinical practice: A literature. *Studies in Health Technology and Informatics, 231*, 82−90.

Massaad, E., & Cherfan, P. (2020). Social media data analytics on telehealth during the COVID-19 Pandemic. *Cureus, 12*(4), Article e7838. https://doi.org/10.7759/cureus.7838

O'Neill, P., Shandro, B., & Poullis, A. (2020). Patient perspectives on social-media-delivered telemedicine for inflammatory bowel disease. *Future Healthcare Journal, 7*(3), 241–244. https://doi.org/10.7861/fhj.2020-0094

Pershad, Y., Hangge, P. T., Albadawi, H., & Oklu, R. (2018). Social medicine: Twitter in healthcare. *Journal of Clinical Medicine, 7*(6), 121. https://doi.org/10.3390/jcm7060121

Petruzzi, M., & De Benedittis, M. (2016). WhatsApp: A telemedicine platform for facilitating remote oral medicine consultation and improving clinical examinations. *Oral Surgery, Oral Medicine, Oral Pathology and Oral Radiology, 121*(3), 248–254. https://doi.org/10.1016/j.oooo.2015.11.005

Pourmand, A., Ghassemi, M., Sumon, K., Amini, S. B., Hood, C., Sikka, N. (2021). Lack of telemedicine training in academic medicine: Are we preparing the next generation? *Telemed Journal and E Health, 27*(1):62–67. https://doi.org/10.1089/tmj.2019.0287. PMID: 32294025.

Ramdurg, P., Naveen, S., Mendigeri, V., Sande, A., & Sali, K. (2016). Smart app for smart diagnosis: Whatsapp a bliss for oral physician and radiologist. *Journal of Oral Medicine, Oral Surgery, Oral Pathology and Oral Radiology, 2*(4), 219–225. https://www.joooo.org/article-details/3501.

Rocha, H. M., Savatt, J. M., Riggs, E. R., Wagner, J. K., Faucett, W. A., & Martin, C. L. (2018). Incorporating social media into your support tool box: Points to consider from genetics-based communities. *Journal of Genetic Counseling, 27*(2), 470–480. https://doi.org/10.1007/s10897-017-0170-z

Terrasse, M., Gorin, M., & Sisti, D. (2019). Social media, e-health, and medical ethics. *The Hastings Center Report, 49*(1), 24–33. https://doi.org/10.1002/hast.975

Ventola, C. L. (2014). Social media and health care professionals: Benefits, risks, and best practices. *P & T, 39*(7), 491–520.

Versteeg, K. M., Knopf, J. M., Posluszny, S., Vockell, A. L., & Britto, M. T. (2009). Teenagers wanting medical advice: Is MySpace the answer? *Archives of Pediatrics & Adolescent Medicine, 163*(1), 91–92. https://doi.org/10.1001/archpediatrics.2008.503

Whitten, P., Holtz, B., & Laplante, C. (2010). Telemedicine: What have we learned? *Applied Clinical Informatics, 1*(2), 132–141. https://doi.org/10.4338/ACI-2009-12-R-0020

Further reading

Dalia, Y. M., Kheir, E. I., Razan, Z. A., Alamri, R. A., & Razan, A. (2022). Social media and medical applications in the healthcare context: Adoption by medical interns. *Saudi Journal of Health Systems Research, 2*, 32–41. https://doi.org/10.1159/000521635

Deema, F., Martinez-Menchaca, H. R., Ahmed, M., & Farsi, N. (2022). Social media and health care (Part II): Narrative review of social media use by patient. *Journal of Medical Internet Research, 24*(1), e30379. https://doi.org/10.2196/30379

Mohammad, S. J., Landman, A., & Gordon, W. J. (2021). Telemedicine, privacy, and information security in the age of COVID-19. *Journal of the American Medical Informatics Association, 28*(3), 671–672. https://doi.org/10.1093/jamia/ocaa310

Section VI

Epilogue

Chapter 16

Innovative uses of social media in public health and future applications

Manoj Sharma
Department of Internal Medicine, Kirk Kerkorian School of Medicine at UNLV, University of Nevada, Las Vegas, NV, United States

In this book, we have discussed types of social media and their current applications in public health and healthcare, gender and age-specific use of social media, implications of social media related to mental health outcomes among diverse populations, applications of social media in public health research, and relation of social media with public health communication and pedagogy. In this final epilog, we will talk about innovative uses of social media as these are emerging in recent years with the aim of projecting the future trajectory of social media use in public health and healthcare. The growing trends of innovative and future uses of social media are summarized in Table 16.1 below.

The growing use of social media among public health and healthcare professionals

There will be a trend of growing use of social media in public health and healthcare at the global level as it has increased the speed of communication between professionals, and this will gain even more prominence in years to come (Catto, 2020; Conrad et al., 2020; Staccini & Fernandez-Luque, 2017). The use of visuals and graphics is likely to enhance the appeal of social media in near future. However, more caution would be needed in interpreting social media messages and its relevance. For example, retweeting is not equivalent to evidence-based practice (Catto, 2020), and this will need to be clarified through various forums to different segments of populations. Also, the time commitment devoted toward social media may reduce research productivity and creativity among public health and healthcare professionals.

Keller et al. (2014) conducted a survey of faculty at Johns Hopkins University and found that at that time, among the respondents, nearly 54% had

TABLE 16.1 Innovative and future uses of social media in public health and healthcare.

No.	Innovative and future uses of social media
1.	The growing use of social media among public health and healthcare professionals
2.	The upward trend of the use of social media in public health research
3.	Social media and crowdsourcing
4.	Artificial intelligence and social media
5.	Rise in the use of social media in teaching, training, and workforce development of public health and healthcare professionals
6.	Increasing use of social media in disease surveillance
7.	Rising use of social media in public health campaigns and messaging
8.	Greater use of social media by local health departments
9.	Social media surveillance in K-12 settings
10.	Counteracting threats from the use of social media
11.	Expanding the use of social media for health education and health promotion interventions
12.	Use of social media for equity and social justice
13.	Use of social media for quality assessment and safety of healthcare services
14.	Use of social media in policy change interventions

used YouTube, 30% had read blogs, 46% had used Facebook, and approximately 7% had used Twitter. If this survey was to be conducted today at universities among faculty of public health and healthcare, the number would be much higher. The number of users is likely to increase to an overwhelming majority or close to 100% users in the future.

Breland et al. (2017) advocate for the use of social media by public health researchers to disseminate their work, influence policy, and build professional networking. However, the authors also provide a word of caution to separate public and private comments while exercising sound judgment in avoiding posting controversial and inappropriate messages in the public domain. Rolls et al. (2016) in an integrative review also found that health professionals are using social media to build virtual communities. We foresee this trend of the use of social media for enhancing networking to continue and gain momentum in the future.

Also, in academic public health, traditionally for tenure and promotion of faculty within university settings, the model relies on accomplishments in teaching, research, and service. With the advent of social media, faculty are reaching out to broader audiences creatively utilizing multiple outlets to share their scholarly work and activities. Acquaviva et al. (2020) have developed a set of guidelines for documenting social media engagement as part of scholarship comprising aspects such as the number of tweets, posts, videos (live and recorded), blogs, innovations shared, educational efforts, podcasts, infographics, etc. We see this trend to be growing in future years where public health faculty will be using these modalities and documenting their accomplishments for greater recognition within academia in *schools and programs of public health*.

The upward trend of the use of social media in public health research

In terms of research with social media, Sinnenberg et al. (2017) examined 137 studies that had used Twitter with over five billion tweets and found that majority of those articles looked at the content analysis of the tweets. Some uses also pertained to recruitment for interventions. The authors recommended the future use of standardized reporting guidelines of Twitter studies. This trend of standardization of guidelines will apply to all forms of social media in the future.

Young et al. (2021) conducted a content analysis of social media images available on Instagram as a method to monitor COVID-19 policy adherence in New York. Such analysis of visuals on Instagram will be undertaken by more public health researchers in the future.

Social media and crowdsourcing

The use of crowdsourcing is likely to increase in the prospective years in the context of social media. Crowdsourcing is a method of procuring data about a task or project by registering the input from a large number of people via the internet and using it for purposes such as problem-solving, surveying, surveillance, etc. For example, it has been used for testing public health messages and materials, developing apps such as Waze, developing educational tools, and mapping geographic information during disasters (Conrad et al., 2020). Social media will serve as a major source for gathering data for crowdsourcing and subsequent applications.

Artificial intelligence and social media

Undoubtedly, the COVID-19 pandemic had a tremendous impact on the popularity of social media and the emergence of unique applications, which is likely to be burgeoning even more. For example, Cresswell et al. (2021) used artificial intelligence (AI)-enabled social media (Facebook and Twitter) analyses to examine contact-tracing apps' use in the United Kingdom. Han et al. (2020) analyzed Weibo texts (a microblogging system in China) to identify COVID-19 public perceptions. Likewise, Tan et al. (2021) also analyzed Weibo posts about COVID-19 sentiments. Golinelli et al. (2020) in a systematic review and found the use of AI-based algorithms being applied to social media data gaining popularity amidst the COVID-19 pandemic. The coupling of AI methodology with social media will increase in public health and healthcare fields.

Rise in the use of social media in teaching, training, and workforce development of public health and healthcare professionals

Social media has been used in teaching students in public health and healthcare, and this trend is likely to increase over the years. For example, Crilly and Kayyali (2020) examined social media's putative application with pharmacy students. The students in their study had to deliver a social media public health campaign on a chosen topic and then make a class presentation using a poster. They found that a majority of the students in their limited sample found this to be an effective strategy. Ramakrishnan et al. (2020) used social media in the training of nephrology interns. Likewise, in a systematic review, Dedeilia and colleagues (2020) found numerous applications of social media in medical and surgical education. Ovaere et al. (2018) have also examined applications of social media in surgical education. Social media will continue to evolve in the training and education of public health and healthcare professionals around the world and will bridge the gaps in training across cultures and continents.

Increasing use of social media in disease surveillance

Another application of social media that is emerging in healthcare is that of disease surveillance (Velasco et al., 2014). Eggleston and Weitzman (2014) describe the use of social media in diabetes surveillance. They describe both a passive approach where the patients do not know that their information is being used for research and an active approach where patients volunteer to give information. We have discussed in this book the ethical aspects of the use of social media, and especially for passive surveillance, these aspects will come to the forefront in years to come. Likewise, active surveillance is likely to enhance the quality of information gathered and develop targeted precise intervention efforts for specific diseases to take place.

Rising use of social media in public health campaigns and messaging

There is a growing use of social media in the general public for procuring health-related content (Van de Belt et al., 2013). Hence, social media has been used for public health campaigns and messaging (Merchant et al., 2021). For instance, in the United Kingdom, Jawad et al. (2015) used social media (Facebook, Twitter, and YouTube) for propagating the harmful effects of waterpipe smoking through a campaign called "*ShishAware*." Vyas et al. (2012) used social media and a short messaging system (SMS) to decrease the sexual risk-taking behaviors among Latino youth in the United States. Such campaigns have also been used in developing countries. For example, Harding et al. (2020) implemented the *Breastfeed4Ghana* campaign using Facebook and Twitter among Ghanaian adults. While this campaign was feasible and had a good outreach, the messages were not retained by those receiving them. There is a need to link these social media campaigns with sound behavioral theories (Sharma, 2022). In the future, a greater dovetailing of behavioral theories with social media campaigns all over the world, both in the developed as well as developing nations, will need to take place. Also, important will be the messengers for these messages. For example, a studyby Solnick et al, 2021 showed that the social media tweets based on personal narratives by emergency physicians regarding COVID-19 recommendations were more effective than those by federal agency officials. This is once again supported by behavioral theories that advocate having credible role models for public health messaging (Sharma, 2022).

Greater use of social media by local health departments

In a 2013 study, Harris and colleagues found that 24% of local health departments in the United States had Facebook, 8% had Twitter, and 7% had both. We envisage that a large majority or even 100% of local health departments in the future will have both these and more. These are also growing in popularity among public health and healthcare universities and will become universal outlets for all agencies.

Social media surveillance in K-12 settings

In K-12 school settings, we have seen in this book that cyberbullying, unhealthy comparisons, and negative mental health outcomes among school students are on the rise due to social media. A trend has been observed where schools and school districts are engaging in social media surveillance (Burke & Bloss, 2020). The authors note that some school districts have gone far ahead with such surveillance. For example, in Florida, a private technology company, FivePoint Solutions, has been hired to screen and scrutinize data from the Department of Children and Families, Florida's Department of Education, Department of Juvenile Justice, Department of Law Enforcement, and local law enforcement, along with posts by students on social media. There are obvious pros and cons of such efforts. In the wake of growing school violence concomitant with an inappropriate use of social media, adverse mental health consequences, such trend of social media surveillance will rise and be a continual point of debate in the near future.

Counteracting threats from the use of social media

Lau et al. (2012) allude to several potential threats arising out of the use of social media, for example, presenting harmful messages that are in direct conflict with public health messages such as those about tobacco use or modeling of other negative behaviors, defiance of public health directives, distorting policy and research findings, and so on. In the future, such negative measures will also gain popularity and would need greater counter combating by public health and healthcare professionals.

Expanding the use of social media for health education and health promotion interventions

Pal (2014) presents an application of social media to diabetes education programs. Zhang et al. (2019) used a social media complaint platform in Beijing to document smoke-free policy violations by the general public. Based on the data, volunteers helped improve compliance in zones where more violations were reported. Social media is now routinely being used in health education and health promotion interventions where blogging, use of Facebook, Twitter, YouTube, etc., are commonly used. Social media has been used for conducting health education and health promotion interventions, and its use will continue to grow in years ahead.

Use of social media for equity and social justice

Merchant et al. (2021) suggest the use of social media to combat racial and structural inequities in the United States. This is an emerging area of concern both politically as well as in public health. The future will see a greater emphasis on social media campaigns and messaging to address social inequities across the United States and globally.

Use of social media for quality assessment and safety of healthcare services

In a Dutch study by Van de Belt et al. (2015), social media sources like Twitter, Facebook, and healthcare rating sites were used to identify the responses of the regulatory body in the Netherlands to incidents reported by individuals and risk-based supervision. The study underscored the potential role that social media could play in such quality assessment. We envisage such applications of social media in other parts of the world in near future.

Use of social media in policy change interventions

Yeung (2018) points out the role of social media in influencing policy change interventions in public health and healthcare. In the past, social media has not really been used for bringing about social change, and this is an unchartered activity that is gaining importance in recent years among the lay public especially with movements such as "Black Lives Matter" and others. We see a greater focus on such applications by public health and healthcare professionals in this regard.

Conclusion

As described in this book, we have seen the multifarious applications of social media in public health and healthcare sectors. Social media is here to stay and grow by leaps and bounds in the coming years. All public health and healthcare professionals must continually strive to update their knowledge about the latest advances related to this field and augment their skills. We will have to build on the positives offered by social media and counteract the negatives offered by this advent in technology. There will be more sophisticated and extensive applications of social media that will emerge in the future about which we can not even speculate. Therefore, we must be ready for accepting and modulating ourselves with change.

References

Acquaviva, K. D., Mugele, J., Abadilla, N., Adamson, T., Bernstein, S. L., Bhayani, R. K., Büchi, A. E., Burbage, D., Carroll, C. L., Davis, S. P., Dhawan, N., Eaton, A., English, K., Grier, J. T., Gurney, M. K., Hahn, E. S., Haq, H., Huang, B., Jain, S., Jun, J., … Trudell, A. M. (2020). Documenting social media engagement as scholarship: A new model for assessing academic accomplishment for the health professions. *Journal of Medical Internet Research, 22*(12), e25070. https://doi.org/10.2196/25070

Breland, J. Y., Quintiliani, L. M., Schneider, K. L., May, C. N., & Pagoto, S. (2017). Social media as a tool to increase the impact of public health research. *American Journal of Public Health, 107*(12), 1890−1891. https://doi.org/10.2105/AJPH.2017.304098

Burke, C., & Bloss, C. (2020). Social media surveillance in schools: Rethinking public health interventions in the digital age. *Journal of Medical Internet Research, 22*(11), e22612. https://doi.org/10.2196/22612

Catto, J. (2020). Is social media worth the risk for health care professionals? *European Urology Focus, 6*(3), 427−429. https://doi.org/10.1016/j.euf.2019.06.003

Conrad, E. J., Becker, M., Powell, B., & Hall, K. C. (2020). Improving health promotion through the integration of technology, crowdsourcing, and social media. *Health Promotion Practice, 21*(2), 228−237. https://doi.org/10.1177/1524839918811152

Cresswell, K., Tahir, A., Sheikh, Z., Hussain, Z., Domínguez Hernández, A., Harrison, E., Williams, R., Sheikh, A., & Hussain, A. (2021). Understanding public perceptions of COVID-19 contact tracing apps: Artificial intelligence-enabled social media analysis. *Journal of Medical Internet Research, 23*(5), e26618. https://doi.org/10.2196/26618

Crilly, P., & Kayyali, R. (2020). The use of social media as a tool to educate United Kingdom undergraduate pharmacy students about public health. *Currents in Pharmacy Teaching and Learning, 12*(2), 181–188. https://doi.org/10.1016/j.cptl.2019.11.012

Dedeilia, A., Sotiropoulos, M. G., Hanrahan, J. G., Janga, D., Dedeilias, P., & Sideris, M. (2020). Medical and surgical education challenges and innovations in the COVID-19 era: A systematic review. *Vivo (Athens, Greece), 34*(3 Suppl. 1), 1603–1611. https://doi.org/10.21873/invivo.11950

Eggleston, E. M., & Weitzman, E. R. (2014). Innovative uses of electronic health records and social media for public health surveillance. *Current Diabetes Reports, 14*(3), 468. https://doi.org/10.1007/s11892-013-0468-7

Golinelli, D., Boetto, E., Carullo, G., Nuzzolese, A. G., Landini, M. P., & Fantini, M. P. (2020). Adoption of digital technologies in health care during the COVID-19 pandemic: Systematic review of early scientific literature. *Journal of Medical Internet Research, 22*(11), e22280. https://doi.org/10.2196/22280

Han, X., Wang, J., Zhang, M., & Wang, X. (2020). Using social media to mine and analyze public opinion related to COVID-19 in China. *International Journal of Environmental Research and Public Health, 17*(8), 2788. https://doi.org/10.3390/ijerph17082788

Harding, K., Aryeetey, R., Carroll, G., Lasisi, O., Pérez-Escamilla, R., & Young, M. (2020). Breastfeed4Ghana: Design and evaluation of an innovative social media campaign. *Maternal and Child Nutrition, 16*(2), e12909. https://doi.org/10.1111/mcn.12909

Jawad, M., Abass, J., Hariri, A., & Akl, E. A. (2015). Social media use for public health campaigning in a low resource setting: The case of waterpipe tobacco smoking. *BioMed Research International, 2015*, 562586. https://doi.org/10.1155/2015/562586

Keller, B., Labrique, A., Jain, K. M., Pekosz, A., & Levine, O. (2014). Mind the gap: Social media engagement by public health researchers. *Journal of Medical Internet Research, 16*(1), e8. https://doi.org/10.2196/jmir.2982

Lau, A. Y., Gabarron, E., Fernandez-Luque, L., & Armayones, M. (2012). Social media in health—what are the safety concerns for health consumers? *Health Information Management, 41*(2), 30–35. https://doi.org/10.1177/183335831204100204

Merchant, R. M., South, E. C., & Lurie, N. (2021). Public health messaging in an era of social media. *JAMA, 325*(3), 223–224. https://doi.org/10.1001/jama.2020.24514

Ovaere, S., Zimmerman, D., & Brady, R. R. (2018). Social media in surgical training: Opportunities and risks. *Journal of Surgical Education, 75*(6), 1423–1429. https://doi.org/10.1016/j.jsurg.2018.04.004

Pal, B. R. (2014). Social media for diabetes health education - inclusive or exclusive? *Current Diabetes Reviews, 10*(5), 284–290. https://doi.org/10.2174/1573399810666141015094316

Ramakrishnan, M., Sparks, M. A., & Farouk, S. S. (2020). Training the public physician: The nephrology social media collective internship. *Seminars in Nephrology, 40*(3), 320–327. https://doi.org/10.1016/j.semnephrol.2020.04.012

Rolls, K., Hansen, M., Jackson, D., & Elliott, D. (2016). How health care professionals use social media to create virtual communities: An integrative review. *Journal of Medical Internet Research, 18*(6), e166. https://doi.org/10.2196/jmir.5312

Sharma, M. (2022). *Theoretical foundations of health education and health promotion* (4th ed.). Jones and Bartlett Learning.

Sinnenberg, L., Buttenheim, A. M., Padrez, K., Mancheno, C., Ungar, L., & Merchant, R. M. (2017). Twitter as a tool for health research: A systematic review. *American Journal of Public Health, 107*(1), e1–e8. https://doi.org/10.2105/AJPH.2016.303512

Solnick, R. E., Chao, G., Ross, R. D., Kraft-Todd, G. T., & Kocher, K. E. (2021). Emergency physicians and personal narratives improve the perceived effectiveness of COVID-19 public health recommendations on social media: A randomized experiment. *Academic Emergency Medicine, 28*(2), 172–183. https://doi.org/10.1111/acem.14188

Staccini, P., & Fernandez-Luque, L. (2017). Secondary use of recorded or self-expressed personal data: Consumer health informatics and education in the era of social media and health Apps. *Yearbook of Medical Informatics, 26*(1), 172–177. https://doi.org/10.15265/IY-2017-037

Tan, H., Peng, S. L., Zhu, C. P., You, Z., Miao, M. C., & Kuai, S. G. (2021). Long-term effects of the COVID-19 pandemic on public sentiments in mainland China: Sentiment analysis of social media posts. *Journal of Medical Internet Research, 23*(8), e29150. https://doi.org/10.2196/29150

Van de Belt, T. H., Engelen, L. J., Berben, S. A., Teerenstra, S., Samsom, M., & Schoonhoven, L. (2013). Internet and social media for health-related information and communication in health care: Preferences of the Dutch general population. *Journal of Medical Internet Research, 15*(10), e220. https://doi.org/10.2196/jmir.2607

Van de Belt, T. H., Engelen, L. J., Verhoef, L. M., van der Weide, M. J., Schoonhoven, L., & Kool, R. B. (2015). Using patient experiences on Dutch social media to supervise health care services: Exploratory study. *Journal of Medical Internet Research, 17*(1), e7. https://doi.org/10.2196/jmir.3906

Velasco, E., Agheneza, T., Denecke, K., Kirchner, G., & Eckmanns, T. (2014). Social media and internet-based data in global systems for public health surveillance: A systematic review. *The Milbank Quarterly, 92*(1), 7–33. https://doi.org/10.1111/1468-0009.12038

Vyas, A. N., Landry, M., Schnider, M., Rojas, A. M., & Wood, S. F. (2012). Public health interventions: Reaching Latino adolescents via short message service and social media. *Journal of Medical Internet Research, 14*(4), e99. https://doi.org/10.2196/jmir.2178

Yeung, D. (2018). Social media as a catalyst for policy action and social change for health and well-being: Viewpoint. *Journal of Medical Internet Research, 20*(3), e94. https://doi.org/10.2196/jmir.8508

Young, S., Zheng, Q., Zeng, D. D., Zhan, Y., & Cumberland, W. (2021). Social media images as an emerging tool to monitor adherence to COVID-19 public health guidelines: A content analysis. *Journal of Medical Internet Research, 10*, 2196/24787. https://doi.org/10.2196/24787. Advance online publication.

Zhang, J., Cui, X., Liu, H., Han, H., Cao, R., Sebrie, E. M., & Yin, X. (2019). Public mobilisation in implementation of smoke-free beijing: A social media complaint platform. *Tobacco Control, 28*(6), 705–711. https://doi.org/10.1136/tobaccocontrol-2018-054534

Further reading

Harris, J. K., Mueller, N. L., & Snider, D. (2013). Social media adoption in local health departments nationwide. *American Journal of Public Health, 103*(9), 1700–1707. https://doi.org/10.2105/AJPH.2012.301166

Qian, W., Lam, T. T., Lam, H., Li, C. K., & Cheung, Y. T. (2019). Telehealth interventions for improving self-management in patients with hemophilia: Scoping review of clinical studies. *Journal of Medical Internet Research, 21*(7), e12340. https://doi.org/10.2196/12340

Index

Note: 'Page numbers followed by "f" indicate figures, "t" indicate tables, and "b" indicate boxes'.

A

Acquired diversity, 126
Active analysis, 196
Addiction, 88
Adolescent, 88
"A dose of truth" campaign, 219—220
Age-specific usage
    adolescence
        habits, 26
        and health information, 28
        and self-esteem, 27
    adults
        habits, 28—29
        vulnerability, 29—30
    children
        elementary-aged children, 25
        parental influence, 24—26
    older adults
        COVID pandemic, 31—32
        media communication, 30
Analytical infodemiological studies, 255—256
Antivaccination beliefs, 222
Appearance anxiety, 54
Application program interface (API), 176
Approval anxiety, 52
Artificial intelligence, 343—344
Asynchronous (real-time) telemedicine, 318
Attachment theory, 88
Attitude, older adults, 109
Autonomy, 102
Availability demands, 52

B

Behavioral addiction, 71t—74t, 88—89
Behavioral confidence, 114t—115t
Behavioral theories, 12—13
Behavior change communication, 280—281
Belongingness theory, 89
Beneficence, 147—148
Bioethics, 334
Black Lives Matter Movement, 133—134
Blackout Twitter, 176—177
Blogging networks, 11
Blogs, 302
Body dysmorphic disorder, 54
Breastfeed4Ghana campaign, 345
Bullying, 49

C

Caplan's social skill model, 89
Centers for Disease Control and Prevention, 248
Children's Online Privacy Protection Act [COPPA], 152
Citizen-centric schemes, 194—195
Classical conditioning, 89
Cognitive skills, older adults, 109
Cognitive theory, 102—104
Collaborative social media, 194—195
Communication, 66
Community Toolbox, 248
Competence, 102
Computer-mediated tools, 3—4
Confidentiality, 155—157
    data confidentiality, 156
    public health studies, 159—161
Confirmation bias, 269
Connection overload, 52
Content analysis, 195
Content production, 334
Control expertise, 110—111
COVID-19 pandemics
    health information sharing, 6—7
    infodemiology. *See* Infodemiology
    lack of socialization, 68—69
    older adult and social media, 31—32
    publications, 261—262
    social media, 68—69, 257
    and social media
        misinformation spread, 8—9, 16
        publications, 7—8, 8f
    Twitter and Facebook, 5
    virtual space, 137

CREDIBLE guidelines, 255–256, 256f
Crowdsourcing, 303, 343
Cyberbullying, 48–49, 84–86

D

Data confidentiality, 156
Deontology, 147
Depression, young adults, 54–55
Descriptive infodemiological studies, 255–256
Dialogic theory and engagement, 245–246
Digital dissemination, 219
Digital equity, 125–126, 136–137
Digital inclusion, 126
Digital stress, 52
Discussions forums, 11
Disease surveillance, 344
Disinformation, 270
Diversity, equity, and inclusion (DEI)
   acquired diversity, 126
   inherent diversity, 126
   mental health
      Black Americans, 133–134
      COVID-19, 132–133
      culture, 129–130
      depression, 129–130
      health insurance, 128
      inclusive policies, 131
      indicators, 131
      personal and environmental experiences, 130
      religion, 129
      resource access, 128, 137
      service utilization, barriers, 128–134
      social factors, 131–132
      social media, 134–135, 137–138
      socioeconomic factors, 127–128
      stigma, 128–129
      trauma Black boys experience, 131
      workplace, 129–130
   psychological climate, 127
   public health practice, 135–136, 136t
   structural diversity, 126–127
Dual system theory, 89

E

Electronic communication, 3–4, 23–24
Emotional loneliness, 102–104
Empirical studies, 4–5
Enterprise social media, 224–225
Epigenetics, 69, 89

Equity, 125
Equity and social justice, 346
Ethical issues, 15–16
   confidentiality, 158
   privacy, 157
Ethics
   beneficence, 147–148
   justice, 148
   vs. morality, 146–147
   nonmaleficence, 148
   professionalism, 147
   respect for persons, 148
   societal effects, 146
   theories, 147
Experimental studies, 180–181, 187
Expressive social media, 194–195

F

Facebook
   audience engagement, 237f
   Congressional members usage, 236f
Facebook alcohol content posting behavior, 183
Facebook-based research study, 153b
Factoid, 270
Fake news (FN) triangle, 259–260, 261f
Fear of missing out (FOMO), 52, 53t
Flow theory, 89

G

Gender-specific usage
   cisgender
      females, 35–36
      males, 36–37
      social media and social support, 33–35
      social media platforms, 34t
      stereotypical media access, 32–33

H

Hashtag, 178
Health, 100–101
Health awareness, 310
Health belief model, 112
Healthcare, 296–297
Healthcare and promotion programs
   community building purpose, 297
   eHealth and public health integration, 297
   healthcare, 296–297
   health education specialists, 297–298
   health institutions, 298

social media
  breaches of patient privacy, 306−307
  data quality, 305
  doctor-patient communication, 301
  general public, 304
  health education and awareness, 304
  health facts circulation, 299−300
  health inventions delivery, 300
  health-related research, 301
  medical education, 303
  platforms for doctors, 301−303
  professional development, 301
  professional recommendations, 305
  skilled image damage, 306
 social network interventions, 297
Healthcare disparities, 126
Healthcare information, 3−4
Healthcare management and disease control, 6
Healthcare practitioners (HCPs), 296
Healthcare services, quality assessment and safety, 346−347
Health communication, 280−281, 288
Health education
  and awareness, 304
  research, 4−5
Health Education Specialist Practice Analysis II (HESPA II 2020), 297−298, 298t
Health equity, 126
  guiding principles, 281
Health literacy, 223−224
Health promotion, 284−287
Healthy People 2030, 126−127

I

Incentive-sensitization theory, 89
Inclusion, 126
Inclusive communication, 281, 288
Industrial media, 3−4
Influencers, 187
Infodemiology, 239, 249, 311
  applications, 254
  combat effects, 265−266
  concept/information incidence rate, 256−257
  COVID-19 battle, 257−259
  crisis events management, 255
  demand-based methods, 256−257
  disinformation, 258−259
  expanded framework, 256−257
  fake news (FN) triangle, 259−260, 261f
  false beliefs, 263t
  historical context, 255
  infodemic knowledge, 262−264
  malinformation, 258−259
  management, 267−268
  misinformation, 258−259, 263−264
  prebunking and debunking interventions, 266−267
  pro-social/antisocial effects, 264−265
  quality assessment, 258−259
  research landscape, 255−256
  supply-based method, 256−257
  volume and velocity, 257−258, 258t
Information access equity course, 226
Infoveillance, 239, 249
Inherent diversity, 126
Innovative action model, 286f, 287t
Internet, 66−67

J

Journal publications, 4
Justice, 148

K

Knowledge information aggregation, 334
Knowledge management systems, 224−225
K-12 school settings, 345−346

L

Linguistic Inquiry and Word Count, 196−197
Literature and social media, 17−18
Local health departments, 345
Loneliness, 102−104

M

Media interventions, 233−234
Media literacy, 223−224, 227
Media-related dimension, 169−170
Media sharing, 334
Medical Directors Forum, 302
Medical education, 303
Mental health, 49
mHealth, 326−329
Microblogs, 302−303
Mini-Mental State Exam (MMSE), 105−106
Misinformation dangers, 8−9
Mismatched social norms, 110

Mixed methods research, 187. *See also* Quantitative and mixed methods research
Morality, 146–147
Multi-theory model (MTM), health behavior change, 112–113, 113f, 114t–115t

N

National Commission for Health Education Credentialing, Inc. (NCHEC), 297–298
"Need to Belong" theory, 102–104
Nonmaleficence, 148

O

Older adults
  depressive symptoms, 100–101
  health issues, 100
  mental health issues, 100
  negative thoughts, 100–101
  population, 100
  social media
    barriers, 111, 111f
    benefits, 105–107
    digital platform, 104
    disadvantages, 104–105
    financial challenges, 108–109
    harmful consequences, 111–112
    literacy-related challenges, 107, 108f
    physical challenges, 108
    psychological challenges, 109–113
    public health theories, 112
  social support and connectedness
    categories, 103t
    life experiences, 101
    loneliness, 102–104
    policy approaches, 106
    psychological needs, 102
    self-determination theory, 102
    structural support, 102
Online learning approach, 288
  communication programs, 279
  health organizations, 279
  learning environments, 278–279
Operant conditioning, 90

P

Participatory dialogue, 114t–115t
Passive analysis, 196
Patient communication techniques, 296
Peele's model of addiction, 70
Person-affect cognition-execution model, 89
Pharmacists, 302
Phishing attacks, 84–86, 90
Platform-swinging practices, 170–171, 187
PlumX Metrics, 218–219
Policy campaigns
  COVID-19 pandemic, 234, 235f
  dialogic theory and engagement, 245–246
  health promotion messaging, 239–242
    e-cigarettes, 241
    nonself-presentation, 241–242
    self-presentation, 241–242
    social media communication campaign, 240–241
    substance abuse, 241
    Twitter, 239–240
    yomevacuno network, 240
  infodemiology, 239
  public health policy, 242–243
  public opinion, 243–246
  social media
    check-ins at public places, 234
    congressional members, 234–238
    crises, 245
    democrat policymakers, 234–238
    policy development, 238–239
Policy change interventions, 347
Post-truth, 270
Prebunking and debunking interventions, 266–267
Prevalence, social media use, 67–68
Privacy
  Facebook, 148, 150, 152
  higher education faculty, 151
  Institutional Review Boards (IRBs), 149
  internet-based information, 149
  investigator, 149
  Periscope, 151
  public health studies, 159–161
  public information, 150
  social media applications, 148
  Twitter, 148–151
  YouTube, 148, 152
Professional ethics, 147
Professional networking, 311, 334
Professional recommendations, 305
Program for Monitoring Emerging Diseases (ProMED-mail), 255
Psychological climate, 127
Psychological health, 90
Psychosocial health, 285–286, 288

Public health
  campaigns and messaging, 345
  and healthcare professionals
    academic, 343
    diabetes education programs, 346
    health education, 346
    integrative review, 342−343
    Johns Hopkins University survey, 341−342
    research, 343
    training and workforce development, 344
  interventions, 4−5, 11−13
  peer-reviewed article publications, 6−7, 7f

Q
Qualitative research
  online interactions, 193−194
  and social media
    active analysis, 196
    analytical tools, 196−198
    citizen-centric schemes, 194−195
    collaborative social media, 194−195
    content analysis, 195
    data acquisition, 197t
    data reliability questionable, 205
    ethical use guidelines, 206−208
    expressive social media, 194−195
    health care professionals, 199−201
    health concerns, 206
    information-rich opportunity, 195−196
    licensing revoked, 206
    passive analysis, 196
    platforms, 194−195
    privacy concerns, 206
    public, 201−202
    public health areas, 198−199
    public health organizations, 203−205, 205f
    research self-identification, 196
    user-generated data and sources, 196t
  social networking services (SNSs), 193−194
Quantitative and mixed methods research, 187
  alcohol-related content, 179
  challenges, 184
  content
    images, 174−175
    text-based, 176−177
    videos, 175−176
  deductive research approach, 170
  engagement, 173−177
  experimental designs, 180−181
  live streaming, 185
  motivations for posting, 178
  platform choice, 170
  platform swinging, 170−171
  reach, 177−178
  real-time data, 181−182
  relationships, 172
  self-report *vs.* actual content, 182−183
  self *vs.* other generated content, 171−172
  10-item scale, 183

R
Rational addiction theory, 90
Real-time data, 181−182
Relatedness, 102
Research dissemination
  academic promotion and tenure process, 219
  challenges and advantages, 221−222
  digital dissemination, 219
  health and media literacy, 223−224
  nonopen access journals, 217−218
  novel publication metric trends, 218−219
  social media
    dissemination avenues, 220
    enterprise social media, 224−225
    health and science communication, 221
    health communication campaigns, 222
    health information, 219−220
    health outcomes, 222−223
    policy makers, 224
    public health practice, 225−226
    web-based platforms, 224−225
  visual abstracts, 220−221

S
Self-cheer leader, 110
Self-determination theory, 90, 102
Self-escape theory, 90
Self-generated content, 171−172
Self-monitoring (real-time) telemedicine, 318
Self-presentation, 169−170
SERMO, 302
Sexting problem, 84−86, 90
Sexual and gender minorities (SGMs), 34
Skill building discussion questions, 283
Slacktivism, 249
Social bots, 270
Social capital, 90

Social capital (*Continued*)
  model, 90
Social cognitive perspective theory, 90
Social cognitive theory, 12–13, 90, 112
Social Determinants of Health, 286–287
Social ecological model, 13–14
Social influence theory, 90
Social learning theory, 90
Social loneliness, 102–104
Social marketing strategies, 281, 283–284, 288
Social media activities, 319
Social media addiction
  behavioral addiction, 71t–74t
  problematic consequences, 78t–84t
  substance-related issues, 70
  theoretical basis, 70–86
  theories/models, 75t–77t
Social media analytics, 311
Social media growth and access, 23–24
Social media image, 110
Social media types, 9–11
Social network analysis, 186
Social networking, 90, 319, 335
Social networking services (SNSs), 193–194
Social networking sites, 9–11
Society for Public Health Education (SOPHE), 297–298
Software-based digital technologies, 277–278
Stimulus-response reinforcement framework, 90
Stressors, 52, 53t
Structural diversity, 126–127
Substance Abuse and Mental Health Services Administration (SAMHSA), 91
Synchronous (real-time) telemedicine, 318

T

Technology acceptance model, 91
Telemedicine
  classifications, 318
  mHealth, 326–329
  social media
    activities, 319
    communication capabilities, 320
    and curriculum, 329–330
    different uses, 326
    Facebook, 319
    fake news, 319–320
    guidelines, 324–325
    in healthcare, 321–323
  healthcare personnel, 320
  medicine and health disciplines, 320–321
  positive and negative points, 331–332
  recommendations, 332–334
  selection, 321
  smartphones, 323
  specialties/branches, 325
  younger generation, 320
  telecommunication technology, 318
Teleology, 147
Telepharmacy, 330, 335
Text-based social media data, 176–177
Theory of planned behavior, 91, 112
Theory of Reasoned Behavior, 32–33
Theory of technology frame, 91
TikTok, 323
Time perspective theory, 90
Twitter
  audience engagement, 237f
  Congressional members usage, 236f
  in healthcare, 322
Twitter-based research study, 154b

U

U.S. Department of Health and Human Services, 249
US adults, 4
Uses and gratification theory, 91

V

Video-based social media, 175–176
Video hosting platforms, 11
Visual abstracts, 220–221

W

Web 2.0, 319
WhatsApp, 322

Y

Young adults
  mental health and counseling trends
    anxiety and stress, 52–54
    autism, 57
    body dysmorphic disorder, 54
    COVID-19 pandemic, 49–50
    daily activity, 57–58
    depression, 52–55
    developmental milestones, 51

digital stress, 52
incidence and prevalence, 50
LGBTQIA + population, 56
machine learning, 49–50
mental disorder, 49
online schooling, 50–51
social comparison, 55–56
transgender and gender-diverse adolescents, 56–57
research-based findings, 46
social media, 14–15
empirical research, 46–48
influencers, 45–46
statistics, 46

Printed in the United States
by Baker & Taylor Publisher Services